21018
21/11/77

Pneumatic Structures
A Handbook for the Architect and Engineer

Thomas Herzog

Pneumatic Structures

A Handbook for the Architect and Engineer

with contributions by
Gernot Minke and Hans Eggers

Crosby Lockwood Staples   London

Granada Publishing Limited
First published in the Federal
Republic of Germany
1976 by Verlag Gerd Hatje, Stuttgart
English translation first published
in Great Britain
1977 by Crosby Lockwood Staples
Frogmore St Albans Hertfordshire AL2 2NF and
3 Upper James Street London W1R 4BP

Copyright © 1976 by Verlag Gerd Hatje, Stuttgart

Translated by Sheila Bacon

ISBN 0 258 97049 9

Contents

Preface

The first volume of Frei Otto's work *Tensile Structures* appeared in 1962. It contained a long chapter on a field of construction which had hitherto received little attention: pneumatic structures. These are based on a physical principle that is frequently found in nature, both organic and inorganic, and has been applied in the field of technology for centuries.

The initiative given by Otto's detailed description of the present state of development was taken up by many people from the end of the 1960s, especially after the first International Colloquium on Pneumatic Structures which took place in Stuttgart in 1967. In England, the Netherlands, Austria and the western United States especially there were countless experiments within and on the periphery of the pop scene; the motive behind them was less that of scientifically orientated interest than of pleasure in the possibility of personally creating "environments" with a minimum of material expenditure in a short period of time. The pneumatic buildings at Expo '70 in Osaka represented a temporary high point in the history of air structures; the pioneer buildings of earlier exhibitions, such as the Crystal Palace, the Eiffel Tower, the Barcelona Pavilion and the German Pavilion in Montreal found in them successors of equal rank.

In the general application of pneumatic structures, architecture – as has happened so often since the industrial revolution – again lags a good way behind other technical fields. In the field of balloon construction, for example, and, somewhat later, of the construction of airships, there have been fundamental developments with regard to membrane material, its cutting and surface protection since the end of the 18th century. These developments, however, have only recently had any influence on architecture. Significantly – apart from a few exceptions – pneumatic structures were until the last few years seen as pure "engineering structures" and the majority of architects, true to their tradition since the 19th century, ignored the whole subject. But, on the other hand, pneumatic forms offer enormous possibilities in shape and colour with a great new design potential.

For some decades – as a result of the increasing mechanisation of the building process – architecture has been dominated by plane, mostly orthogonal forms with hard, cold, machine-produced surfaces. While nearly all previous attempts to oppose this with a sensuous plastic world have meant a negation of the technical/structural dimension of architecture, building with pneumatic structures offers the possibility of a synthesis. They employ forms that are technically highly developed, using soft, flexible, movable, roundly spanned, "organic" shapes, which can be of great sensous beauty when sensibly used.

For the production of a constant internal climate that differs from the external climate, energy must be supplied to buildings. In the case of large buildings the internal climate is determined primarily by the regulating effect of the outer walls. The energy-based heating need only have a limited additional effect in adjusting temperature. If the mass of the outer walls is reduced, as perhaps in the skeleton structure, the interior climate is dependent primarily on the partial of full air conditioning installations which are then necessary.

In the case of pneumatic structures an additional feature is that the stability and variability of the form is directly related to the supply of energy. The pneumatic principle will therefore doubtless play an important role in the realisation of "kinetic architecture" which is increasingly widely discussed.

This book is a report on the state of development. It begins with a general introduction. This is followed by a detailed documentation of completed and projected structures, in which not only pneumatic structures related to architecture are shown, so that a better overall picture of the subject can be obtained and the forms that are relevant for use in the building field can be better located in the spectrum of what is basically possible. The technical descriptions following the picture section show in detail how pneumatic structures hitherto have been constructed. By summarising the details the various solutions to a special problem can be visualised. The two last chapters are made up by excursuses on structural problems with tables for calculating air supported structures, as well as on pneumatic structures used as form work for shell structures.

In summer 1972 the manuscript was in its first draft and underwent supplements and corrections until early 1974. I obtained the time and opportunity for my work through my position at the Institute of Construction at the University of Stuttgart and by means of a scholarship for a period of study in the German Academy of Villa Massimo in Rome.

I thank Gernot Minke and Giuseppe Morabito for their expert advice, Axel Menges for the editorial work on the text and Ursel Prehn for her laborious work in collecting the material.

Special thanks are also due to Rainer Hascher, Claudia Häfele, Erna Herzog and above all to my wife, Verena Herzog-Loibl, without whose co-operation this book would never have been completed.

Thomas Herzog

1. The phenomenon of pneumatic structures

1.1. Introduction

If a flexible *membrane* which is only capable of supporting tension is stressed by the differential pressure of a gas, normally air, then a pneumatic form (from the Greek "pneuma" breath of air) arises. It is deformed in the direction of the less dense agent until its surface is stable in both position and form.

Each pneumatically stressed membrane is capable of resisting external forces. In making use of this capacity, the stressing medium becomes the *supporting medium* and therefore a structural element. The resulting structure becomes a *pneumatic loadbearing structure.* This can be formed either as a *single* or as a *double loadbearing membrane structure.* The number of membranes between the space to be utilised and the exterior determines whether it is a single or a double membrane structure. Figs. 1 and 2 make the principal difference clear. Figs. 3–6 give further schematic examples, Figs. 3 and 4 representing single membrane structures and Figs. 5 and 6 double membrane structures.

In the case of Figs. 1, 3 and 4 the supporting medium must be air and must have a physiologically harmless density. In the case of Figs. 2, 5 and 6 a different gas or high grade compressed air can also be used. In both cases it is necessary to make the pressure area as airtight as possible. These are termed *closed* pneumatic structures (this includes all pneumatic buildings and parts of buildings so far known). However it is possible for the *membrane* itself – apart from small openings for regulating pressure – to be *closed* or to be only one part in the formation of an externally closed cavity. In the latter case the *membrane* is *open* but the *pneumatic building* is *closed* (Figs. 1–4).

If a pneumatically stressed membrane does not form a closed cavity and is not part of the formation of such a cavity, then it is termed an *open pneumatic structure.* The membrane is purposely formed thus so that it benefits from part of the energy of the applied air pressure (sails, parachutes, kites).

1.2. Design rules for pneumatic structures

As pneumatic structures in nearly all technical applications until now have used membranes which have only slight elasticity, their final form in the non-inflated state must be generated by suitable cutting patterns. For determining this form soap film models with a thickness of 0.1 to 1 μ have proved useful. They have an outer and inner liquid surface – as opposed to droplets with only an outer liquid surface.

With regard to their surface all shapes produced with soap bubbles can be thought of as "ideal" pneumatic forms since, because of the fluidity of their film, forms always occur in which there are equal membrane stresses at every point on the surface. Within the prescribed boundary conditions the largest possible volumes and the smallest possible surface areas always form. One refers to *minimal surface* areas. Thus an optimisation of form in relation to use of material takes place. The dead weight and the resulting deformation are so small in the case of soap film models with a span of less than 10 cm that they can generally be ignored.

The most important of the design rules for boundary surfaces are described briefly below (detailed explanations in Bibl. 22; Bibl. 119; Bibl. 167). In any case deformations which arise from outside forces are ignored.

If a soap bubble is suspended freely in space, it is not bound by any boundary conditions. It is affected only by the intermolecular cohesive powers of the soap film and the inner relative pressure. The lamella forms a spherical surface as the only finite surface of constant curvature which is free from singularities (Fig. 7). It conforms to the general equation that defines the relationships of any stable fluid surface (Bibl. 167, p.15) and also applies to all soap bubble tests described below:

$$p = \sigma \left(\frac{1}{r_1} + \frac{1}{r_2} \right), \qquad (1)$$

in the specific form of the sphere:

$$p = \frac{2\sigma}{r} = \text{constant, as} \qquad (2)$$

$$r_1 = r_2 \text{ or}$$

$$\sigma = \frac{p \cdot r}{2}, \qquad (2')$$

where:
p = the pressure in the bubble,
σ = the surface tension,
r_1 = the largest radius of curvature of the surface,
r_2 = the smallest radius of curvature of the surface.

Moreover the two radii of curvature in the case of the sphere are equal and therefore describe a surface which is *doubly curved in the same direction* or *synclastic.*

From the interaction shown in equation 2 it follows that, as a result of the different radii of curvature in bubbles of different sizes, the internal pressure in the smaller bubbles must be larger. This is confirmed by the fact that when two soap bubbles are brought into contact with each other the smaller one inflates the larger (Fig. 8; Bibl. 167, p.11).

Cylindrical surfaces also fulfil the equilibrium condition in accordance with equation 1, but cannot be produced as unbounded surfaces and therefore need sealing at both ends. If one uses coils of wire, places a cylindrical lamella between them and closes the structure by means of spherical segments, then according to equation 2 the radius of the sphere is twice as large as the radius of the cylinder (Fig. 9). The surface of the cylinder is regarded as *singly curved.*

If one uses spherical segments to terminate the cylindrical surface and selects the same radius for both cylinder and spherical segments, then at the contact circles the "cylinder generatrices" pass over into the tangents of the spherical segments (Fig. 10).

The membrane tension σ_1 in the spherical skins continues in the same degree in the longitudinal tensions of the cylinder skin. It follows therefore that the transverse or "ring" tensions σ_2 become twice as large as σ_1, which can lead to the collapse of buildings under extreme wind loading. For cylinder ring tensions

$$\sigma = p \cdot r. \qquad (3)$$

If a soap bubble floats on a liquid surface, then it adjoins on that surface a denser medium, into which the tensions of the lamella surface are introduced at the edge (Fig. 11). The higher internal pressure compresses the floor of the bubble down a little. The bubble forms a hemisphere with synclastically curved lamellae and its cross-section forms a circle with the fluid upper surface. By the introduction of further boundary conditions, cross-sections that differ from the circle are formed.

For example, if one uses a three- or four-angled wire frame, then in the area of the angles surfaces are formed which are *doubly curved in opposite directions* or *anticlastic;* they are also called saddle surfaces (Figs. 12, 13).

Also, depending on the size of the bubble, more or less strongly defined anticlastic surfaces are formed at the areas of constriction by means of "restrictive" wire loops (Figs. 14, 15). Numerous forms of pneumatically stressed soap films can be found conforming to the sectional shapes used (Fig. 16).

Several bubbles always have the tendency to pile up together, as the outer skin of an agglomeration of bubbles also tends towards a circular shape, i.e. tends to occupy the smallest possible surface in relation to its volume.

Up to four cohesive bubbles can, if they are the same size and have the same internal pressure, form identical spherical sections in which all internal lamellae are planar.

If the bubbles are of different sizes, then the internal lamellae are arched in the direction of the larger bubble. There is a direct relation for the radius of this arching to the radii of the bubbles concerned:

$$r_3 = r_1 \cdot r_2 \, (r_1 - r_2). \qquad (4)$$

At one edge only three lamellae and in one point only four edges, or four bubbles, can meet together. The lamellae always meet at an angle of 120°. Four edges always form an angle of 109°28' (Morandi angle).

Fig. 17 gives an illustration of simple bubble combinations. Larger three dimensional agglomerations are called foam. If the bubbles are of equal size, then all inner bubbles have plane contact surfaces and form polyhedrons (Figs. 18, 19). Only the outer lamellae are curved. If the bubbles have different volumes, then the lamella framework inside consists of more or less strongly arched concave and convex single lamellae according to the size of the bubble. This means that all pneumatic structures whose surfaces represent adjacent spherical sections can be constructed. Moreover, by means of additional inner stresses the total size of the structure, which is otherwise limited in its dimensions by the strength of the membrane, can in theory be extended indefinitely (analogous foam).

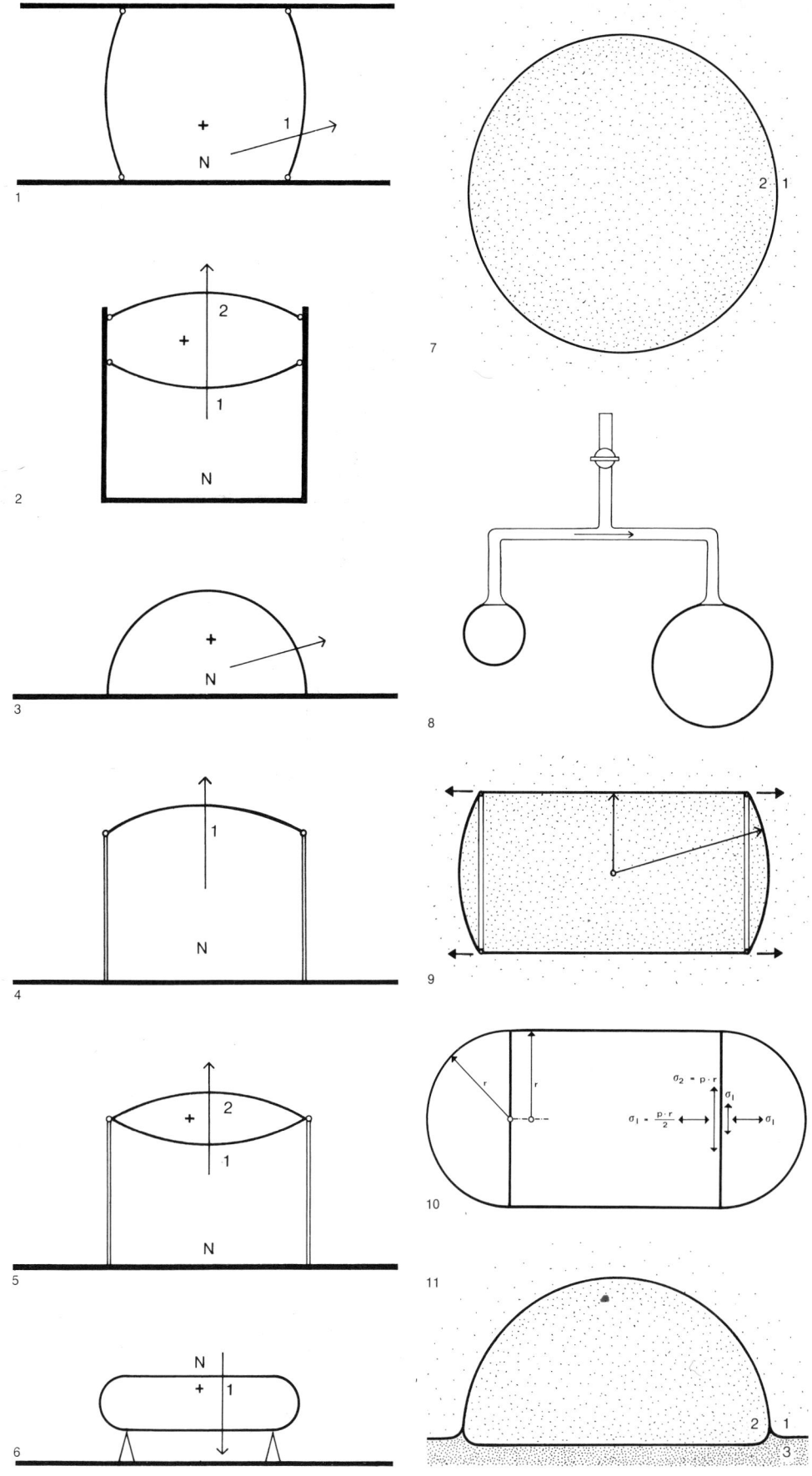

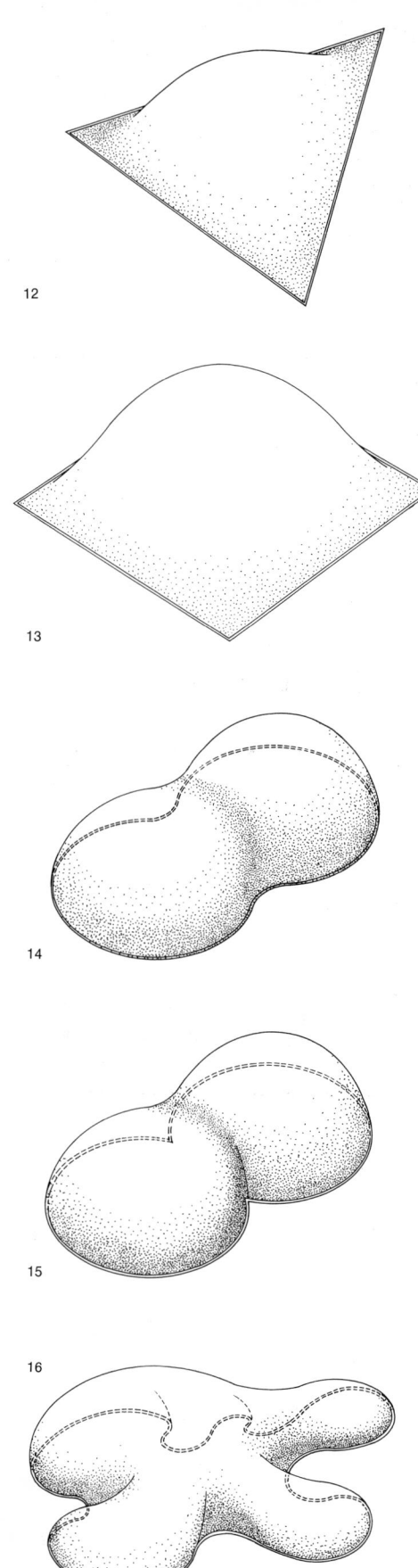

12

13

14

15

16

For the study of pneumatic forms in which the membrane tensions are not the same at every point on the surface and which therefore cannot be represented by soap bubble models, resilient rubber membranes with a thickness of 0.1 to 0.4 mm are particularly suitable when used within their linear area of elasticity.

With the aid of such membranes, for example, it is possible to mould any shape in which spheres, arranged in a straight or curved line, can be inscribed without leaving any excess volume (Figs. 20, 21; Bibl. 119). This can give rise to unilaterally curved, synclastic and anticlastic surface areas.

At places where the radii of curvature are small, zones of lesser stability occur due to a corresponding decrease in the membrane tensions.

If membranes are stressed by means of negative pressure then the imaginary spheres describing the upper surface are found on the outside, for in this case the membrane is stressed by the relative positive pressure of the surrounding atmosphere. However, additional stabilising elements are always necessary, as it is not possible to have a "free" negative pressure structure consisting of only one membrane without boundary conditions (Figs. 24–27). Here too the total of the radii of curvature is always decisive for the evaluation of the surface tension.

Where in the case of soap bubbles wire frameworks have to be used for producing the shape, special boundary conditions are also necessary in the case of pneumatic structures made of non-flexible or slightly flexible materials. The tensions at various points on the membrane are otherwise so different that wrinkles can occur. For example, a structure with the basic shape of a square cushion tends to take on the form of a sphere (Fig. 28). Then there is no tensile stress in the opposite direction to the wrinkle. In order to still produce the tensile stresses necessary to resist the external forces, additional support elements must be introduced (Fig. 29).

In the case of entrances, windows, connection points for fans, etc., there are openings in the membrane which also are subject to specific design laws.

If one places a loop of thread into a plane soap lamella spanned inside a wire ring and pierces the lamella within the loop, the loop forms a circle as a result of the overall equal tensile stresses (Figs. 30, 31).

If one fastens a slack thread in two places on the wire ring and pierces the lamella as shown in Figs. 32 and 33, the thread forms an arc of a circle (Bibl. 167).

This fact should be taken into account in the edge formation of membranes when using tensile stressed cables for the transmission of force.

If in a membrane the tensile stresses prevailing in different directions are not equal (as e.g. in the case of cylindrical forms with hemispherical ends) the shape of the opening should – to avoid wrinkles – be adapted to these forces.

As the membranes of pneumatic structures, apart from special forms, do not form plane surfaces, the curves of openings are correspondingly distorted in three dimensions.

1.3. Morphological classification of pneumatic structures

In the following classification (Table 1) pneumatic structures are described and compared by means of formal characteristic properties. Phenomenological variables other than those used here are certainly possible, but the author has limited himself to categories already in use, as it seemed more sensible to him to create method in the framework of practised terminology than to introduce additional criteria and formulations.

It was stated that pneumatic *structures* could be *open* or *closed*. Equally *membranes* can be *open* or *closed*.

As explained in the chapter on "design rules", the sphere and its sections represent the optimum pneumatic form on account of the equal stress in their surface. If one recognises the sphere as a structure which has the same proportions in three dimensions, then it is natural to classify deviation from this form on the basis of *differing proportions*.

Differentiation should therefore be made between structures with:

– two dimensions of similar size and one larger dimension, e.g. "tubes", "masts", "columns", "towers";

– two dimensions of similar size and one smaller dimension, e.g. "cushions", "lenses", "discuses", "mattresses";

– three dimensions of similar size, e.g. "balloons", "balls", "spheres", "bubbles".

Naturally there are borderline cases between these alternatives. In assessment, therefore, the main directions of extension are compared in their relationship one to another; if a structure is twice as long as it is wide but its height only amounts to a third of its width, then the relationship between width and height is the decisive factor in its classification. The absolute dimensions play no role in this differentiation factor.

The form can be further classified according to the *types of curvature of the outer surface*.

As already shown, it can be:

– singly curved,

– doubly curved in the same direction or synclastic,

– doubly curved in opposite directions or anticlastic.

Plane membrane sections can also occur in pneumatic structures when there are interior skins whose edge is attached to the outer membrane (similar to the inner lamella in the case of equal sized soap bubbles). However, these skins are not stabilised in plane, because there is no difference in pressure. They obtain their tension through the tensile forces acting at the edge.

A further aspect of classification is the establishment of whether an object represents an individual pneumatic structure or comprises several pneumatic structures, and whether these structures are the *same* as each other or *different*.

If the structures are bound together so that they are *not separable* from each other then it is a *combination*. This is the case when adjacent elements possess common membranes so that one element must be destroyed in order to separate

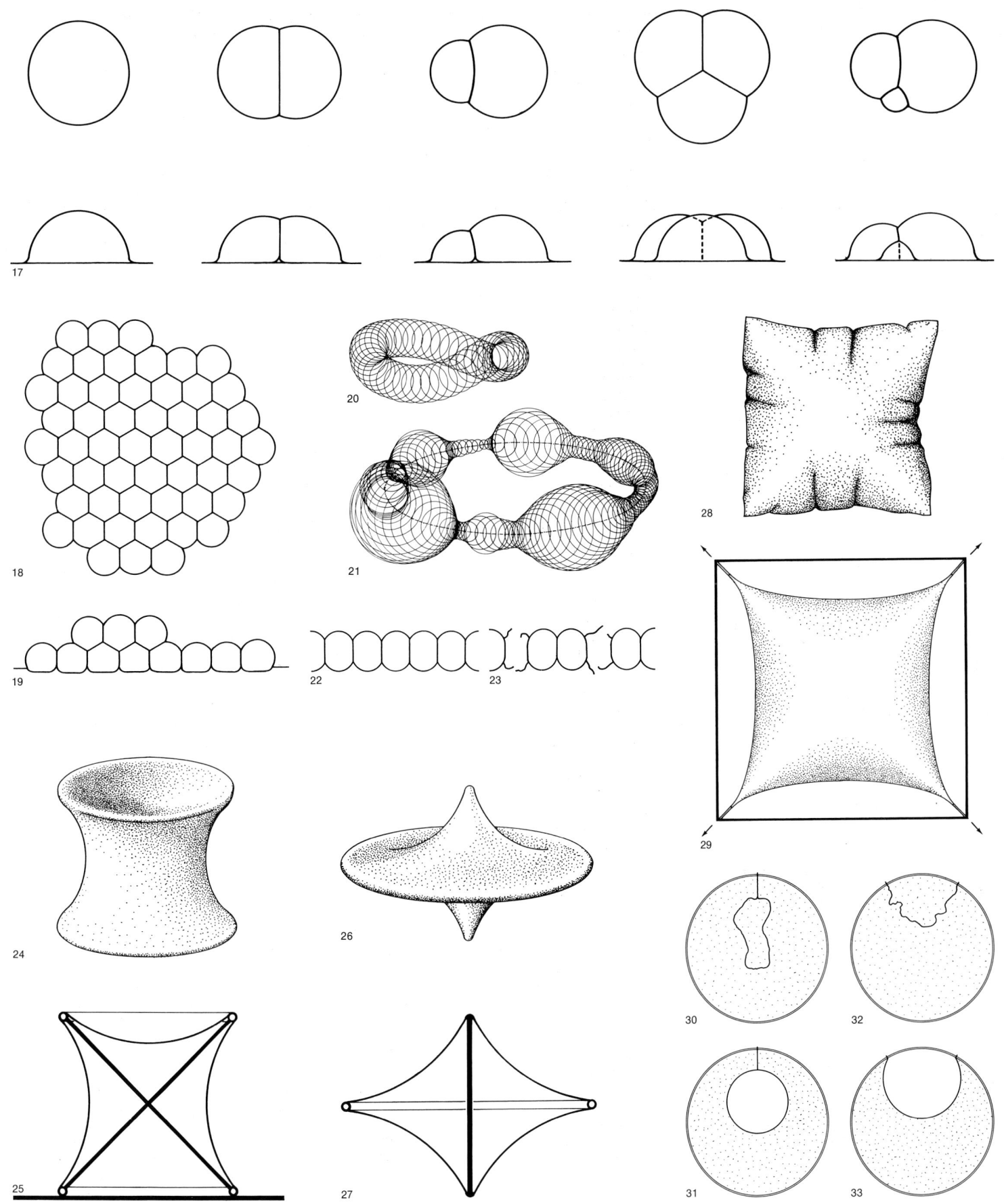

17

18

19

20

21

22   23

24

25

26

27

28

29

30   32

31   33

Table 1. Morphological classification of pneumatic objects

feature	alternatives		
type of membrane **type of structure** 6 alternatives	membrane open structure open	membrane closed structure closed	membrane open structure closed
proportion of structure 7 alternatives	one dominant dimension	two dominant dimensions	three dimensions of similar size
type of curvature 7 alternatives	singly curved	doubly curved in the same direction	double curved in opposite directions
type of connection 7 alternatives	no connection	additions	
		equal structures	unequal structures

12

possible mixtures of forms			
membrane open and membrane closed, structure closed	membrane open, structure open and structure closed	membrane open and membrane closed, structure open and structure closed	
one dominant dimension and two dominant dimensions	one dominant dimension and three dimensions of similar size	two dominant dimensions and three dimensions of similar size	one dominant dimension and two dominant dimensions and three dimensions of similar size
singly curved and doubly curved in the same direction	singly curved and doubly curved in opposite directions	doubly curved in the same direction and doubly curved in opposite directions	singly curved and doubly curved in the same direction and doubly curved in opposite directions
combinations		additions and combinations	
equal structures	unequal structures	equal structures	unequal structures

13

the next but one element from the first (Figs. 22, 23).

If, on the other hand, they are *separable* structures, then they can be separated from each other by releasing the connecting mechanism (the same applies to the assembly). The object that is formed by such individual structures represents an *addition* (additive system). (The terms addition and combination are also used as features for connections in the description of support structures.)

The question of the kind of connection between individual structures is important when considering production, erection, dismantling, transport volume, prefabrication and development of structural details, as well as for replacement of damaged parts.

The morphological features for differentiation of pneumatic structures are shown in the first three lines of Table 1. Each horizontal line gives the alternative characteristics and shows which mixtures of forms are possible within one line of alternatives.

The fourth line describes the possible types of connection of individual pneumatic structures to an object. In the first lines the "type of membrane" and "type of structure" are summarised, as they can only occur in the way shown. (The connection of closed membranes to open structures is not possible.)

The principle employed for this representation is in accordance with the morphological box (Bibl. 168). Each alternative of a form variable can be joined to every other alternative of another form variable.

The diagrams of examples should illustrate the actual feature. They are sketches of simple elements which also possess features from every other line of alternatives, but in which the feature by which they are classified is especially marked.

Within this classification there are about 2,000 different possibilities for describing a pneumatic object according to morphological viewpoints. Some of the lines of connection might seem at first to make no sense. However, certainly none of the possible solutions given in the schedule should be excluded, as so far only a small number of these have been realised or projected, and the schedule should also help to find new solutions by means of new forms of alternative connections.

It is not possible to schematise all the single forms possible within the rules of formation. Therefore a criterion such as the geometry of the structure (cylinder, sphere, cone, etc.) must be abandoned at this point, although such a procedure is obvious at first and is found in the bibliography, albeit with clear limitation of the possible spectrum of forms. The fact that geometrically simple structures have previously been preferred in use is consistent with the favourable exploitation of membrane webs with regard to models, the simplified production and calculation process and the possibility of standardisation. However, the standard types represent only a fraction of the possible forms.

2. Pneumatic buildings – structural design alternatives in pneumatically stabilised membrane structures

Gernot Minke

2.1. Introduction

2.1.1. Function and aim of the investigation

Buildings supported by air are among the lightest structures known. With a weight of 1 to 2 kp/m² in relation to the covered ground surface it is possible, for example, to achieve spans of 100 m.

The principle of the membrane stabilised by positive pressure has been used by mankind for thousands of years, but in building technology it was introduced only about 25 years ago, so that the associated problems are still relatively unexplored.

Before one can make a well founded judgement on the possible uses of pneumatically stabilised structures or loadbearing elements, a basic investigation of the structural design alternatives is necessary. For this reason, with the help of the following systematic representation of the different systems and types of pneumatically stabilised membrane structures, a survey is given of the many possible formations of these lightweight structures.

In this survey the different systems are fully discussed; the individual types – which represent the specific characteristics of the system – are partially discussed by example only.

2.1.2. Definition of terms

A loadbearing structure has the function of transferring certain forces within prescribed boundary conditions. It can be allotted to a specific support system and to a specific support form.

The *structural system* is a static system that is neither formalised nor materialised. If additional (secondary) stabilising elements occur in a loadbearing system then it seems logical to divide the system, according to type and formation of the secondary elements, into sub-systems which are called *structural types*. One should refer to a *structural form* when the longitudinal and cross-sectional proportions are known; thus a structural form is a building which is certainly formalised but not materialised. One should refer to a loadbearing *structure* if material and size as well as system and form are given, that is when it is a materialised building. If the loadbearing structure is not yet finally defined or if a specific quantity of structures is referred to, then the term *structural kind* is used henceforth.

If the consideration of pneumatic structures is restricted to their function as a loadbearing structure, then pneumatic structures are *loadbearing structures stabilised by differential pressure*. The differential pressure is necessary for the loadbearing function and consequently the stabilising medium is part of the structure; by this definition sails, hot air balloons and parachutes are not pneumatically *stabilised* but pneumatically *stressed* structures. If one relates the term "pneumatic" to the direct stabilisation, then it can only refer to loadbearing structures made of flexible surface elements – of membranes; loadbearing structures made of curved, stiff surface elements remain shells even when they are pneumatically stressed (high pressure containers) or in addition pneumatically stabilised. Then the specialised term *pneumatically stabilised membrane structure* is used instead of the general term "pneumatic structure".

A membrane structure stabilised in plane does not generally change its form and its loadbearing behaviour significantly if a liquid or granular medium is used to stabilise it instead of a gaseous one. Therefore the term "pneumatic structure" is frequently wrongly used for all three kinds of membrane structure stabilised in plane (Table 1).

The following discussion includes only "buildings" that are accessible, artificially produced structures, fixed to their bases. Furniture, containers and all air controlled structures (air cushioned vehicles, counter-air currents) and structures moved by wind (sails, windmill sails) are not considered.

2.1.3. Criteria of differentiation and alternatives of form

Pneumatically stabilised membrane structures can be sub-divided by a series of different features:
- type of differential pressure,
- degree of differential pressure,
- type of support medium,
- formation of membranes,
- number of membranes,
- dimension of additional stabilising elements,
- formation of additional stabilising elements,
- arrangement of additional stabilising elements,
- type of surface curvature,
- dimension of main directions of expansion,
- magnitude,
- type of usage,
- type of membrane material,
- degree of variability.

For each of the above mentioned features there are different "characteristics". For example two "characteristics" are used to distinguish the type of differential pressure: positive pressure and negative pressure. That is, a distinction is made between positive pressure systems and negative pressure systems.

In Table 2 eleven features (criteria of differentiation) of pneumatically stabilised membranes are shown with two to nine different characteristics in the form of a "morphological box".[1] In contrast to other methods, this method of classification, made famous by Zwicky (Bibl. 168) makes possible a simple and clear representation of very many features and characteristics and is therefore especially suitable as an aid to produce variety. However, it is necessary in the use of this method for the main features of the problem area (here, pneumatically stabilised membrane structures) to be defined in sufficient detail. In Table 2 a feature with its respective characteristics is shown on every line. The combination of every single characteristic from every line gives

[1]Notes see p. 29.

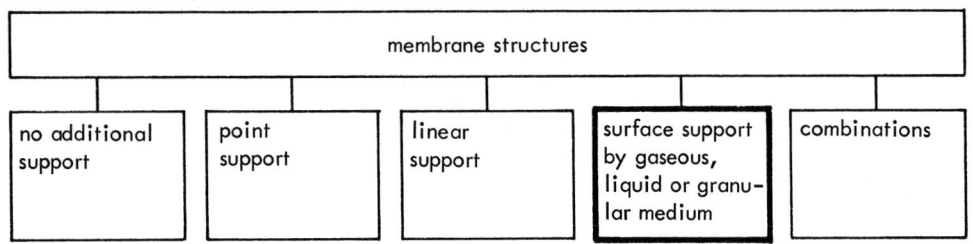

Table 1. Classification of membrane structures.

number	feature	group	characteristics	variety
1	formation of membranes	structural system	a) single b) double	2
2	kind of pressure		a) positive b) negative	4
3	kind of additional support		a) no b) point c) linear	12
4	formation of additional stabilizing elements	structural type	3 b: a) rosette b) ring c) bulged surface 3 c: d) cable e) truss f) arch	28

5	arrangement of additional stabilizing elements		3 b: a) single b) in row c) crossed
			          d) radial e) irregular
			3 c: f) single g) one way h) radial
			          i) tangential k) two-way
			          l) three- (and more) way m) irregular
				148

6	formation of tertiary support		3 b, c: a) no b) tension c) compression d) bending	580
7	dimension of main directions of expansion	structural form	a) one b) two c) three	1 740
8	kind of curvature		a) single b) synclastic c) anticlastic	5 220
9	kind of membrane material	structural kind	a) elastic b) thermoplastic c) non-elastic/ adjustable d) non-elastic/rigid	20 880
10	degree of span		a) up to 20 m b) 20 - 100 m c) more than 100 m	62 640
11	kind of addition		a) no b) one direction c) two directions d) three and more directions	250 560

Table 2. Classification of pneumatically stabilised membrane structures in the form of a "morphological box".

1,492,992 different alternatives. However, as some of the combinations are mutually exclusive ("incompatible") there are only 250,560 genuine ("compatible") alternatives. In this classification survey the characteristics of the first three features decide the *structural system*. For the classification of the low pressure system it is logical to mention also the combination of point (b) and linear (c) support in the type of secondary support.[2] Thus 16 different structural systems are defined. These are represented in Table 3 and explained in Section 2.2. Within these morphological boxes, however, no combinations of different characteristics of the same feature are shown; these are basically feasible, but would in this case strongly detract from the clarity of the chosen classification scheme. If all possible combinations of characteristics were considered then the number of possible combinations would greatly increase.

The characteristics of features 4 to 6 decide the *structural type,* but it should be noted that this differentiation occurs only in the case of structural systems with additional (secondary) support. The structural types are therefore to be regarded only as their sub-systems. These are explained in greater detail in Section 2.2.4.

The characteristics of features 7 and 8 describe the *structural form,* while the characteristics of features 9 to 11 deal with the loadbearing structure as a materialised building, i.e. they describe the *structural kind.*

With these eleven features and their characteristics 250,560 different kinds of loadbearing structures can be distinguished and described.

The development of classification in the form of the illustrated morphological box may encourage, by means of a systematic procedure, the discovery of new kinds of support structures or the compilation of description schedules or lists of criteria for the description and assessment of different kinds of loadbearing structures.

However suited this form of representation is to clearly illustrating a great number of different kinds of structures, it is not very suitable for plain visualisation of the individual systems and types of pneumatically stabilised membrane structures. For this the matrix seems to be the most suitable form of representation. The following sections give a survey of the possibilities for the structural formation of pneumatically stabilised membrane structure. Here only features 1 to 6 (Table 2), which define the different structural systems and types, are considered. Features 7 and 8, which define the structural form, have already been explained in Section 1.3 in the phenomenological description of pneumatic structures.

Features 9 to 11, which define the kind of loadbearing structure, are not considered in any more detail here.

2.2. Low pressure systems

2.2.1. General characteristics

In the surveys of the different systems of pneumatically stabilised membrane structures, two primary systems can be differentiated: low pressure and high pressure systems.

In the case of *low pressure systems* (Table 3) the differential pressure of the media divided by the membrane generally amounts to 10 to 100 mm of water pressure. The membrane is thus stressed by a normal pressure of 10 to 100 kp/m². In the case of *high pressure systems* the differential pressure generally amounts to 2,000 to 70,000 mm of water pressure. This means that the membrane is exposed to a differential pressure of 2,000 to 70,000 kp/m². The variable static function of these two groups of systems is explained in the description of the high pressure system in Section 2.3.1.

2.2.2. Single membrane systems, double membrane systems

According to the formation of the membrane the low pressure systems can be divided into *single* and *double* membrane structures. In single membrane structures an (accessible) space under positive or negative pressure is formed or closed by one membrane. In double membrane structures, however, the (accessible) space formed or closed by the membrane structure is not under positive or negative pressure. In this case the parts of the membrane surrounding the support medium are always curved in opposite directions to each other. Double membrane structures can, by reason of their appearance, also be called "cushion structures". As some authors also use this term for high pressure tubular structures (see Section 2.3) it will not be used here.

2.2.3. Negative pressure systems, positive pressure systems

The second division of low pressure systems into *negative pressure systems* and *positive pressure systems,* according to the kind of internal pressure, may seem insignificant at first, as in the stabilisation of the structure only the differential pressure of the media divided by the membrane is important. As the diagrams in Table 3 indicate, the kind of pressure differential has, however, a great influence on the structural form and hence also on the structure of the boundary formation, on the static and dynamic loadings of the membrane and on the spatial form.

In positive pressure systems the membrane is always curved outwards (convex), whereas in negative pressure systems the membrane is always curved inwards (concave) – except in the area of secondary support. As a result, in negative pressure systems snow and water pockets can very easily occur in the roof as well as instabilities of form due to aerodynamic loading in

low pressure systems				
I single membrane structures				
	0 no additional support	P additional point support	L additional linear support	P+L additional point and linear support
I n negative pressure	I n 0	I n P	I n L	I n P+L
I p positive pressure	I p 0	I p P	I p L	I p P+L
II double membrane structures (inflated)				
	0 no additional support	P additional point support	L additional linear support	P+L additional point and linear support
II n negative pressure	II n 0	II n P	II n L	II n P+L
II p positive pressure	II p 0	II p P	II p L	II p P+L

Table 3. Classification of low pressure systems.

wind – disadvantages that are not usually present in positive pressure systems. Moreover negative pressure systems usually require high supports at the edge or in the centre. This means relatively expensive secondary structures. These disadvantages have meant that negative pressure systems have so far hardly been used. However, it must be mentioned here that these systems, in combination with positive pressure systems and high pressure tubular structures, can be very suitable for some purposes.

2.2.4. Systems with additional support

2.2.4.1. Type and application of the principle

The third feature of the low pressure system is the kind of additional, i.e. secondary, stabilisation of the membrane used.

As the types of air supported buildings with large spans which have been realised so far lead one to conclude that the significance of the principle of additional support has not yet been properly comprehended, the characteristics and application of this principle will be investigated in greater detail in this section on structural support systems and in the following sections on the corresponding structural support types.

Additional support within the membrane surface (for example by means of cables) is generally suitable in the case of large spans, in order to decrease the radius of curvature and thereby reduce tension in the membrane. By this means it is possible to use conventional flexible membrane materials, which are simple to manufacture and to work on, even for the largest spans.

The following consideration should serve to clarify this principle (Fig. 1): if one assumes the same air pressure, then in an air supported building with a semicircular cross-section, when the span is doubled, the stress in the membrane doubles. However, as the total height also doubles, the stress from wind forces increases. In order to achieve the same stability as in case A the internal pressure must be increased in case B, which would cause another increase in membrane stress. A further disadvantage in case B is the greater relative volume of the interior in relation to the floor space. Because of this, energy costs for heating and pressurisation increase.

If these two disadvantages are to be avoided then, reducing the height of the envelope, at an identical maximum height of C and A and the same internal pressure, the stress in the membrane in case C is 2.5 times larger than in case A. Thus if one wishes to increase the span without obtaining a greater volume and greater membrane tension, this is only possible by reducing the radii of curvature with the aid of additional stabilising elements, as in case C.

If the reduction of the radii of curvature is achieved by means of channel cables, care must be taken that the tension in a cable is directly proportional to its radius of curvature (Fig. 3, 4, 5). If the cables show the same curvature as the total form of the air supported building, then no reduction of the total tensile stress can be obtained (Figs. 3, 6). Thus in the formation of structures in this system it is important to curve the cables as strongly as possible, which can be achieved through the development of boundary restrictions and through tertiary internal anchorings (Figs. 4, 5). As Fig. 2 shows, any large surfaces can be bridged by such internal anchorings.

In the past it was frequently ignored that for pneumatically stabilised membrane structures, the dynamic load and not the internal pressure is of primary significance for the dimensioning of the membranes, the secondary stabilising elements and the anchorages. The crucial factor with flat structures is the tensile stresses caused by wind suction on the side facing the wind below the vertex (see Section 6.3.3). As it is possible for the stress in the membrane due to wind suction or to instabilities of form (oscillation in wind) to be many times higher than the statics calculation from the normal differential pressure shows, an optimisation of form is necessary in the case of larger spans, above all in respect of the aerodynamic loading. Quite generally it can be said: the flatter the total form, the less is the effect of wind suction and of wind pressure. In the case of very flat structures the occurrence of wind pressure can be almost completely avoided.

If theoretically maximum spans of 2 or 3 kilometres can be achieved with aerodynamically favourable forms, such structures are not practicable without reduction of the radii of curvature with the aid of strongly curved channel cables, as they are much too expensive in terms of material (Fig. 3).

If low membrane tensions, distribution of space and low air volume are determining factors for the choice of system, then in a span of 30 m a reduction in radii of curvature by means of additional stabilising elements can be economical (Figs. 4, 6); for buildings with a short design life and spans less than 10 m this also applies where

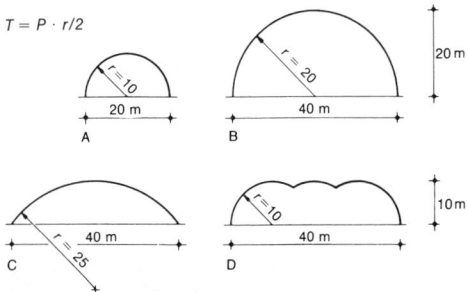

$T = P \cdot r/2$

1. Relationship between tensile stress (T), internal pressure (p) and radius of curvature (r).

membranes are made of thin inexpensive plastic foils (see p. 131).

For additional stabilisation of the membrane two basically different kinds of support can be distinguished: point or linear. The different formations are described in more detail in Sections 2.2.4.2 and 2.2.4.3.

Additional point supports are usually suitable for average and large spans, but the development of support points is generally complicated and expensive in terms of material, so that it is normally more economic to make use of the principle of additional linear support.

With certain boundary formations and with some fields of use it can be practical to combine the principle of point and linear support, i.e. if it is thereby possible to produce sufficient curvature in all areas of the membrane or if, for example, funnel-shaped areas of larger diameter are required in the membrane surface for the drainage of rainwater or snow.

Thus there are four separate characteristics for the feature of additional stabilisation by means of secondary elements. These, together with the two previously mentioned classification features for the group of low pressure systems, define 16 different systems, which are presented in Table 3 in system diagrams and are provided with a mnemo-technically constructed code.

2.2.4.2. Additional point supports

On the principle of point supports

If one attempts to strengthen the curvature of the surface in a balloon made of an elastic rubber membrane by pressing with a pencil point on the membrane, it becomes clear that at the "support point" very high tensions occur in relation to the rest of the membrane, which can very easily lead to the skin splitting. On the other hand, if one expands these individual support points to a circular band or a rounded-off bulged surface, then it is possible, if this expanded support surface is large enough in relation to the whole membrane, for the membrane to show the same tension distribution at every point and thus the danger of the membrane breaking at the support points is no greater than at any other point.

The proof for this, i.e. that with compatible formation of the support elements there is equal tension distribution, can be achieved with the help of soap membranes (Figs. 7–10; Bibl. 120). The relationship between the size of the point support areas, the extent of deformation and the size of the total membrane can be defined with analogous soap bubble models.

If one introduces a circular support element into a flat membrane from below, then the membrane becomes anticlastically deformed (Fig. 10). A membrane of this kind thus forms a minimum surface with equal stress distribution.

The support ring can, however, only be raised to a certain height before the membrane tears. The larger the ring in relationship to the membrane surface, the higher it can be raised and the larger the possible deformation of the originally flat surface.

On the development of point supports

With point supports three different formations can be defined: "rosette", "ring" and "bulged surface" (Table 4). The term *rosette* is used only when a tensioned element is concerned in which the forces from the membrane are collected into edge cables and from there into tertiary support elements (Figs. 7, 8, 11). It is also possible to use only a single cable, which is then referred to as a "loop". This loop forms a curve of the same curvature in space. As each section of the loop lies in the plane of the membrane immediately adjacent and the membrane tensions are the same at all points, the radii of curvature of the loop – each measured in the tangent plane of the membrane – must also be the same.

The term *ring* is used for a circular element under bending stress as well as tensile or compressive stress (Fig. 10). The bending stress, which usually predominates, results from the angle between membrane plane and ring plane as well as from the support of the ring by tertiary elements.

The term *bulged surface* is used when there is a spherical rounded-off support surface which is mainly under compressive stress. The bulged surface can be a balloon (Fig. 12) which adapts to increasing curvature and directs forces away from the membrane. It can, however, be formed from flexible star-shaped lamellae or clover leaf shaped discs under bending tension or, in special cases, be resolved in individual disc-like surfaces – formations which are already well known from tent construction (Figs. 13, 14).

In the case of ring and rosette supports in pneumatically stabilised membrane structures, the surface within the support element must be closed. This is most easily achieved by means of a low tensioned membrane, which at high points becomes a "dome" (Fig. 11) and at low points a "funnel".

Survey of the different types of loadbearing structure

In Table 4 on the horizontal axis the three basic forms of additional point stabilisation are defined, while on the vertical axis five of their typical arrangements are given. The combination gives 15 different alternatives, which are defined as structural types. As these alternatives are the same for positive and negative pressure systems as well as for single membrane and double membrane systems, 60 different types of pneumatically stabilised membrane structures with additional point support are defined in all.

In this survey the types are given only with their code. Thus the first place gives the number of membranes (I = single membrane, II = double membrane), the second place the kind of internal pressure (p = positive pressure, n = negative pressure), the third place the type of additional support (P = point, L = linear), the fourth place the formation of the additional loadbearing elements (ro = rosette, or loop, ri = ring, bu = bulged surface) and the fifth place the five possible arrangements of the individual supports, as shown in the first column. The first three places of the code thus define the structural system, the following places the structural type.

2

3

4

5

6

2. Greenhouse, study project. Design: Frei Otto and Conrad Roland, 1959.

3. City in the Arctic, project. Design: Atelier Warmbronn (Frei Otto and Ewald Bubner), 1970/71.

4, 5. Study projects. Design; Seminar Pneumatische Konstruktionen, Institut für Umweltplanung, Ulm, under the direction of Gernot Minke, 1971.

6. Multi-purpose hall on the English South Coast, study project. Design: Croucher, Minke and Salt, 1969.

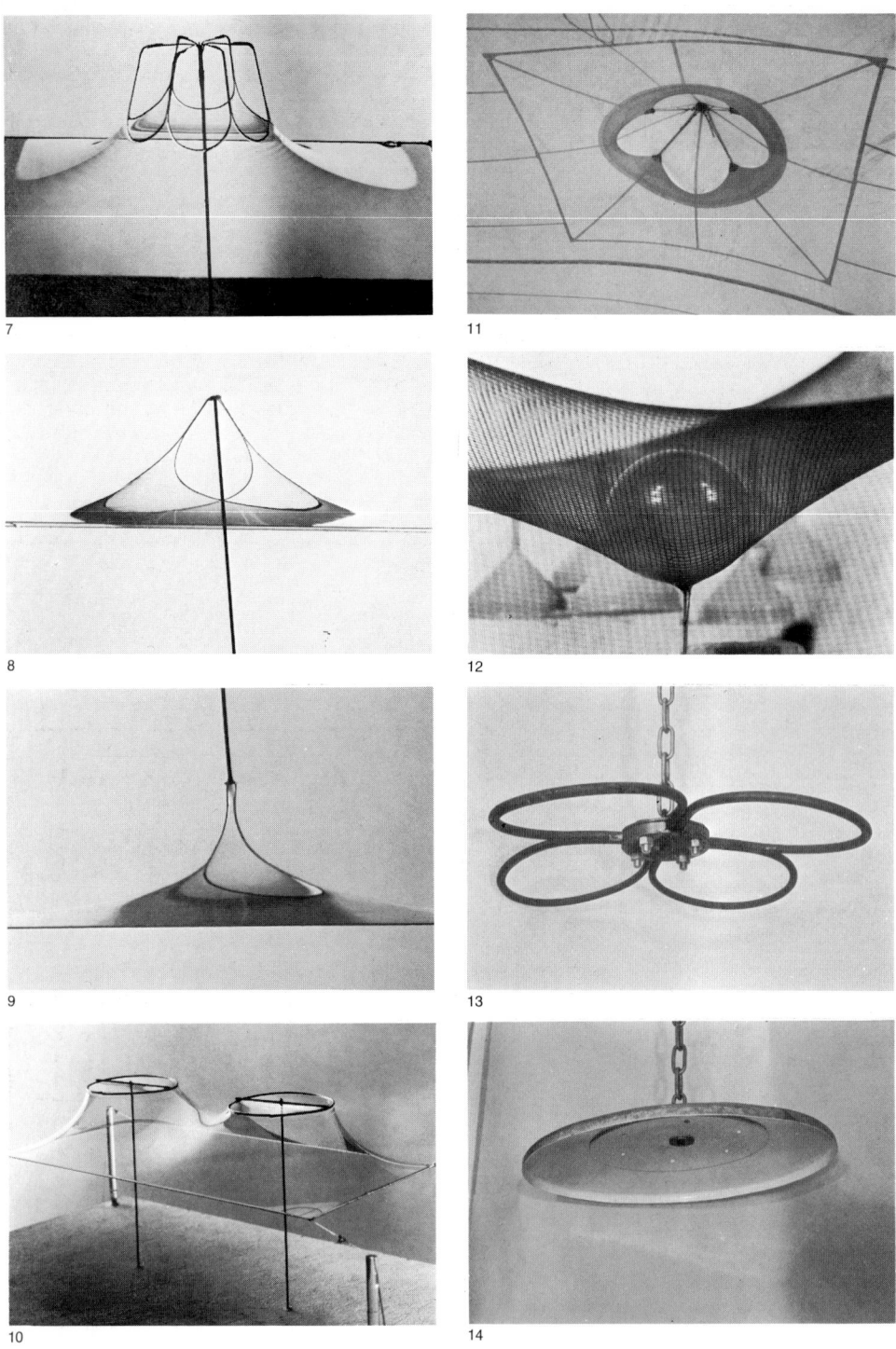

7

8

9

10

11

12

13

14

Moreover, if one includes the kind of tertiary support, then 120 different types of structure in this group of systems can be distinguished in all. If one ignores tertiary elements under bending tension which are in general not practicable, then in the case of positive pressure systems internal tertiary supports are under tensile stress, and external ones under compressive stress, while it is the other way round with negative pressure elements. For this reason the symbols "T" for tensile stress and "C" for compressive stress were introduced into the code.

As the *formation* of the tertiary support is in this case more significant for determining the characteristics of the structural type than the *arrangement* of the point secondary supports, the arrangement of the secondary elements is not considered in the diagrammatic representation of the structural types in Table 5. Thus in these diagrams only the 24 basic types of pneumatically stabilised membrane structures with additional point support are shown.

Examples of alternatives of design and usage
As Figs. 15 and 16 show, systems stabilised by negative pressure with point supports seem to differ very little in appearance from non-pneumatically stabilised membrane structures with point supports. However, if one looks at the curvatures of the surface, these systems can be clearly identified: while in the case of pneumatically stabilised membrane structures there is a predominance of curvatures with two sides in the same direction (synclastic) – curvatures with two sides in opposite directions (anticlastic) can only appear in boundary and support areas – in simple membrane structures without pneumatic stabilisation synclastic curvatures never occur, but only anticlastic ones.

The synclastic surfaces, which in the case of negative pressure systems are always curved inwards, can be dangerous because of the formation of water or snow pockets. It is therefore necessary in the formal development of these systems to take care that at every place on the membrane there is sufficient slope for water drainage.

In the case of the multi-purpose hall shown in Fig. 15, the point support was achieved by high pressure balloons. The shape of the hall was determined with a subsequently stiffened rubber membrane deformed by negative pressure; this resulted in an almost uniform tension at all points on the surface. The project shown in Fig. 16 is a double membrane system in which the inner skin forms a part of a spherical surface. For normal wind loadings a negative pressure of only 0.001 excess atmospheric pressure gives sufficient stabilisation in this case.

In the case of additional point stabilisation of positive pressure systems, the development of these secondary support elements requires a considerable amount of structural expenditure. However, structural systems of this kind offer a relatively large variety of forms. In the projects shown in Figs. 17 and 18 the low points are fixed by means of a guyed ring; the membrane continues in funnel-shaped pipes in which rain and snow can be drained away. In the project shown

7–10. Deformation of an originally flat soap membrane by means of individual supports. Investigation by the Entwicklungsstätte für den Leichtbau under the direction of Frei Otto.
11. Roof covering for the swimming pool in Boulevard Carnot, Paris. Design: Robert Taillibert and Frei Otto, 1967. Rosette shaped support of the membrane.

12. Bird cage in Hanover, study project. Design: Thomas Klumpp, 1970. Support of the membrane by a balloon.
13, 14. German Pavilion, Expo '67, Montreal. Design: Frei Otto and Rolf Gutbrod with Hermann Kendel, Hermann Kiess and Larry Medlin. Proposals for the support of the roof membrane.

in Fig. 19 the low points are produced by a "bulged surface", whereby the membrane curvatures are shaped so that rainwater can be drained away outside even in the area of the bulged surface.

2.2.4.3. Additional linear supports

On the type and development of linear supports
Additional linear support in a pneumatically stabilised membrane structure acts exactly like point support in increasing the curvature in the membrane surface and thus reducing the tension in the membrane. However, while it is relatively difficult in the case of a point support to maintain a uniform distribution of stress to the membrane, this is generally easy to achieve with linear support. In the case of additional linear support of the membrane, three different formations can be defined: "cable", "beam" and "arch".

The term *cable* should be applied as a general term for all linear flexible elements which are only under tensile stress. This also includes, for example, chains, threads, wires and cords. The cable appears in positive pressure systems as a "channel cable" and in negative pressure systems as a "ridge cable".

With a relatively small reduction in the radius of curvature it is practicable, instead of a cable, to use a plaited or woven cord of metal, natural fibres or synthetic fibres. In special cases it is even possible to obtain the same effect just by a multiplication of the membrane layers in a sort of banded area. The kind of formation is therefore primarily dependent on the extent of the desired reduction in the radius of curvature (Figs. 4, 5). However, production, transport (packaging) and erection, as well as the possibility for simple connection to tertiary guying elements, can also be decisive factors for the kind of formation of these secondary stabilising elements. An especially interesting formation of a linear support under tensile stress, for which the overall term "cable" was introduced, is the "membrane rib".

By "membrane rib" is understood a flat membrane which, just like a cable, produces an extra linear guying of the pneumatically stressed membrane. Using membrane ribs it is possible for the resultant channels to show only a very small curvature or even to run in a straight line (Figs. 20–27). In order to keep the lowest possible forces in the boundary cables of the membrane ribs, these must be relatively strongly curved. (The tension in the cable is directly proportional to its radius of curvature and the surface tension from the membrane rib.)
Membrane ribs are especially suitable for positive pressure double membrane systems (Figs. 26, 27), particularly as the enclosed volume can be greatly reduced by this.

Table 4. Kind of formation and arrangement of additional point support.

The membrane rib can also simply help to divert forces from the pneumatically stabilised membrane into a cable net lying below it and running parallel to it (Fig. 25). As the example of the American pavilion in Osaka has shown (see pp. 116, 117), a combination of roof membrane and cable net can be created, which is very simple in terms of production and erection and which also offers the advantage that the cable net lies below the roof membrane and is protected from the weather.

The term *beam* was chosen for all linear support elements mainly under bending stress.
The beam is a stiff, primarily straight element, which is only introduced for extra support in exceptional cases. Such an exception can occur where a very small span is concerned or if the beam also takes on further functions at the same time, for example that of track for a crane runway.

The term *arch* is used for all linear support elements mainly under compressive stress.
The arch is a curved stiff element that is predominantly under compressive stress. The use of arches for additional support will only come into question for small and average spans, as they are limited by the dead weight of the element. As a rule the use of the arch will only be practicable for negative pressure systems.

Survey of different types of structure
The differentiation of structural systems into structural types results from the above mentioned *formation* and *arrangement* of the additional support elements. There are seven different kinds of arrangements shown in Table 6.

formation / arrangement	rosette/loop ro	ring ri	bulged surface bu
single 1	I p P ro 1 I n P ro 1 II p P ro 1 II n P ro 1	I p P ri 1 I n P ri 1 II p P ri 1 II n P ri 1	I p P bu 1 I n P bu 1 II p P bu 1 II n P bu 1
row 2	I p P ro 2 I n P ro 2 II p P ro 2 II n P ro 2	I p P ri 2 I n P ri 2 II p P ri 2 II n P ri 2	I p P bu 2 I n P bu 2 II p P bu 2 II n P bu 2
two (and more) way 3	I p P ro 3 I n P ro 3 II p P ro 3 II n P ro 3	I p P ri 3 I n P ri 3 II p P ri 3 II n P ri 3	I p P bu 3 I n P bu 3 II p P bu 3 II n P bu 3
radial 4	I p P ro 4 I n P ro 4 II p P ro 4 II n P ro 4	I p P ri 4 I n P ri 4 II p P ri 4 II n P ri 4	I p P bu 4 I n P bu 4 II p P bu 4 II n P bu 4
irregular 5	I p P ro 5 I n P ro 5 II p P ro 5 II n P ro 5	I p P ri 5 I n P ri 5 II p P ri 5 II n P ri 5	I p P bu 5 I n P bu 5 II p P bu 5 II n P bu 5

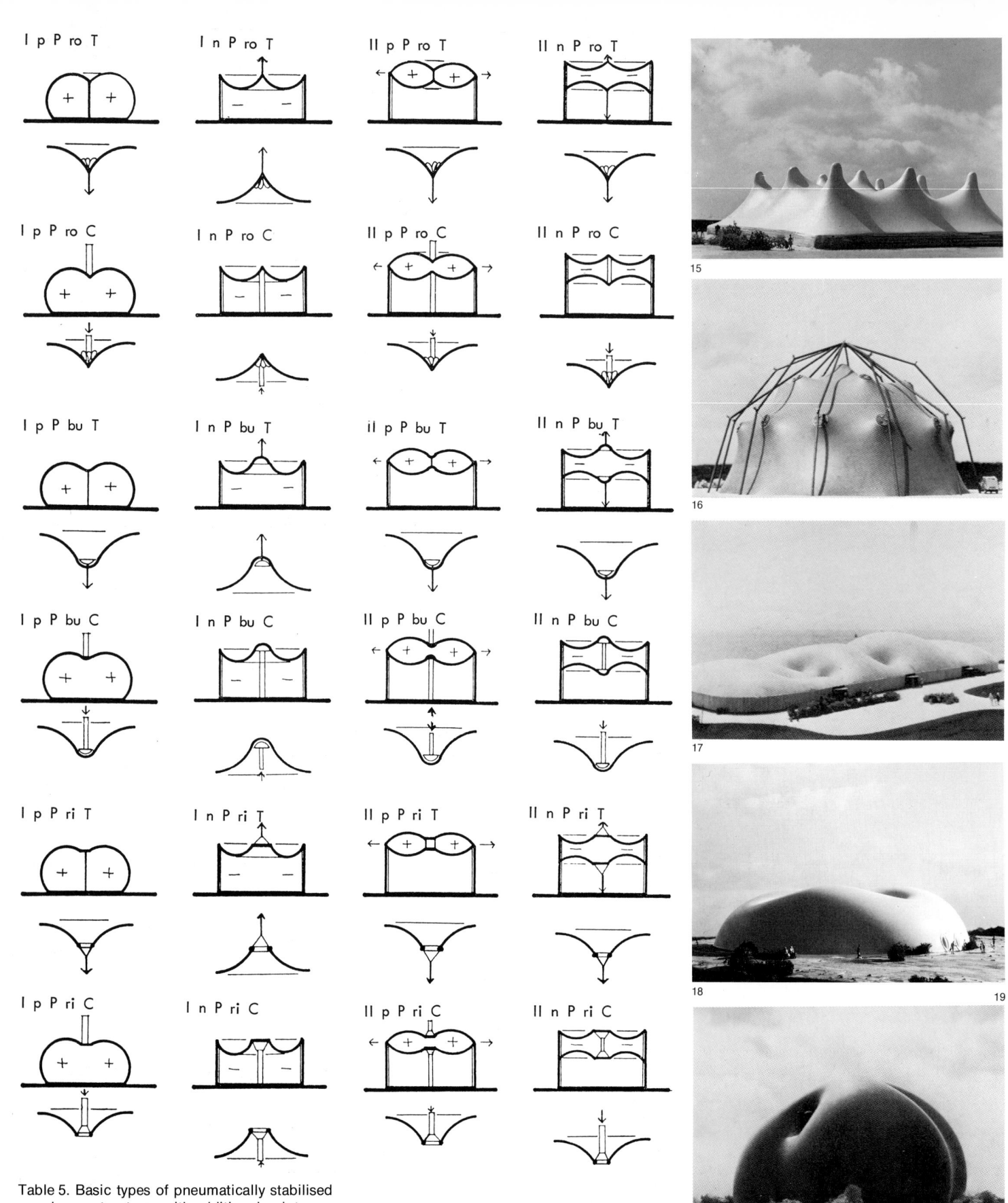

Table 5. Basic types of pneumatically stabilised membrane structures with additional point support.

In this survey on the horizontal axis the three basic forms of additional linear support (cable, beam, arch), in this case for single membrane positive pressure systems, are shown; on the vertical axis seven typical arrangements of these supports are given (single, parallel, radial, tangential, two-way, three [and more]-way, irregular). If one disregards the combinations of the three different kinds of support then there are 21 different type definitions in all. As these apply to positive and negative pressure systems as well as to single membrane and double membrane structures, a total of 84 different structural types of pneumatically stabilised membrane structures with additional linear support are defined.

In Table 7 the twelve type groups, for which the seven different possible arrangements are defined, are represented in the form of a mnemo-technical code and in the form of diagrams.

Table 8 shows the 84 defined types of single- and double-membrane structures with linear support.

The division of the 84 types described here systematically should serve as inspiration and/or working aid for analytical and synthetical subprocesses in the design of such structures. In a representation of all 84 types, it becomes clear that one series of types is evidently less practicable or structurally less efficient for application than another. Some types can even be regarded as useless solutions. As, however, an assessment of a structure is not possible without data on size, material and usage, this fact is purposely ignored in this systematic representation.

Examples for possible design and use

It can be recognised from the negative pressure systems shown in Figs. 28 and 29 that with additional linear support significantly different forms occur than with non-pneumatically stabilised membrane structures. The membrane elements of the examples shown would not be stable against wind suction and wind pressure without pneumatic stabilisation, as their corners lie on a plane. The additional linear support can, as is clear from these examples, offer an advantage over point support in that all membrane parts show the same cutting pattern.

Fig. 30 on the other hand shows a project that can be developed in a similar form as a non-pneumatically stabilised membrane structure or as a cable structure. In these two cases the areas round the crown of the arch are only curved very slightly to form a saddle so that normally additional measures against deformation caused by wind must be taken there. However, in the case of negative pressure systems there is generally sufficient curvature and thus also adequate stability against deformation. The surface of these areas is synclastically curved.

In the case of positive pressure systems channels always arise through additional linear support. In the examples shown in Figs. 31 to 37 various possible arrangements are illustrated: parallel (Fig. 31), tangential (Fig. 32) and radially arranged cables (Fig. 33) as well as regular (Figs. 34 and 35) and irregular (Figs. 36 and 37) nets.

15. Multi-purpose hall, study project. Design: Seminar Pneumatische Konstruktionen, Institut für Umweltplanung, Ulm, under the direction of Gernot Minke. Single membrane negative pressure system.

16. Demountable exhibition hall, study project. Design: Seminar Pneumatische Konstruktionen, Institut für Umweltplanung, Ulm, under the direction of Gernot Minke, 1971. Double membrane negative pressure system.

17. Multi-purpose hall on the English South Coast, study project. Design: Croucher, Minke and Salt, 1969. Single membrane positive pressure system.

18. Exhibition hall in Delft, study project. Design: Gernot Minke with students of the Technische Hogeschool Delft, 1971. Single membrane positive pressure system.

19. Exhibition hall, study project. Design: Seminar Pneumatische Konstruktionen, Institut für Umweltplanung, Ulm, under the direction of Gernot Minke, 1971. Single membrane positive pressure system.

20–27. Linear support by means of membrane ribs.

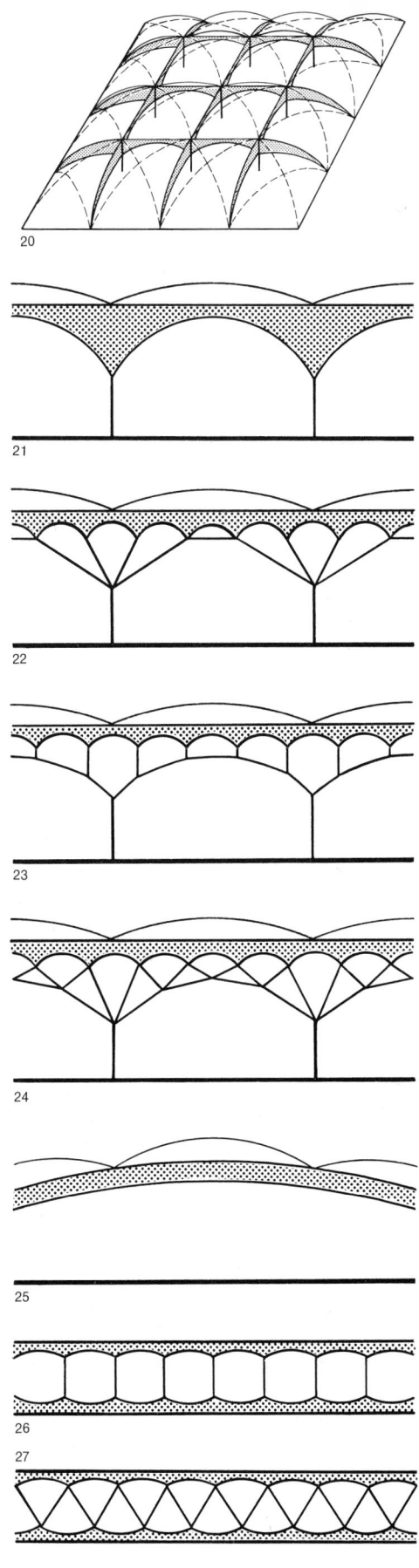

20
21
22
23
24
25
26
27

stress / arrangement	tension (cable) ca	bending (truss) tr	compression (arch) ar	combinations
single (1)	l p L ca 1	l p L tr 1	l p L ar 1	ca 1 + tr 1 ca 1 + ar 1 tr 1 + ar 1 ca 1 + tr 1 + ar 1
parallel (2)	l p L ca 2	l p L tr 2	l p L ar 2	ca 2 + tr 2 ca 2 + ar 2 tr 2 + ar 2 ca 2 + tr 2 + ar 2
radial (3)	l p L ca 3	l p L tr 3	l p L ar 3	ca 3 + tr 3 ca 3 + ar 3 tr 3 + ar 3 ca 3 + tr 3 + ar 3
tangential (4)	l p L ca 4	l p L tr 4	l p L ar 4	ca 4 + tr 4 ca 4 + ar 4 tr 4 + ar 4 ca 4 + tr 4 + tr 4
two-way (5)	l p L ca 5	l p L tr 5	l p L ar 5	ca 5 + tr 5 ca 5 + ar 5 tr 5 + ar 5 ca 5 + tr 5 + ar 5
3-(and more) way (6)	l p L ca 6	l p L tr 6	l p L ar 6	ca 6 + tr 6 ca 6 + ar 6 tr 6 + ar 6 ca 6 + tr 6 + ar 6
irregular (7)	l p L ca 7	l p L tr 7	l p L ar 7	ca 7 + tr 7 ca 7 + ar 7 tr 7 + ar 7 ca 7 + tr 7 + ar 7

Table 6. Kind of acting force and arrangement of additional linear support.

28. Pendopneu, study project for an exhibition hall. Design: Gernot Minke with students of the Technische Hogeschool Delft, 1971. Single membrane negative pressure system.

29. Hanover fair stall, study project. Design: Gernot Minke with students of the Technische Universität (TU) Hannover, 1970. Double membrane negative pressure system.

30. Sports hall, study project. Design: Seminar Pneumatische Konstruktionen, Institut für Umweltplanung, Ulm, under the direction of Gernot Minke, 1972. Single membrane negative pressure system.

31. Exhibition hall, study project. Design: Gernot Minke with students of the Technische Hogeschool Delft, 1971. Single membrane positive pressure system.

32. Exhibition hall on the English South Coast, study project. Design: Minke, Stevens and Warne, 1969. Single membrane positive pressure system.

33. Roof covering of the Alpamare swimming pool, Bad Tölz. Design: Gernot Minke, 1971. Single membrane positive pressure system.

34. Factory in Delft, study project. Design: Gernot Minke with students of the Technische Hogeschool Delft, 1971. Single membrane positive pressure system.

35. Demountable roof structure for a touring exhibition, study project. Design: Hirst, Kamel and Minke, 1969. Double membrane positive pressure system.

36, 37. Exhibition hall, study project. Design: Seminar Pneumatische Konstruktionen, Institut für Umweltplanung, Ulm, under the direction of Gernot Minke, 1971. Single membrane positive pressure system.

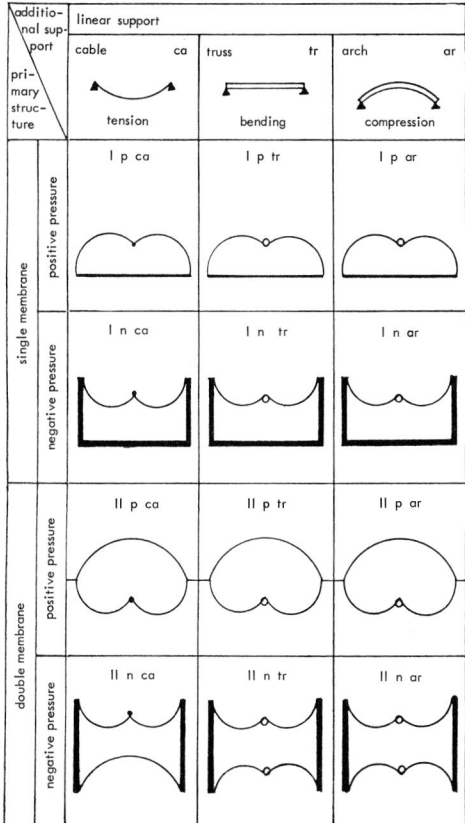

additional support	linear support		
primary structure	cable ca ⌣ tension	truss tr ▭ bending	arch ar ⌢ compression
single membrane — positive pressure	I p ca	I p tr	I p ar
single membrane — negative pressure	I n ca	I n tr	I n ar
double membrane — positive pressure	II p ca	II p tr	II p ar
double membrane — negative pressure	II n ca	II n tr	II n ar

Table 7. Basic types of pneumatically stabilised membrane structures with additional linear support.

28

29

30

31

32

33

34

35

36

37

25

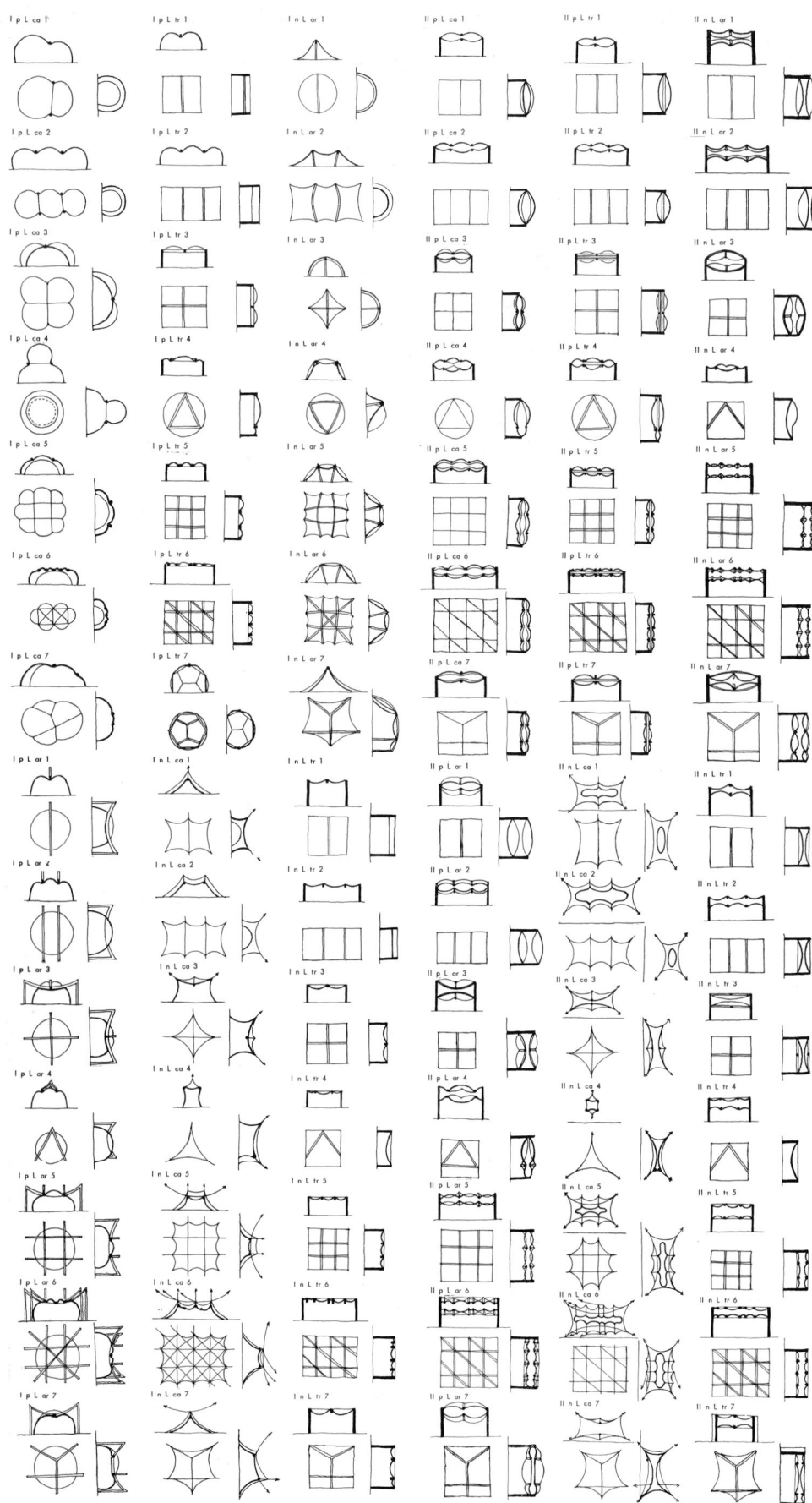

Table 8. The 84 possible types of single- and double-membrane structures with additional linear support.

Table 9. Combinations of negative and positive ▷ pressure systems.

In the case of the project shown in Fig. 32 the central zone was formed by a stronger membrane. As its radius of curvature is three times as large as the radius of curvature of the edge zones, a membrane tension three times as large occurs in the central zone.

Due to the channel cables in the roof of the Alpamare swimming pool in Bad Tölz (Fig. 33) a favourable distribution of internal space could be created and the internal pressure required could be reduced to 0.002 excess atmospheric pressure. Because of the low internal pressure the glass walls needed no additional reinforcement.

2.2.5. Combinations, additions

Although pneumatically stabilised membrane structures have so far mostly been realised as individual structures, a practical economic application for a number of systems seems to be made possible through combination with other pneumatically stabilised systems.

The term *combination* is used if two or more structures of the same or different kinds are directly joined together to form a new structure. If two or more structures are joined together with the help of connection points then this is called an *addition*. In this case each structure is maintained as an autonomous support element.

By means of a logical combination it can be possible to cancel out the disadvantages of individual systems. Thus negative pressure systems show a series of disadvantages (formation of water pockets, aerodynamically unfavourable form), which in certain circumstances can be avoided by combination with positive pressure systems.

The diagrams in Table 9 show examples of the 64 possible simple combinations of the eight negative pressure systems in Table 3; in some cases a vertical as well as a horizontal combination is illustrated.

This representation by means of examples of some of the possible combinations should show how new kinds, forms or types of structures can be found by means of synthesising processes with the help of classification. It should also demonstrate that the spectrum of structural and design possibilities is still by no means fully comprehended and that as yet only a very small number of the alternatives offered to us in the field of loadbearing structures have been realised. Whether the synthetically determined alternatives are practicable for specific uses can only be put to the test when there are constructive requirements related to a specific project.

In the case of additions the formation of the secondary support elements at the edges and the type of connection are generally decisive for the economy of the whole structure. In the possible applications of a system of prefabricated units made of four double membrane standard elements shown in Figs. 38 and 40, the elements are self-supporting by means of an internal compression stressed structure and connected by means of movable connection units so that a secondary support structure can be omitted. In

the test structure shown in Fig. 41 the membrane is stressed by means of flexible, compressed, thermal insulating material.

2.3. High pressure systems

2.3.1. Type and formation

High pressure systems under pneumatically stabilised membrane structures consist of tube-like elements and are therefore termed "tube structures" (in special cases these can also be spherical as Figs. 12 and 44 show). The tube elements show a very strong curvature in one direction, but a very small or no curvature in the other direction, and are able to transfer transverse forces in the direction of low curvature. They can, for example, take on the support function of a beam, an arch, a grid or a lattice shell (Table 10) and thus belong to the group of frame structures. As the membrane of this structure can only take up tensile forces, the compressive forces arising in load application must be compensated for by a corresponding initial stressing of the membrane. The differential pressure required for this generally lies between 2,000 and 70,000 mm of water pressure and this is 100 to 1000 times as large as in low pressure systems. Thus high pressure systems differ from low pressure systems in the differential pressure, the structural function and the type of support action.

The high pressure systems, in comparison with other structures transferring transverse forces, have a relatively low structural efficiency (Bibl. 105, p. 50–53; Bibl. 107) and are thus only used when requirements such as easy erection and dismantling, low weight and low transport volume are decisive factors for the choice of structural system. This is the case, for example, in floating structures for short term use (Fig. 42).

A significant property of tube structures is that their form within certain limits can adapt to stresses through external forces. Thus they are particularly suitable as support elements for cable net and membrane structures which are subject to strong changes in form and tension (Figs. 12, 42–44): if the tension in the tensile stressed skin increases then the support surface increases.

2.3.2. Survey of the various structural support systems

As high pressure systems have different structural functions and characteristics from low pressure systems, the differentiating features for low pressure systems given in Table 2 are not applicable here.

In order to distinguish between high pressure systems the following features should be taken into account:
1. The pattern of the elements,
2. The kind of connection of the elements to each other.

In the case of the first feature the three characteristics "straight", "buckled" and "arched"

38

39

40

41

38–40. Buildings for agricultural use, study projects. Design: Gernot Minke with students of the Technische Hogeschool Delft, 1971.

41. Test structure for student working areas. Design and execution: Seminar Leichtbaukonstruktionen, SHfbK Hamburg, under the direction of Gernot Minke, 1973.

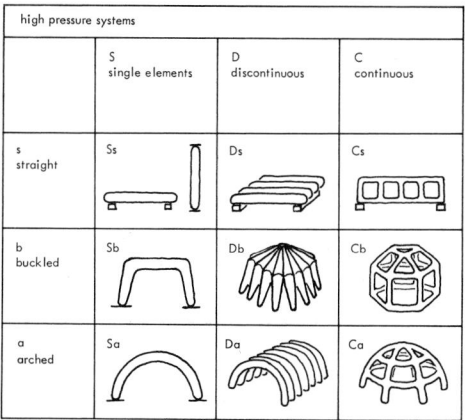

high pressure systems			
	S single elements	D discontinuous	C continuous
s straight	Ss	Ds	Cs
b buckled	Sb	Db	Cb
a arched	Sa	Da	Ca

Table 10. Classification of high pressure systems.

42. Information centre Kenniskapsule, Delft. Design and execution: Gernot Minke and Sean Wellesley-Miller with students of the Technische Hogeschool Delft, 1971.
43, 44. Test structures in Delft. Design and execution: Gernot Minke with students of the Technische Hogeschool Delft and Stevin Labor, 1971.

42

43

44

should be considered; with the second feature the three characteristics are "single elements", "discontinuous" and "continuous".
The nine systems thereby defined are put together in the form of a matrix in Table 10. They are provided with a mnemo-technically derived code and presented as system diagrams.
The terms "straight", "buckled" and "arched" relate to the axis of the element. In the case of system "Ss", with the loadbearing effect of a beam, for example, a small curvature can occur in the longitudinal direction for structural or manufacturing reasons, or the beam can be put together from two truncated cones.
Through addition and combination (see Section 2.2.5) surface structures, which represent a discontinuous or continuous system for transferring forces, can be developed from individual elements.

2.3.3. The different types of loadbearing structures

The nine systems for high pressure structures shown in Table 10 can be divided according to their structural performance into *structural types* and these can if necessary be divided again according to the type of surface curvature and the arrangement of their elements into *structural forms,* as is usual in the case of frame structures.
Thus with system "Ss", for example, the "beam" type of structure can be distinguished from the "column" type and with system "Ds" the "plate" type can be distinguished from the "disc" type. In the case of system "Cs" the "Vierendeel truss" and single layer and multi-layer grid types occur; these types of systems can have one, two or three main directions of expansion and thus a one, two or three dimensional loadbearing effect.
The lattice shell system ("Ca") can be split up according to the kind of surface curvature into unilaterally, synclastically or anticlastically curved structural types, or into rotation, translation and ruled surface types, and further, according to the arrangement of elements for example, into net, radial, lamella and geodesic dome types.
The differentiation of kinds of structure can correspond to the classification of low pressure systems stated in Table 2.
As pneumatic high pressure tube structures have the same systems, types and forms of structure as frame structures and as their use is appropriate in only a very few loadbearing buildings, a detailed description of their structural types is not given here.

Notes

[1] The features were limited for reasons of clarity to low pressure systems, because high pressure systems (as is explained in more detail in Section 2.3.1) take on other structural functions and thus have extensively different characteristics.
[2] The suggested classification was tested in relation to its analytical and synthetical function in the comprehension of loadbearing structures so far built, and in the design of loadbearing structures not yet realised, in the following seminars led by the author: T. U. Hanover, January 1970; T. U. Delft, February and September 1971; I. U. P. Ulm, December 1971. The following systematic diagrams came from the reports on these seminars or are based on the studies carried out there (Bibl. 83; Bibl. 84; Bibl. 123).

3. Examples of pneumatic structures from nature and historical technology

3.1. Examples from nature

Some 62% of all animals can fly. A majority of these use as sail surfaces skins which, when they are not pneumatically stressed, lie folded together on the body or hang down limply. These are all open membranes.

The wings of the blue dragonfly *(Aeschna cyanea)* consist of fine membered ribs and between these are spanned membranes which inflate quite weakly under stress (Fig. 1). The strenght of the skin amounts to approximately 3 u and has an equivalent weight of 3.7 g/m². (Bibl. 74, p. 78, 79, 84.)

The flying frog of the Sunda Islands *(Rhacophorus reinwardtii)* has, between his greatly lengthened toes and fingers, skins which give him a controlled fluttering flight (Fig. 2; Bibl. 132, p.154).

Amongst mammals capable of flight the best known are bats (Fig. 3). There are about 500 different types; the largest have a wing span of 90 cm. (Bibl. 132.)

Organic cell structures are not pneumatic forms. If the cell walls consist of solid material then they are self-supporting compartments or shells – similar to those in industrially manufactured foams made of latticed polyurethanes, polystyrenes or glass. If the cell walls consist of flexible membranes, then they are stabilised in their position and form not by gases but by fluids.

Soap bubbles or bubble agglomerations in the sea foam are, however, genuine pneumatic forms with closed membranes.

The law of different internal pressure in liquid bubbles of different sizes can be seen illustrated in the photograph of the broken egg (Fig. 5). One sees how the inner lamellae curve each to the larger bubble.

An interesting experiment was carried out by the (British) Royal Aircraft Establishment in shooting a bullet through a soap bubble (Figs. 6–9). In Fig. 6 the shot flies towards the soap bubble. In Fig. 7 it is in the centre, and in Fig. 8 it comes out again on the opposite side. A fraction of a second later the bubble breaks (Fig. 9), uniformly and symmetrically, a proof of the equal tension in its surface.

A good example of a pneumatic form with a closed membrane is also offered by the natterjack toad *(Bufo calamita)* with its deafening rasping on warm summer nights and its impressive inflatable sack in its throat (Fig. 4).

3.2. Examples from historical technology

The fact that an open, limp membrane surface is stressed under wind loading, thereby changing its form, and that defined forces occur at its edge is one of the early technical experiences of mankind.

The appearance of a sailing ship is very well known, so only one example is shown here (Fig. 10).

The same principle was used in land vehicles by the lesser known sail carts, which were used primarily on flat land – above all in coastal areas. About 4,000 years ago Amenemhet III drove into

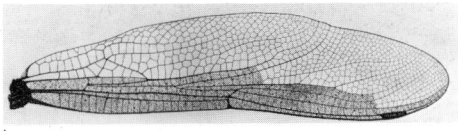

1

2

3

4

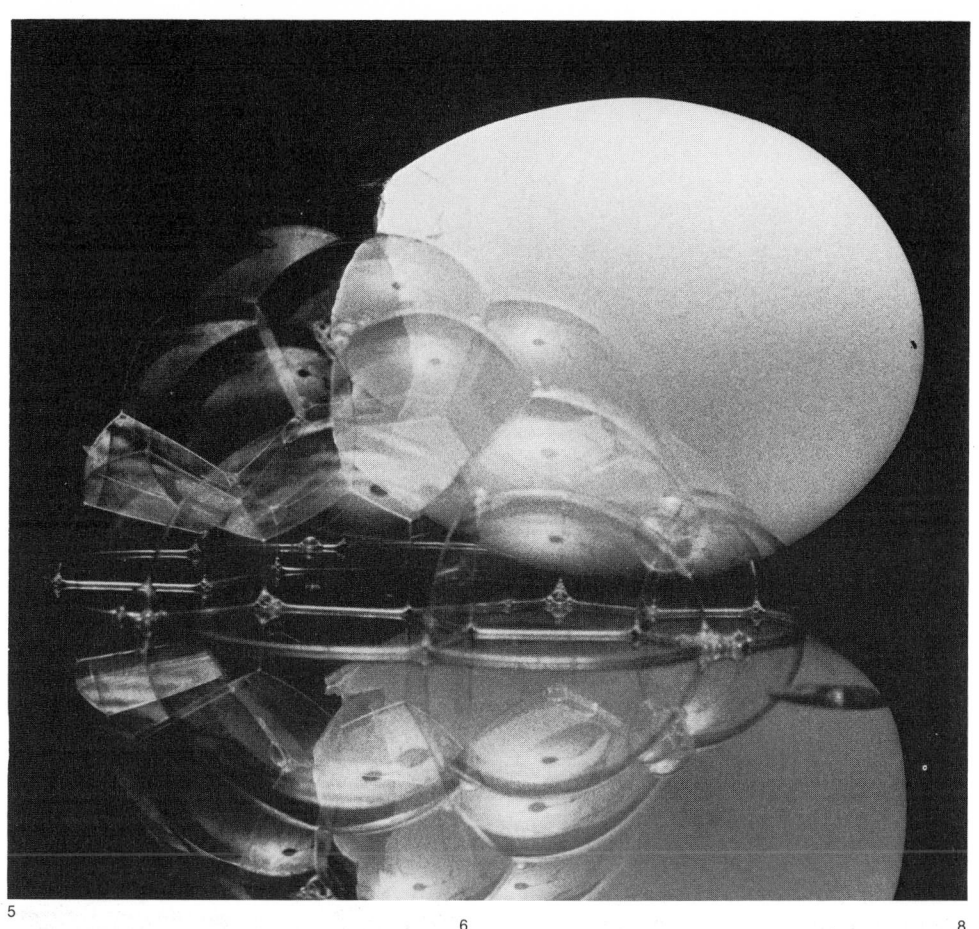

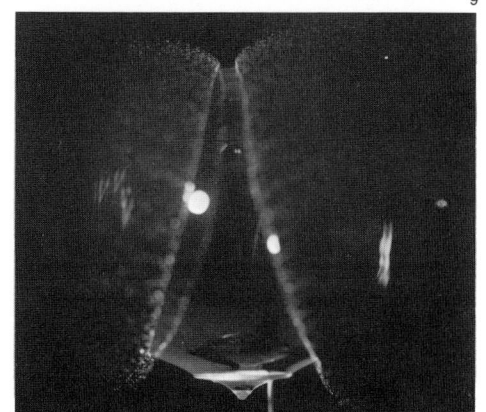

1. Wing of the blue dragonfly (Aeschna cyanea).
2. Flying frog of the Sunda Islands (Rhacophorus reinwardtii).
3. Bats.
4. Natterjack toad (Bufo calamita).
5. Broken egg.
6–9. Different phases of the destruction of a soap bubble damaged by a bullet.
10. Ship of the "invincible Armada", 1588.
11. Vehicle with sails, 1599.
12. "Partenza di Pulcinella per la luna".

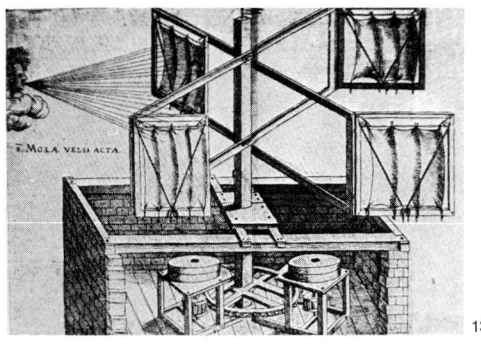

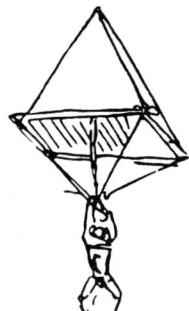

13

14 15

13. Wind wheel with hinged blades for driving two mills, by Fausto Veranzio, circa 1595.

14. Sketch of a parachute by Leonardo da Vinci, circa 1500.

15. Parachute by Fausto Veranzio, circa 1595.

16, 17. Illustrations for the novel *La Découverte Australe,* 1781.

18. Dacians with field kites on Trajan's Column in Rome.

19. Warm air kites from an armoury book in Frankfurt on Main, circa 1490.

18

19

the desert "with shafts and sails". In China the sail was also known as an aid to the movement of carts. The sail cart built in 1599 by Simon Stevin for Moritz, Prince of Orange, shows clearly in the structure of the cart body the relationship to sailing ships (Fig. 11). The vehicle held 28 men and ran behind a flying sheet at a speed of 7 miles per hour. The same prince is also said to have "gone out for recreation" to the Dutch shores after the Siege of Nieuport (1600) with another vehicle of this kind. (Bibl. 58, p. 1270 ff.)

A "space ship", a mixture of sailing boat and sail cart, is shown in the "partenza di Pulcinella per la luna" (Fig. 12), the curious vision of a moon journey which generously neglects several physical laws.

In 1595 the Italian, Fausto Veranzio, designed a wind wheel with hinged blades for driving two mills (Fig. 13). The surfaces, which serve to catch the wind, consist of membranes supported linearly by cables.

Long before their introduction into air travel, parachutes can be found in drawings and engravings. A sketch by Leonardo da Vinci, originating about 1500, shows a "tent roof made of compressed canvas", whose deformation

through wind has, however, been ignored in the illustration (Fig. 14) – in contrast to the parachute from the machine book by Veranzio, for which the latter proposes a structure related to his millwheel (again with linear cable supports) (Fig. 15).

On a copperplate engraving for the novel *Ariane* by Desmarets (Paris 1639) the prisoner flees by means of a parachute jump using a sheet. A somewhat complicated apparatus is used in the abduction of Christine by Victorin in an illustration for the novel *La Découverte Australe* by Restif de la Bretonne (Leipzig 1781). Atop a fantastic flying apparatus a parachute is fixed which is closed in ascent and open in descent (Figs. 16, 17). It is interesting that the illustrator has shown more perception of the form of a pneumatically inflated membrane in the representation of the flight membrane than in that of the parachute, which looks like an umbrella unstressed by air pressure. (Bibl. 58, p. 279–282.)

Sail and box kites are still favourites as toys with children today. The latter were first discovered by. L. Hargrave in 1896.

Sail kites have been known for centuries. Their basic form is an inverted isosceles triangle on top of which is a circle segment or an obtuse

16

17

angled triangle. The membrane is made of material, parchment or paper.

In the times of the Dacians, Scythians, Parthians and Persians – the Romans also since Constantine – mobile kites were used as banners. In the early Middle Ages Christian armies carried them; they were also known in India at that time. (Bibl. 23; Bibl. 58, p. 198 ff.)

On the frieze of Trajan's Column in Rome Dacians with their field kites are depicted (Fig. 18). The kites consisted of open metal caps and membrane tubes fixed to them; these were inflated by air to look like bodies of animals. (Bibl. 121.) Sometimes burning torches were put into the mouths of the kites at night. They gave the impression of beasts spewing forth fire and smoke, moving through the air under their own power. Even in the year 1241a "Christian army" is said to have been put to flight by such a banner carried by the Mongols.

Complete animal bodies with clawed feet and stabilising wings on either side were constructed in the 15th century. In an armoury book from Frankfurt on Main dated 1490 there is a dragon kite which is inflated by hot air, moves forwards by means of a rocket in the hind part of the body and is held only by a long string (Fig. 19).

20. Warm air balloon by the brothers Joseph Michel and Jacques Etienne Montgolfier, 1783.
21. Hydrogen balloon by Jacques Alexandre César Charles, 1783.
22. Balloon with sail by Terzuolo, 1855.
23. Balloon by Giffard, 1852.
24. Airship by August von Parseval, 1906.
25. Stratosphere balloon by Auguste Piccard, 1931.

24

20

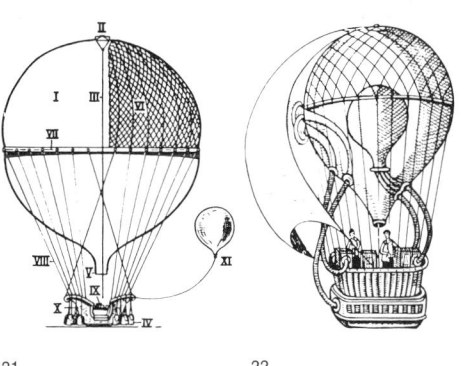

21   22

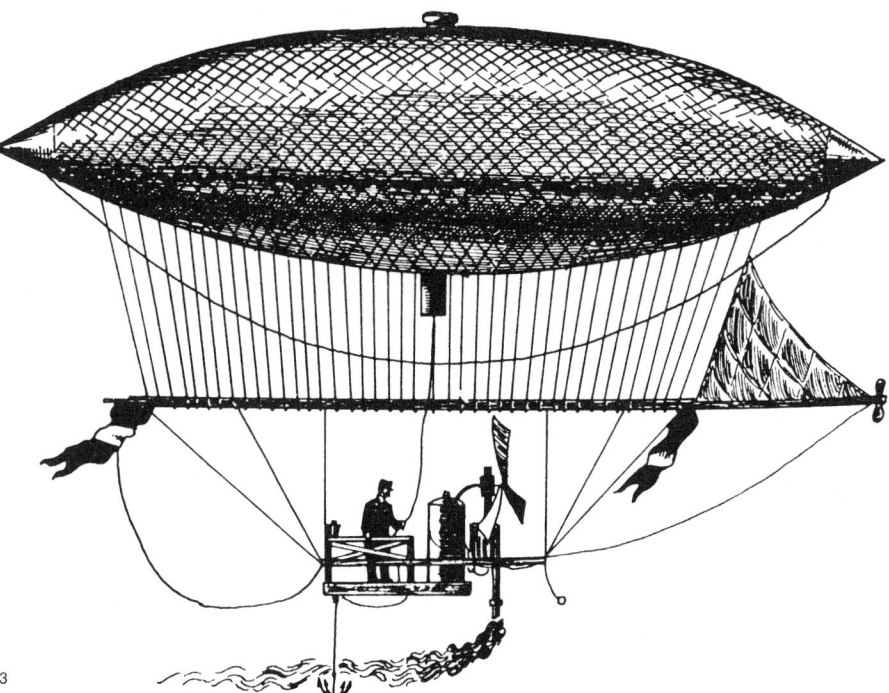

23

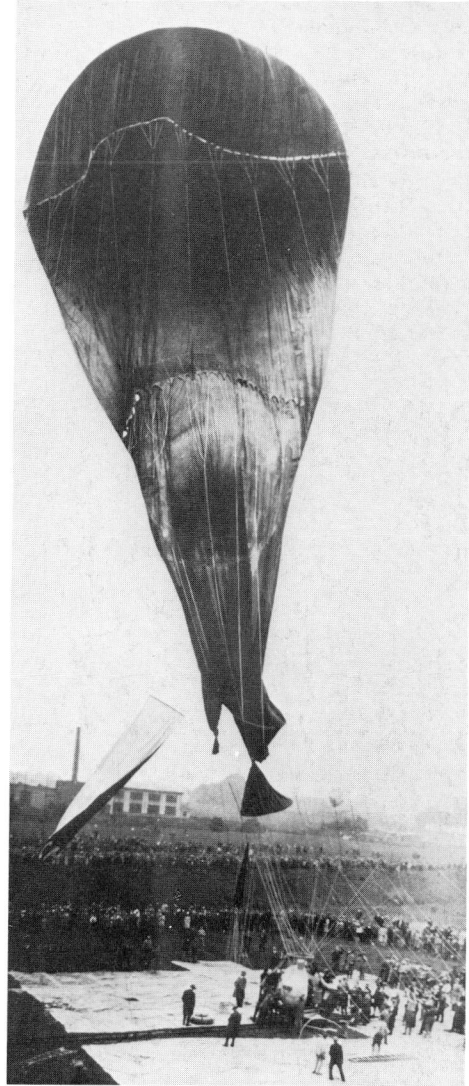

25

119,339

PATENT SPECIFICATION

Application Date, Nov. 20, 1917. No. 17,063/17.
Complete Left, June 20, 1918.
Complete Accepted, Oct. 3, 1918.

PROVISIONAL SPECIFICATION.

An Improved Construction of Tent for Field Hospitals, Depots, and like purposes.

I, FREDERICK WILLIAM LANCHESTER, of 41, Bedford Square, London, W.C. 1, Engineer, do hereby declare the nature of this invention to be as follows:—

The present invention relates to an improved construction of tent for field hospitals, depots, and like purposes.

The present invention has for its object to provide a means of constructing and erecting a tent of large size without the use of poles or supports of any kind. The present invention consists in brief in a construction of tent in which balloon fabric or other material of low air permeability is employed and maintained in the erected state by air pressure and in which ingress and egress is provided for by one or more air locks.

...

COMPLETE SPECIFICATION.

An Improved Construction of Tent for Field Hospitals, Depots, and like purposes.

I, FREDERICK WILLIAM LANCHESTER, of 41, Bedford Square, London, W.C. 1, Engineer, do hereby declare the nature of this invention and in what manner the same is to be performed, to be particularly described and ascertained in and by the following statement:—

...

Having now particularly described and ascertained the nature of my said invention and in what manner the same is to be performed, I declare that what I claim is:—

1. A tent or roof structure supported by internal air pressure with means of access comprising an air lock.

2. A tent or roof structure in accordance with Claim 1, in which a rope network is picketed or anchored to the ground and contains a canvas or fabric envelope of low air permeability provided with an inwardly turned sealing flap and adapted to be supplied by air pressure supplied by a fan or blower or by a wind cowl, or by both means in combination.

3. A tent or roof structure in accordance with Claims 1 and 2, in which the air locks take the form of van bodies adapted for the stowage of the network envelope and anchorage material for the purpose of transport.

4. A tent or roof structure in accordance with Claims 1 and 2 substantially as and for the purposes hereinbefore described.

Dated this the 19th day of June, 1918.

F. W. LANCHESTER.

Redhill: Printed for His Majesty's Stationery Office, by Love & Malcomson, Ltd.—1918.

26. Patent specification *An Unproved Construction of Tent for Field Hospitals, Depots and like purposes* for F. W. Lanchester, 1918.

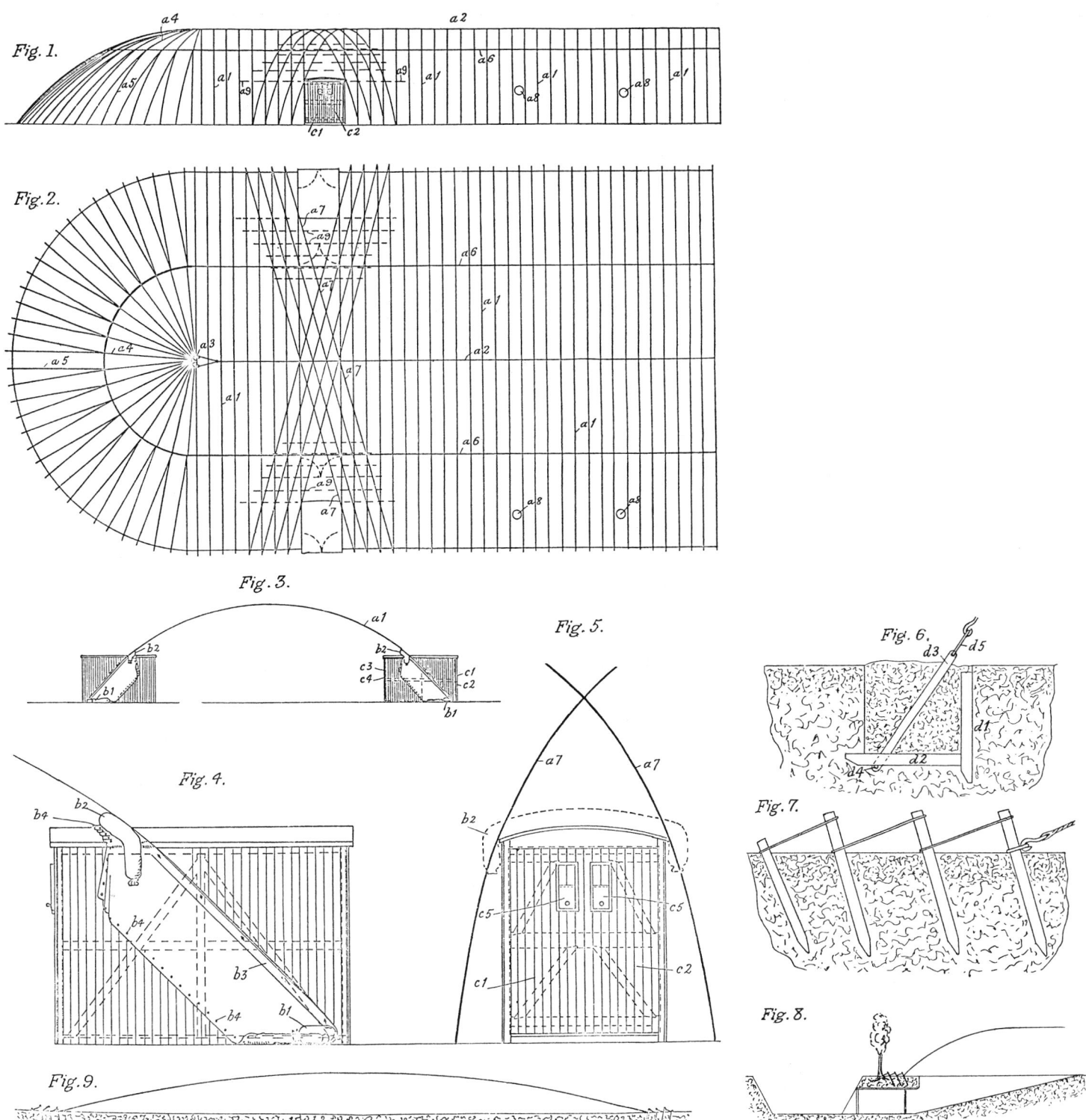

Fig. 1.

Fig. 2.

Fig. 3.

Fig. 4.

Fig. 5.

Fig. 6.

Fig. 7.

Fig. 8.

Fig. 9.

Here we are already dealing with closed membranes which can be viewed as forerunners of the later hot air balloons. (Bibl. 58, p. 653, 658.)

In 1826 G. Pocock built in England a vehicle that was pulled by large kites – the sail carts of 1600 had probably been forgotten.

Kites were first used for scientific purposes by Wildon in 1749 for the measurement of atmospheric temperature and in 1752 by Franklin for the proof of thunderstorm electricity. Kites were very important as research apparatus in meteorology. (Bibl. 23; Bibl. 58, p. 651 ff.)

Free and captive balloons are generally regarded as forerunners of present day pneumatic structures. As we have seen, pneumatic forms have existed for very much longer, yet balloons have remained by far the most important prototypes to date.

Historians today are still not agreed on the intellectual authorship of these hollow membrane spheres. To many the warm air balloon made out of a canvas envelope lined with paper by the brothers Joseph Michel and Jacques Etienne Montgolfier still counts as the first structure of its kind (Fig. 20). The balloon went up on 5th June 1783, at first unoccupied, months later with animals, then with a man aboard.

However, the brothers cannot be considered as the discoverers of the balloon. In his Utopian novel Les Etats et Empires du Soleil Cyrano de Bergerac (1619–1655) describes a smoke-filled balloon that, with the assistance of a sail, carries a cabin in space. A copperplate engraving of this fantasy appeared in 1657. Round about the same time the Jesuit, L. Laurus, also developed the idea of a warm air balloon. On 8th August 1709 in Lisbon the Brazilian cleric, Bartholomew Lourence de Gusmao, finally ascended 200 feet into the air in a hot air balloon. The following extract is taken from a Russian manuscript: "In the year 1731 in Asen an official of the province of Nerechtes Krjakutnoi made a large ball and blew it up with horrible stinking smoke. To it he fastened a loop and sat himself in it, and the evil spirit raised him higher than the birch trees and then threw him against the bell tower, but he grasped the rope used to ring the bells and thus lived". (Bibl. 58, p. 64 ff.)

Much more significant for the future of balloon travel, however, was the discovery by the French physicist, Jacques Alexandre César Charles, who only some two months after the ascent of the "Montgolfière" sent up the first hydrogen balloon (Fig. 21). In its basic structure it remained the prototype for all gas balloons until the present day. "The envelope took the form of a sphere and was tear resistant and airtight. It was made of silk, which was strengthened inside with the help of liquid rubber. On top there was a gas valve which could be opened from the basket with a release cord. In this way one could release gas and prevent an undesired ascent, just as an undesired descent could be prevented by throwing out ballast in the form of sandbags. The inflation tube remained open. Because air pressure forced the gas into the balloon, the gas could not escape on the ground despite the openings. But in the air the open gas cylinder manifold prevented the balloon bursting, for because of the low air pressure the surplus gas, which then expanded, could escape automatically. Naturally the loss of momentum, which would have led to a crash if near the ground, had to be compensated for by throwing out ballast". (Bibl. 99.)

It is also significant that the "Charlière" basket was already fastened to a cable net that fed the forces of its deadweight uniformly into the balloon membrane.

The stratosphere balloon of Auguste Piccard, which in 1931 starting from Augsburg set a new height record (exactly 16,000 m), was the largest balloon in the world at the time. By means of the gradual expansion of the hydrogen it only obtained its spherical shape at the highest point (Fig. 25).

The efforts following the discovery of the balloon as a means of transport were primarily concerned with propulsion and steering. By using the muscles of man working with air oars, of horses working with air blades, and even of tamed eagles, attempts were made to be independent of the wind.

Again and again projects appeared using sails dependent on the wind (Fig. 22), the means of propulsion.

With the arrival of more suitable means of propulsion, in particular internal combustion engines, balloons were given a longer shape which was aerodynamically better (Fig. 23; Bibl. 23; Bibl. 99).

The Parseval airship (Fig. 24) is regarded as the first really practicable dirigible airship in the world, a successor to the famous Parseval-Sigsfeld kite balloon.

The bodies of later airships, in particular the famous Zeppelins, are not pneumatics. They have metal frames which are covered on the outside by a skin. The gas which produces the uplift is in closed chambers inside.

The idea of also using the pneumatic principle in buildings originates from the English engineer F. W. Lanchester; in 1918 he even obtained a patent on it (Fig. 26). The largest of the pneumatics that he designed in the following period was to have a diameter of 650 m! Regrettably he did not live to see the breakthrough of his idea.

4. Examples of pneumatic structures from the technology of today

The following section gives a survey of pneumatic structures of various kinds as they are designed or produced in the field of technology of today. The term "technology" is used in its original sense – as an antithesis to nature rather than to art.

It has already been stated that not all the pneumatic structures shown in the classification system have yet been realised. In the following details, therefore, only those differentiating features have been chosen for which examples can be found.

These are arranged according to:

type of structure – structure open (So), structure closed (Sc);

type of membrane – membrane open (Mo), membrane closed (Mc);

proportion – one dominant dimension (1 di), two dominant dimensions (2 di), three dimensions of similar size (3 di);

additional support – no additional support (0), additional point support (P), additional linear support (L).

Classification according to usage has purposely been omitted, because usage can change and is frequently interchangeable. It is also the result of temporary interpretation.

1

2

1, 2. Parachutes

Today, more than ever before, parachutes are used as a means of transport. Increasing air traffic and their low transport volume make them irreplaceable for both civil and military purposes. The illustrations show how the weight of the live load is transferred to the membrane by means of a great number of cables. These cables at the same time bring about a reduction of the total membrane curvature and correspondingly a reduction of the membrane stresses.

3. Bat glider
Development and manufacture: Ryan

There are problems involved in the use of a parachute in space travel as it falls almost vertically, cannot execute any controlled flight movements and cannot cover any extensive flight path over the earth in a desired direction. Its advantage, however, is that it can be folded up very compactly.

The bat gliders, which were developed by order of NASA, combine the advantages of the parachute with those of the glider plane; they can be folded up and are also manoeuvrable.
(Bibl. 74, p. 85.)

3

4, 5. Heavy Lift Balloon, Model 530K
Design and manufacture: RAVEN Industries, Inc.

This recently developed balloon is intended for heavyweight transport over impassable terrain. In its lifting characteristics it is similar to the helicopter, but considerably cheaper to produce and maintain. It only has to be grounded in wind speeds of 50 km/h and more. The photograph shows that the balloon on the ground has a strong similarity to a pneumatic structure.

6. Mylar – Sphere
Design and manufacture: RAVEN Industries, Inc.

Such a balloon made of approximately 0.1 mm thick foil can carry scientific apparatus over 50,000 m high into the stratosphere. In August 1960 a pneumatically stressed sphere of this kind, whose surface was covered with a vapour deposit of aluminium, was sent as "Echo 1" into an earth orbit. Further reflector bodies followed it. The diameter of Echo I, which was inflated in outer space, was 30 m, yet its packing diameter in the nose of the rocket was less than 70 cm.

5

4

6

7, 8. Information pavilion at Expo '70, Osaka
Design: Taiyo Kogyo Co., Ltd.
Manufacture: as above, 1970

The pavilions invite comparison with the Berlin SPD pavilion (Fig. 9). The structures are of similar size; they are each anchored to a steel ring in the bottom third of the structure (see p. 155) and their intended usage is also similar. In the case of the SPD pavilion the structure is certainly of the single membrane type, but how should the Japanese example be classified? If one regards it as part of the roof, then it is a double membrane structure. This is a border-line case which makes clear the dependence of these differentiating features on the use.

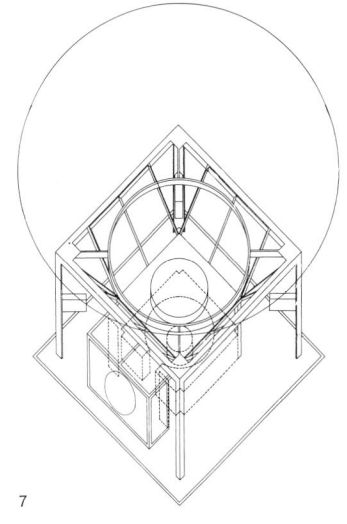

7

8

9

9. SPD publicity pavilion in Berlin
Design: L. Stromeyer & Co. GmbH
Manufacture: as above, 1962

This building is remarkable in that it represents a complete sphere (with a diameter of 12 m), whereas all later spherical single membrane structures consist only of spherical sections. The form relationship with the Field Constable's House in Ledoux is striking.

10, 11. Winter protection for the construction of the TV tower in Dresden
Design: Deutsche Post, project office (H. Rühle, I. Bauer, G. Drechsler, E. Macher)

As in the previous examples a pneumatically stressed envelope is attached to a steel ring with a diameter of 5 m. The retaining ring is connected to the climbing framework of the tower; sealing is effected by a skirt running around the hem. (Bibl. 144, p. 121.)

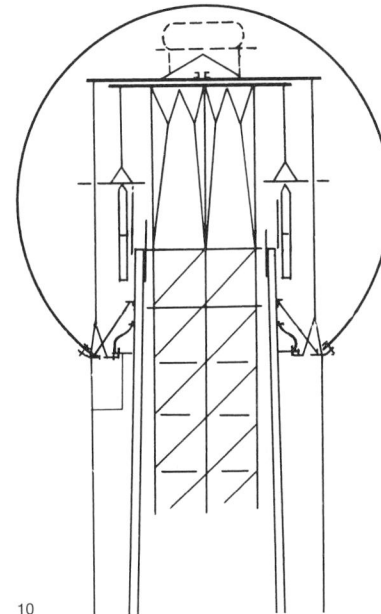

10

11

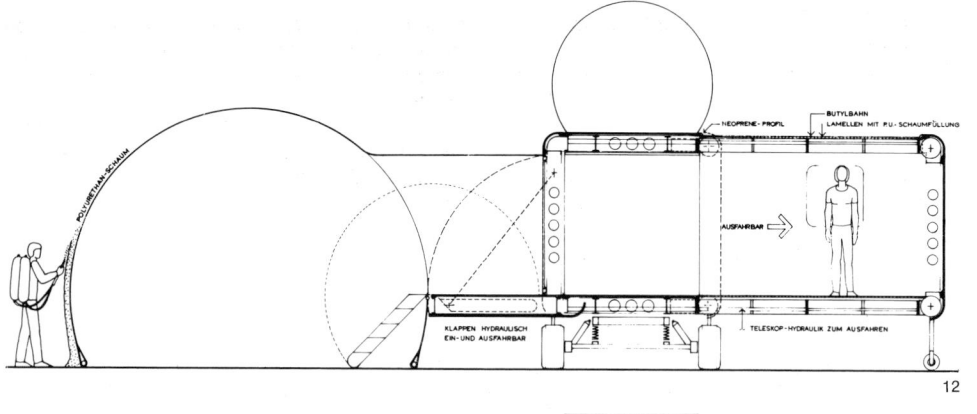

12

12—14. Mobilhaus
Design: Manfred Schiedhelm, 1970

The idea of this design, which was a contribution to a Japanese competition, is to create a mobile "container" house, having a functional area that can be greatly increased by pneumatic components being extended and attached as additional rooms. These alterations can be made quickly and selection of the different volumes, and of their size, is to be left to the user. However, this is an unrealistic proposition in terms of the current availability of suitable structural materials.

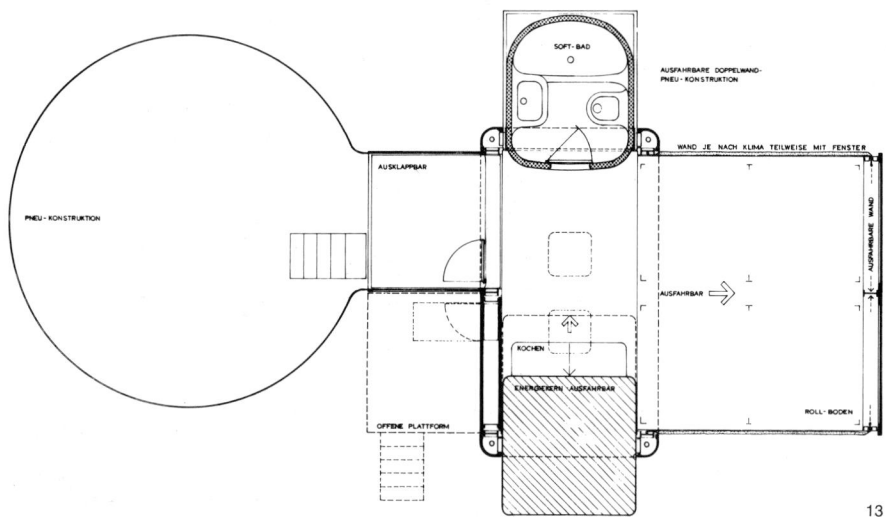

13

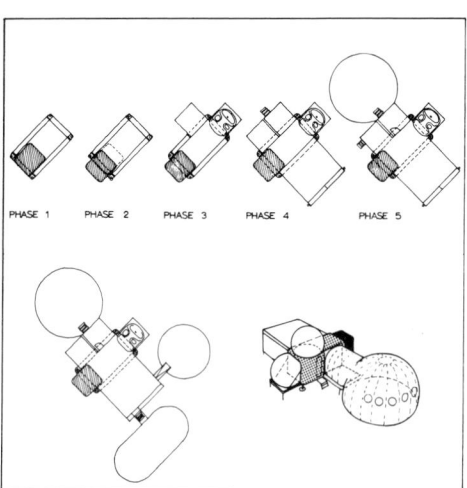

14

15

15. Environment Bubble
Design: François Dallegret, 1965

The sketch appeared as an illustration of an article by Reyner Banham, published in 1965, which indicated the role of the technical utility systems in the development of modern architecture. The pneumatic bubble, equipped with a utility element, should enable the user to lead the life of a modern nomad. (Bibl. 7.)

16

16–18. USA Pavilion for Expo '70, Osaka, initial projects

Design: Davis, Brody, Chermayeff, Geismar, de Harak Ass., 1968

The design team had won the competition for the US Pavilion (Fig. 16) with its first project of a double membrane cube structure made of synclastic panels. Neither the first nor the second (Figs. 17, 18) project was built, because of the cost. (Congress cut the budget from 16 million dollars to 10 million, of which only 4 million remained for the pavilion and all the exhibition items.)

The version finally built (see pp. 116, 117) is far less spectacular but certainly just as suitable for demonstrating the scope of pneumatic structures.

(Bibl. 44, p. 208 ff.)

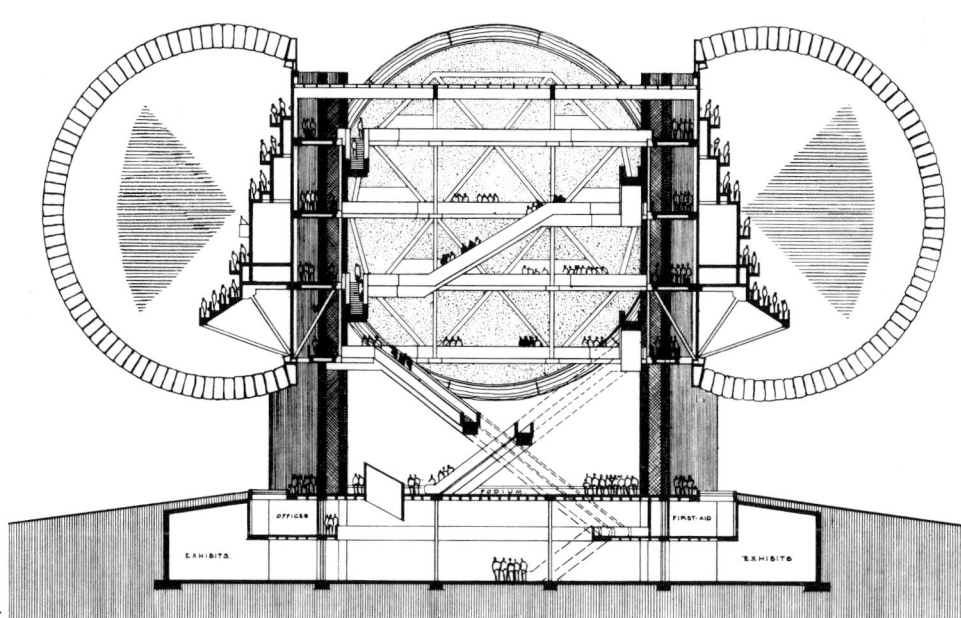

17

18

19

20

19, 20. Water walk
Design and manufacture: Eventstructure
Research Group, 1968

These pneumatic tetrahedrons, which allow walking on water, consist of 0.5 mm thick transparent or translucent PVC foils. Entrance is by airtight and watertight zip fasteners.

21

**22, 23. Die Wolke – Das Haus aus der Dose
(The cloud – house out of a can)**
Design: Coop Himmelblau, 1968

The cloud represents a mobile pneumatic unit of space, equipped with a variable platform which can be used in various ways. The total unit is to be transported packed in a giant can.

Without doubt the proposal has a certain technical appeal, but it is better not to question the relationship of cost to work space, nor the general validity of such a structural system in the face of the present living and environmental problems.

22

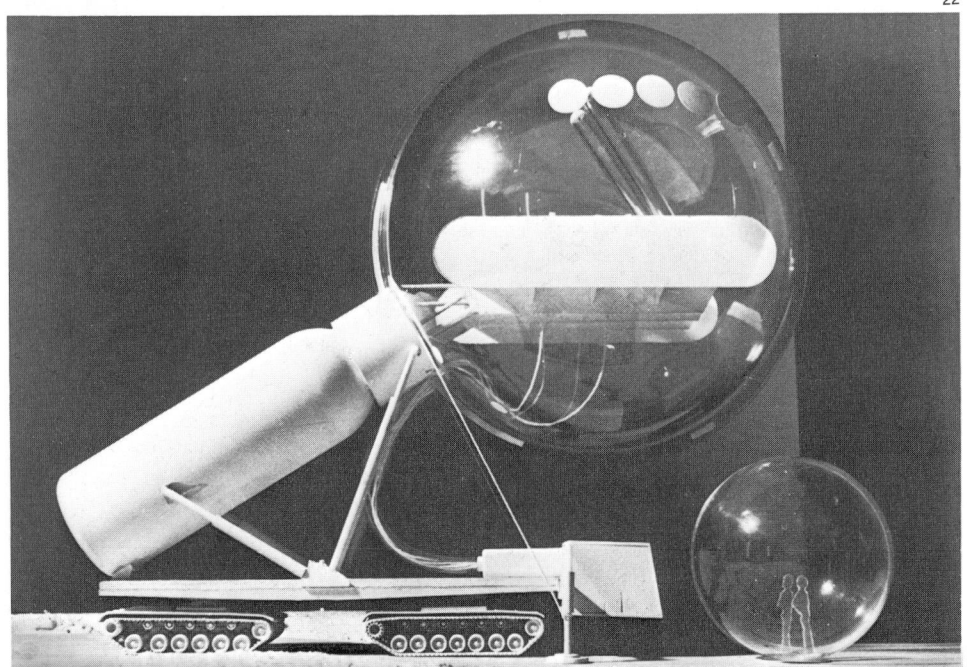

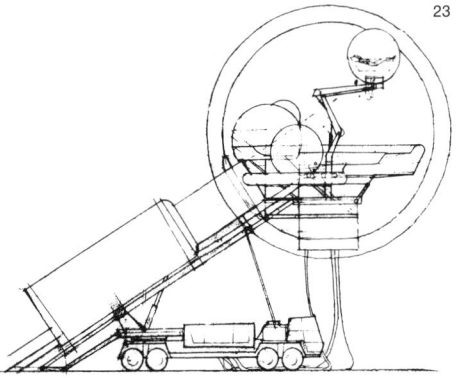

23

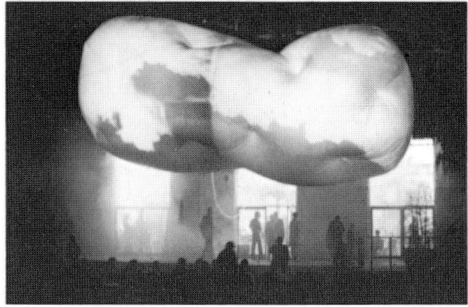

24

24, 25. Cloud, Stedelijk Museum, Amsterdam
Design: Eventstructure Research Group, 1970

The structure presents the rare case of a closed membrane with additional point support. It has no cutting pattern which would anticipate the final shape; thus a large number of wrinkles formed which, however, in this case were not meant to be avoided.

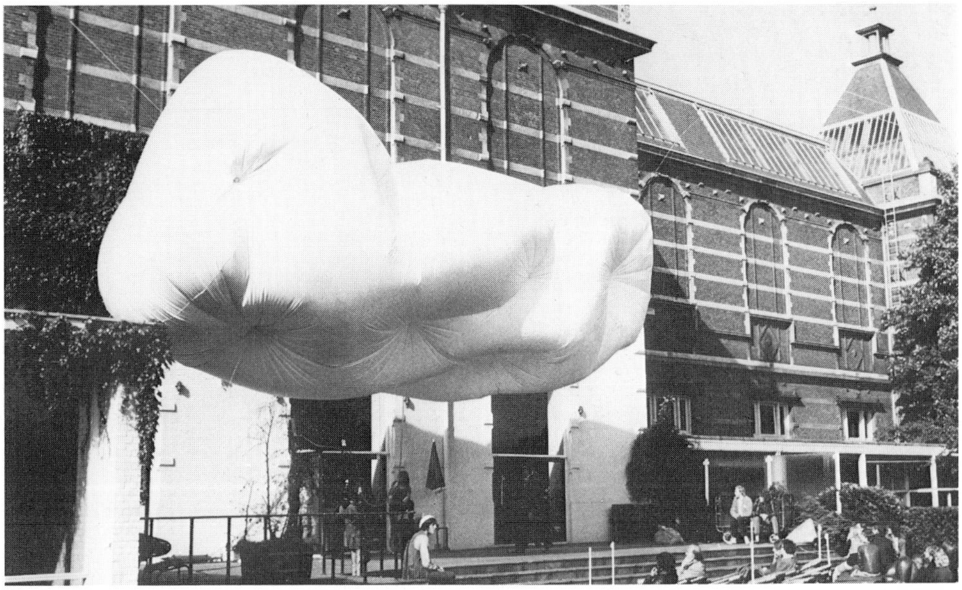

25

26

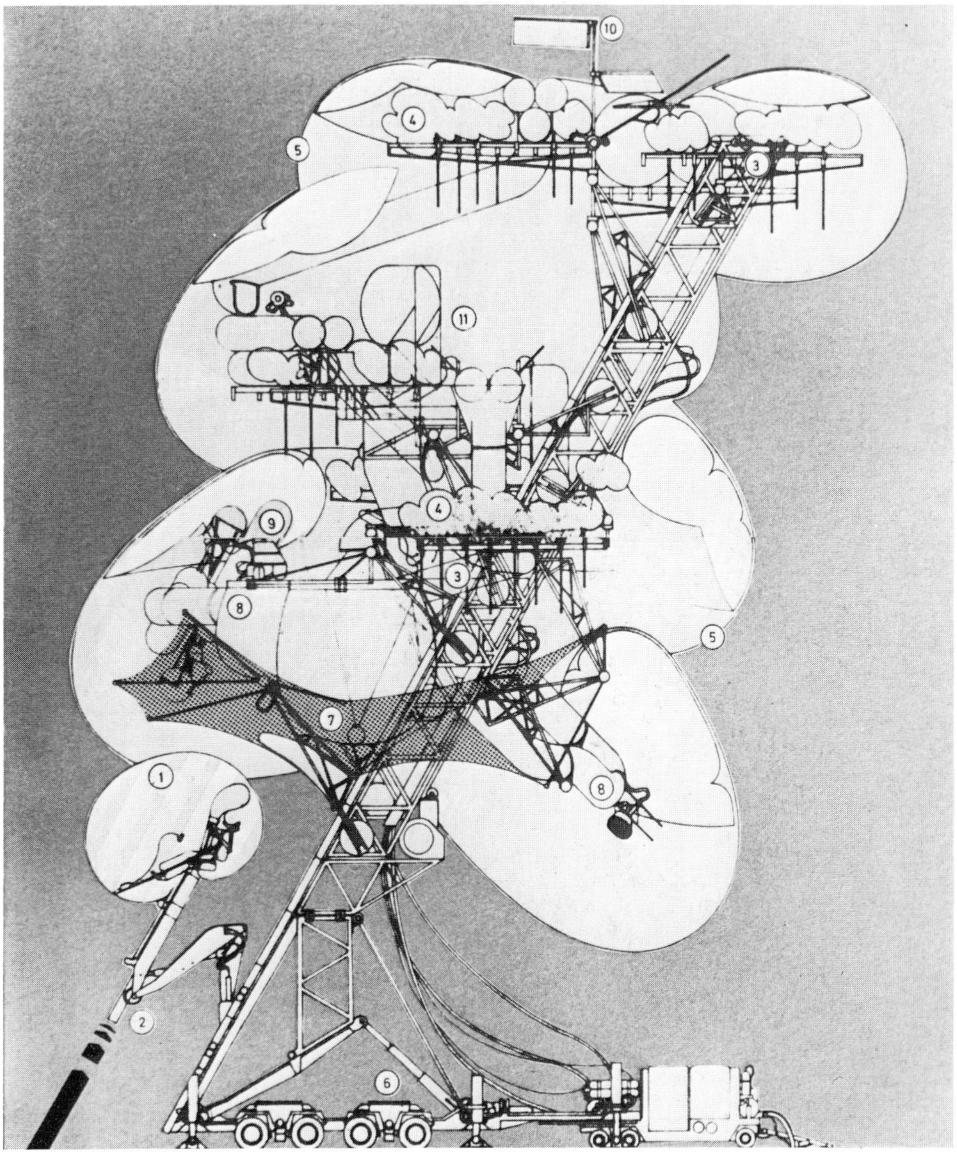

◁ **21. Michelin man**
Design and manufacture: Ballonfabrik — See- und Luftausrüstung GmbH + Co. KG, 1970

This "Pneuman" looks like a predecessor of early space suits. The constrictions were achieved, not by additional stabilising elements, but by the cutting pattern of the material.

26. Wolke – Gruppendynamischer Wohnorganismus (Cloud – group dynamic living organism)
Design: Coop Himmelblau, 1968

"Cloud" is clearly a favourite name for pneumatic structures. The illustration comes from a research programme for future living forms. However, the design laws for pneumatically stressed membranes have not been fully complied with.

27–29. Ricoh Pavilion, Expo '70, Osaka

Design: Nikken Sekkei, Ltd.
Manufacture: Taisei Construction Co., Ltd., and
Goodyear Aerospace Corp., 1970

A light source of 300 electronically controlled
lamps was installed inside the luminous yellow
balloon. Linear anchoring of cables and membrane ribs gave the balloon its characteristic
shape and reduced the radii of curvature. It had
a pipe connection to a "parent balloon" in the
basement, from where the internal pressure was
kept constant. In storms, or for maintenance
purposes, the balloon was drawn in by means of
winches.

28

27

29

30, 31. High tension test station for Felten & Guilleaume Carlswerk AG, Cologne-Mülheim
Design: Frei Otto
Manufacture: L. Stromeyer & Co. GmbH

The double sphere was achieved by means of a centre cable which forms a deep groove.

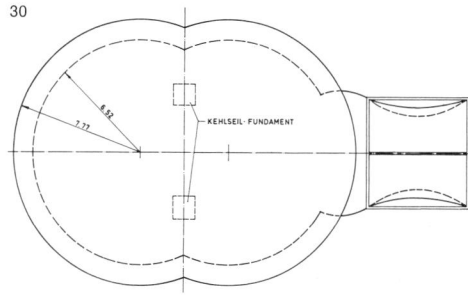

32–34. Brass Rail Restaurants, New York World Fair, 1963–64
Design: Victor Lundy
Structural calculations: Severud-Elstad-Krueger
Manufacture: Birdair Structures, Inc., 1963

Brass Rail Restaurants are a particularly successful example of the early attempts by Lundy to exhaust the possibilities of form in pneumatic structures.

The elegant lightweight structures were 23 m high and had a diameter of 18 m. The impression of piles of spheres was produced by means of guy ropes fastened to a central steel mast.

34

35. Boston Arts Centre Theatre
Design: Carl Koch and Margret Ross
Structural calculations: Paul Weidlinger
Manufacture: Birdair Structures, Inc., 1959

The theatre has 2,000 seats; its diameter is 44 m and the height of the cushion roof in its centre is 7 m. The membrane forces are taken up at the edge by three cables that are anchored to the corners of a polygon made out of steel sections. Thus the edge beams are only loaded for buckling stress (see also p. 157). The internal positive pressure of the cushion is 25 mm of water pressure. Originally it was intended to be used as form work for a concrete shell dome, but it was found that the pneumatic structure functioned very well; it remained completely intact after a bad hurricane in 1960. (Bibl. 119, p. 110.)

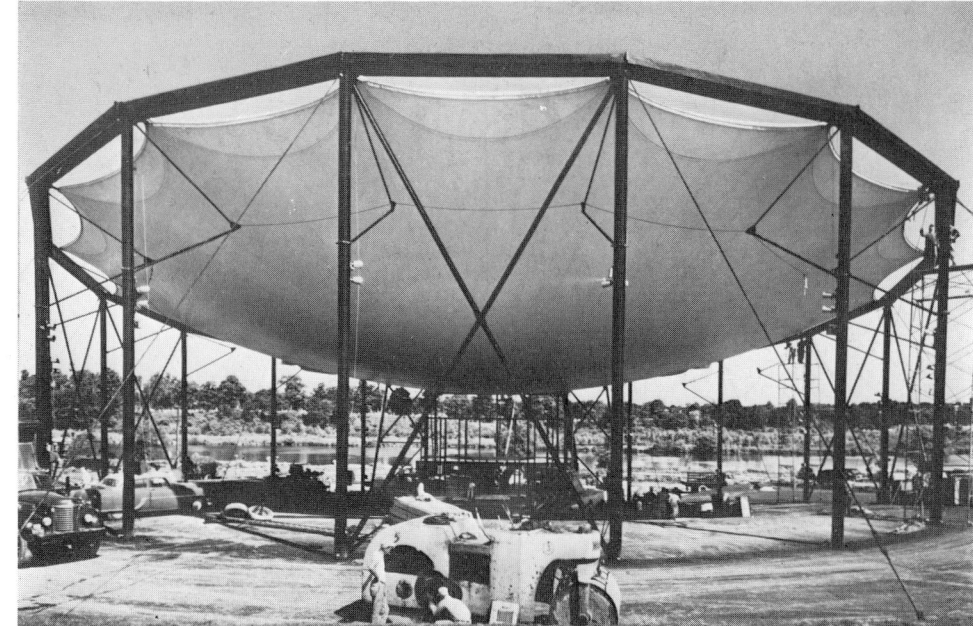

35

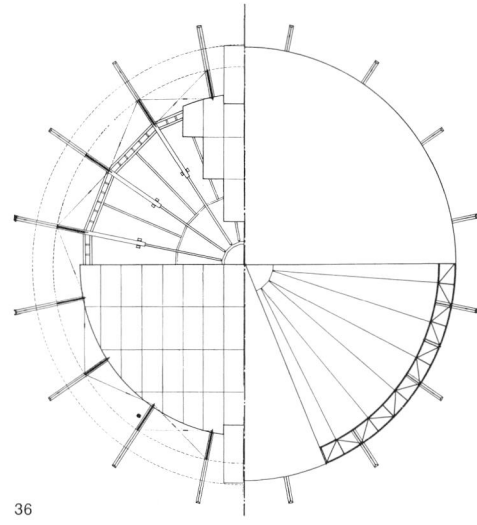

36

37

36–38. Mobile pavillon for RAI (Radiotelevisione Italiana)
Design: Achille and Pier Giacomo Castiglioni
Manufacture: IPI, 1967

The Italians call this pavilion for travelling exhibitions a "flying saucer" (disco volante). The translucent material of the roof and walls consists of PVC coated polyamide fabric. Only the roof is a pneumatic structure. As can be seen from the section, the radius of curvature of the upper membrane is considerably smaller than that of the lower membrane. Because of its relatively flat shape the roof surface is less strongly stressed by wind pressure than by wind suction, so that the upper membrane, which because of its strong curvature can resist higher loadings, actually is more heavily loaded.

38

39

40

39–42. Roof of the Festival Plaza, Expo '70, Osaka

Design: Kenzo Tange
Structural calculations: Y. Tsuboi and M. Kawaguchi
Manufacture: Toray Industries, Inc., 1970

Tange's roof is considered by some people to be magnificent and masterly, by others to be gargantuan.

The steel space framework was set at a height of 30 m over the Festival Plaza; it weighed 5,720,000 kg and measured $108 \times 291.6 \times 7.60$ m on a grid of 10.80×10.80 m.

The structure was rectangular in plan, but there was a "break" formed by a circular opening of approximately 58 m diameter, through which emerged the 60 m high "Tower of the Sun". The pneumatically stressed double membrane elements which were used as covering were light, translucent and unaffected by thermal stress in the steel structure (see also p. 157).

The upper membrane consisted of six different layers of polyester film – a weather protective layer (200 u), a heat reflecting layer (200 u), three "loadbearing" layers (250 u each) and an airsealed coating (50 u). The lower membrane did not have the heat reflecting layer.

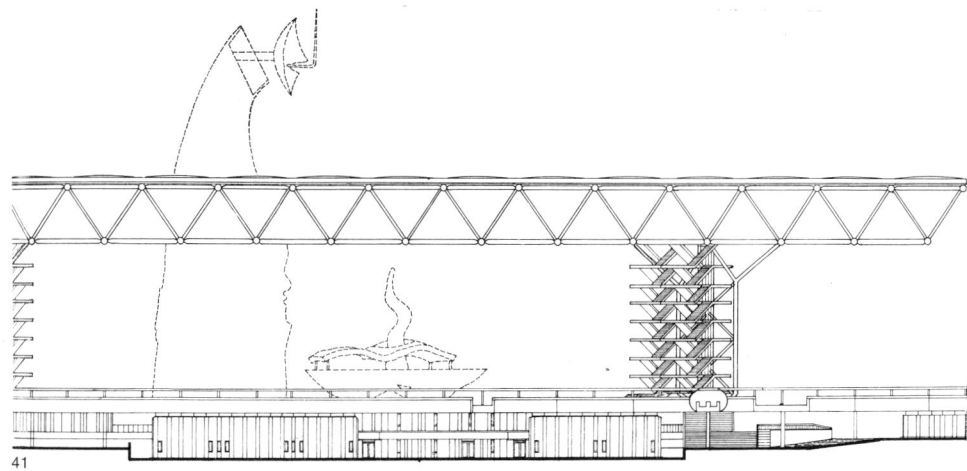

41

42

A life span of six to seven years was assumed. The elements were flame resistant, and if a fire had occurred under the huge roof the sheeting would have melted to form outlets for the smoke and heat. The internal positive pressure was 50 mm of water pressure in calm weather and 100 mm in storm conditions. The structure had been developed to such maturity that it needed only a few control instruments to maintain the positive pressure.

In Fig. 42 two assembly cars for the air cushions can be seen. They ran in grooves which at the same time served as drainage channels.
(Bibl. 88, p.5ff; Bibl. 148.)

43, 44. Leisure centre with alterable air cushion roof at Rülzheim, near Germersheim

Design: Rudolf Kleine, Georg Kuhn, Klaus Richrath and Albert Schieber
Structural calculations: Wulf Witte
Manufacture: Krupp Universalbau, 1974

The air cushion roof consists of two membranes of which the outer one has the shape of a spherical segment while the inner one curves in the opposite direction with two different radii. The latter is supported by a circular gallery. The diameter of the structure is 36 m. Steel cables inserted in the membranes in the form of arches transmit the uplift forces produced by the internal pressure in the cushion roof into 16 anchorages. Hinged steel plates retain and arrest the cable ends as well as the so-called travelling cables. When the structure is closed both membranes are fixed to the concrete foundation. In summer the anchorages are released from the ground plates and, being raised by the internal pressure, the roof glides 4 m upward. It is then held in position by the travelling cables. In case of high wind load the internal pressure of 30 mm of water pressure is being increased.

Being open on all sides in summer the structure serves as coffee shop for the visitors of the neighbouring swimming pool (see p. 133) and is also used for open air theatre performances. During the cold season when the walls are closed the structure, which can be heated, is to be used for exhibitions, lectures, theatre performances and so on.

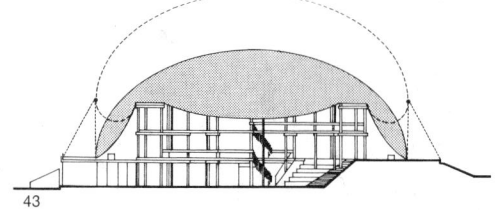

43

44

45, 46, Movil

Design: José Miguel de Prada Poole
Manufacture: Tolder S.A. and students of the Escuela Tecnica Superior de Arquitectura, Madrid

This is a prototype whose potential uses have not yet been thoroughly investigated. The structure becomes mobile by controlling the positive pressure of the cushions, which are independent of one another. It is a solution related to the "Dynamat" (see p. 56) whose further development promises interesting results.

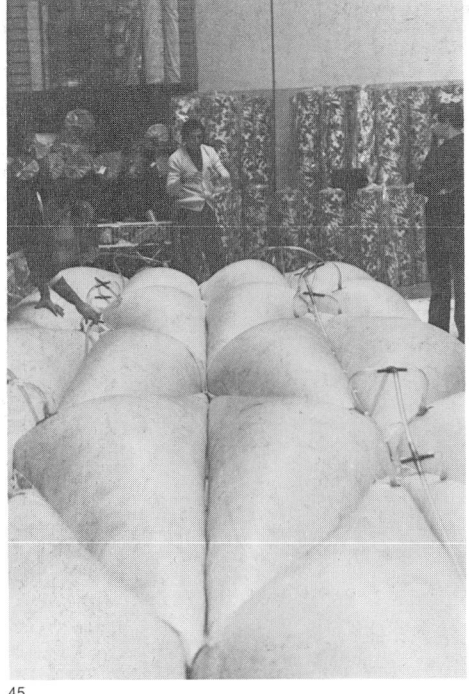

45

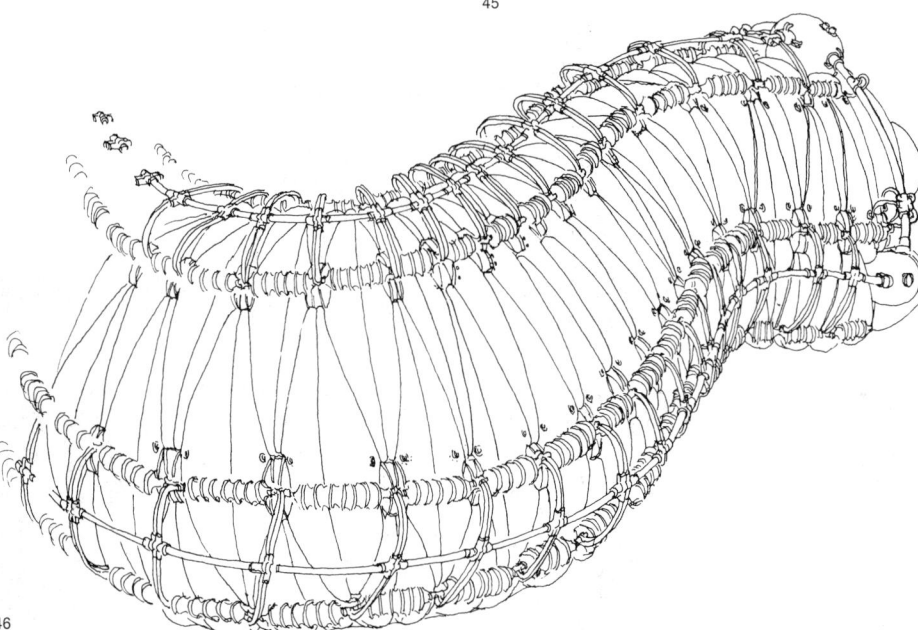

46

47, 48. Discontinuous structure – trial construction for Expoplastica 69

Design: José Miguel de Prada Poole, 1968
Manufacture: Alcudia S.A., 1969

Hexagons made of two layers of plastic foil were attached at their edges and inflated. Fig. 47 shows the model built in 1968, in which each individual cushion is connected to a central fan by means of an external air pipe. The prototype shown at the exhibition was about 5 m high.
The structure is worthy of mention here the attempt was made to erect a dome from low pressure cushions without a supporting framework.

49. Projection screen at the Deutsche Industrieausstellung 1968, Berlin

Design: Frei Otto and Bernd-Friedrich Romberg
Manufacture: L. Stromeyer & Co., GmbH

This structure became known as the first membrane structure stressed by internal negative pressure.
In contrast to positive pressure structures producing an interior cavity that can be utilised in case of low pressure, membranes stressed by negative pressure need additional auxiliary structures – here a steel frame with tension cables – which keep apart the membrane surfaces as they are drawn inwards.
The scope for negative pressure structures in architecture has received little investigation as yet.

50. Modular inflatable cushion structures

Design: Eventstructure Research Group, 1970

Triangular cushions with a lateral length of 8 m are flexibly connected and can be assembled in various different forms. The apexes are held by a central mast with cables.

47

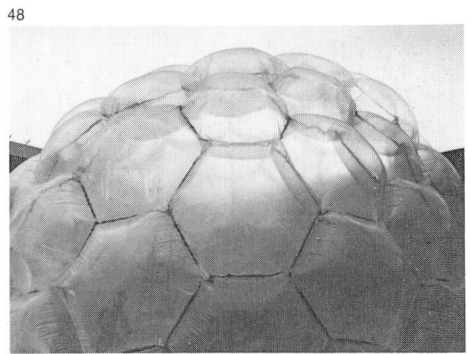

48

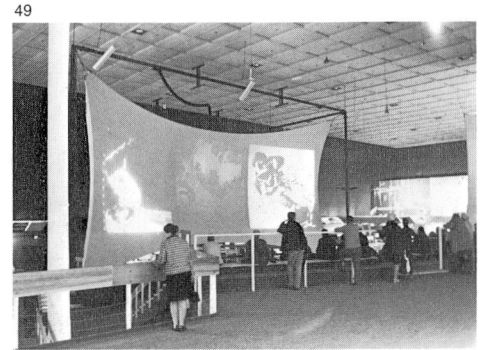

49

51

54

50

These polyhedrons also have flexible connections between their lateral surfaces. They consist of flat pneumatic cushions that are self-supporting and can be added to form structural shapes of all kinds.

57–59. Filling elements for grid shells
Design: Gernot Minke with students of the Technische Hogeschool Delft, 1971

The studies were also carried out with double membrane elements in the form of regular polygons. The intention was not to seek new detail solutions but to show some possibilities existing in the connection of light grid shells and pneumatically stabilised cushions. With flat structures such as these an easily visible anticlastic curvature appears at the corners.

57

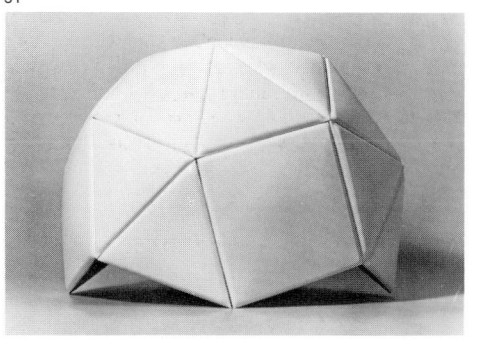

52

55

58

53

56

59

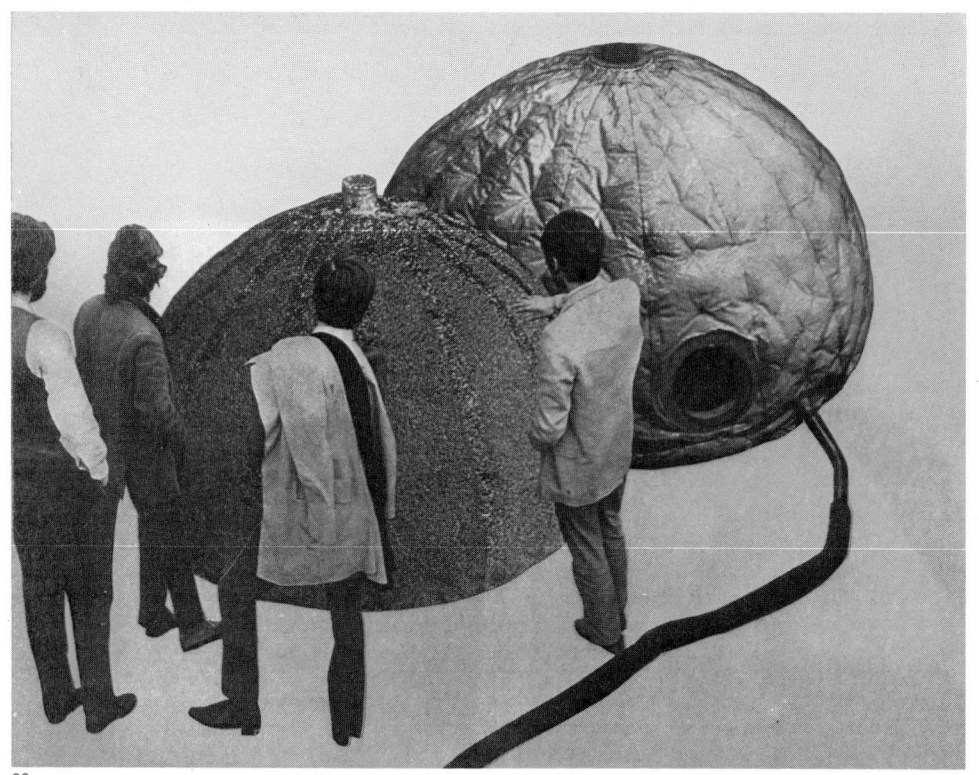

60

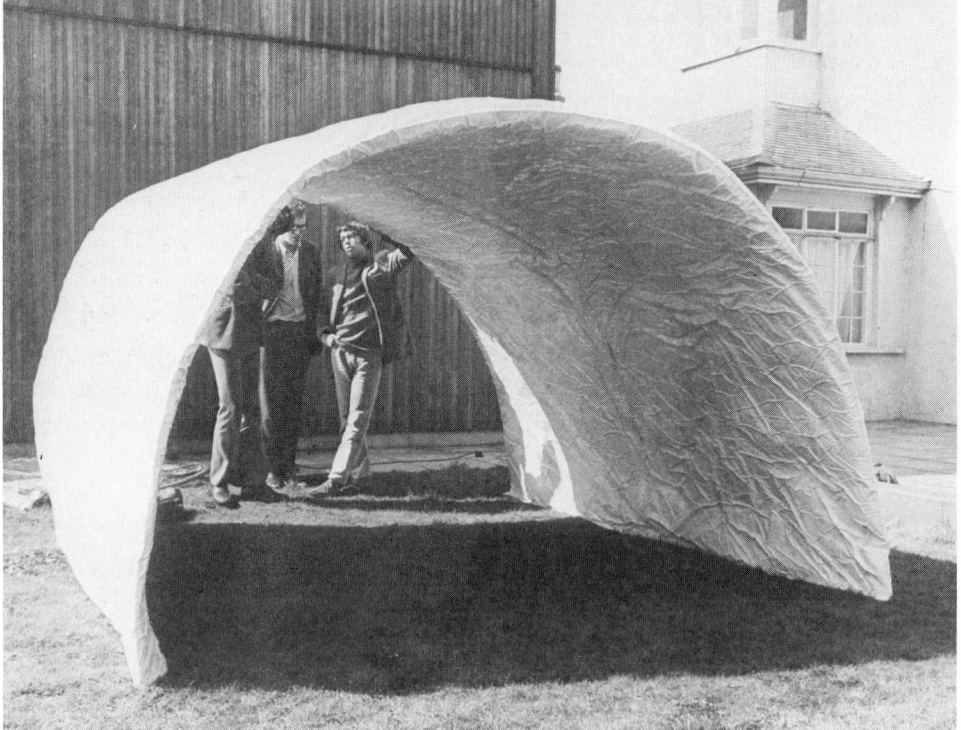

61

63

62

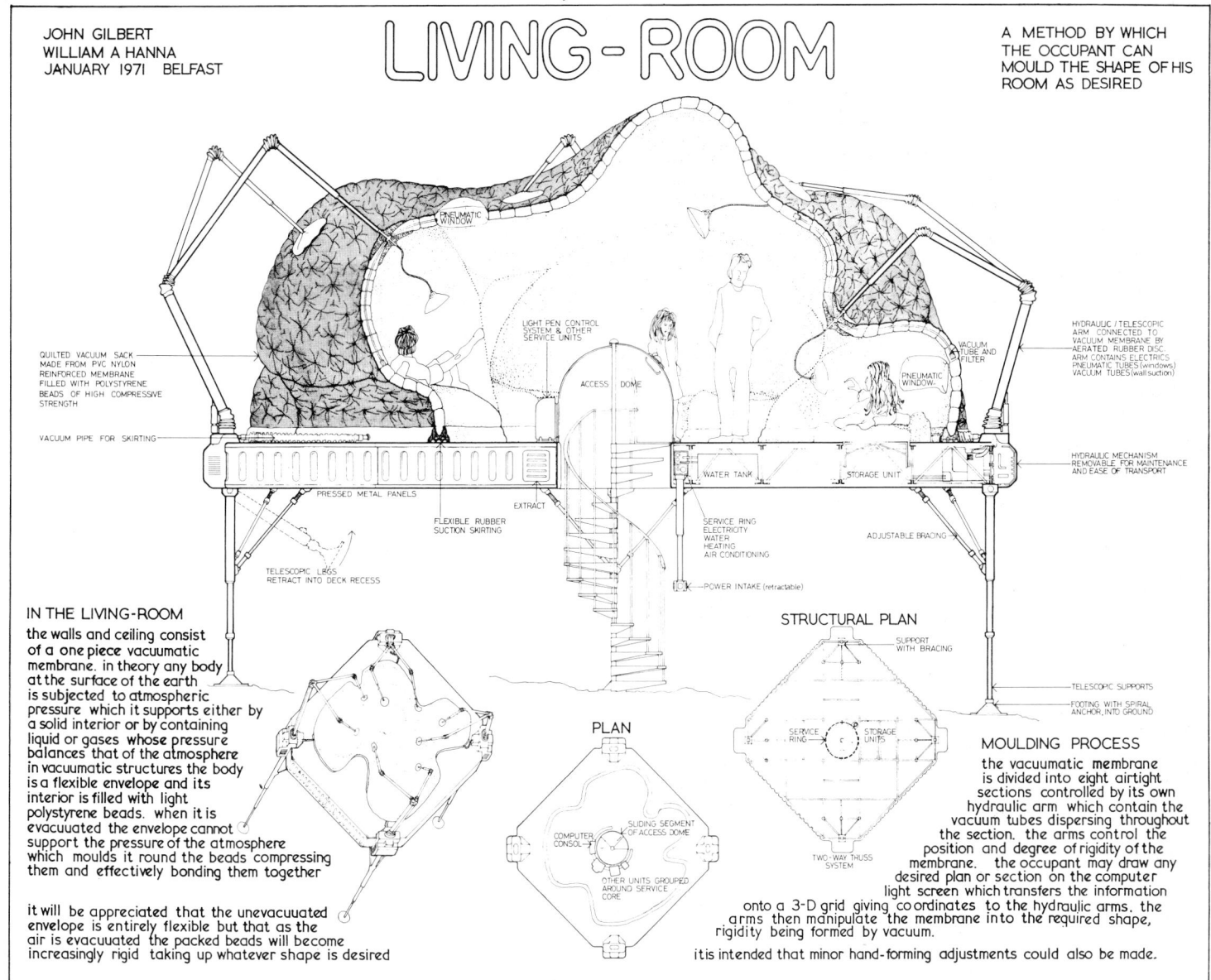

LIVING-ROOM

QUILTED VACUUM SACK
MADE FROM PVC NYLON
REINFORCED MEMBRANE
FILLED WITH POLYSTYRENE
BEADS OF HIGH COMPRESSIVE
STRENGTH

VACUUM PIPE FOR SKIRTING

PNEUMATIC
WINDOW

LIGHT PEN CONTROL
SYSTEM & OTHER
SERVICE UNITS

ACCESS DOME

HYDRAULIC / TELESCOPIC
ARM CONNECTED TO
VACUUM MEMBRANE BY
AERATED RUBBER DISC.
ARM CONTAINS ELECTRICS
PNEUMATIC TUBES (windows)
VACUUM TUBES (wall suction)

VACUUM
TUBE AND
FILTER

PNEUMATIC
WINDOW.

HYDRAULIC MECHANISM
REMOVABLE FOR MAINTENANCE
AND EASE OF TRANSPORT

PRESSED METAL PANELS

FLEXIBLE RUBBER
SUCTION SKIRTING

EXTRACT

WATER TANK

STORAGE UNIT

SERVICE RING
ELECTRICITY
WATER
HEATING
AIR CONDITIONING

ADJUSTABLE BRACING

TELESCOPIC LEGS
RETRACT INTO DECK RECESS

POWER INTAKE (retractable)

IN THE LIVING-ROOM

the walls and ceiling consist
of a one piece vacuumatic
membrane. in theory any body
at the surface of the earth
is subjected to atmospheric
pressure which it supports either by
a solid interior or by containing
liquid or gases whose pressure
balances that of the atmosphere
in vacuumatic structures the body
is a flexible envelope and its
interior is filled with light
polystyrene beads. when it is
evacuated the envelope cannot
support the pressure of the atmosphere
which moulds it round the beads compressing
them and effectively bonding them together

it will be appreciated that the unevacuated
envelope is entirely flexible but that as the
air is evacuated the packed beads will become
increasingly rigid taking up whatever shape is desired

PLAN

COMPUTER
CONSOL

SLIDING SEGMENT
OF ACCESS DOME

OTHER UNITS GROUPED
AROUND SERVICE
CORE

STRUCTURAL PLAN

SUPPORT
WITH BRACING

SERVICE
RING

STORAGE
UNITS

TWO-WAY TRUSS
SYSTEM

TELESCOPIC SUPPORTS

FOOTING WITH SPIRAL
ANCHOR INTO GROUND

MOULDING PROCESS

the vacuumatic membrane
is divided into eight airtight
sections controlled by its own
hydraulic arm which contain the
vacuum tubes dispersing throughout
the section. the arms control the
position and degree of rigidity of the
membrane. the occupant may draw any
desired plan or section on the computer
light screen which transfers the information
onto a 3-D grid giving co-ordinates to the hydraulic arms. the
arms then manipulate the membrane into the required shape,
rigidity being formed by vacuum.

it is intended that minor hand-forming adjustments could also be made.

64

60—64. Vacuumatics

Design and manufacture: John Gilbert, Marcus Patton, Chris Mullen, and Stanley Black, under the direction of Ivan Petrović, 1970

Vacuumatics have developed as a result of a course on space cell construction at Queen's University, Belfast. The novelty of the technical principle involved, the care taken in its working out and the resulting prospects for its use certainly make this structure one of the most significant since Lanchester's time in the field of pneumatic structures.

The vacuumatics consist of two membrane layers between which is a filling of light granular material. When the air between the membranes has been extracted the external atmospheric pressure forces the membrane against the filling material which in turn is more strongly com-pressed. In order to maintain a plane structure the membranes are coupled together by local individual connections. The filling material also provides point support to the membrane.

As the negative pressure increases the resulting shell becomes stiffer. Thus the rigidity of such a "negative pressure mattress" can be increased or decreased as required and the form of the structure can thereby be changed.

Vacuumatics are on the fringe of pneumatic structures. The loadbearing capacity is produced only under negative pressure by a compound material with a porous composition (granulate) and reinforced boundary areas (membranes). Without the availability of this negative pressure the structure would consist only of slack membranes with granular filling. Thus the negative pressure is a structural element. The rounding-off of the membrane, which is characteristic of pneumatic structures, takes place in the micro area, as the section shows.

The significant advantages of "Vacuumatics" over conventional pneumatic structures are the random variability of form produced by a temporary reduction in negative pressure as well as the outstanding properties of thermal insulation inherent in the use of light synthetic globules (e. g. Styropor).

The domes shown in Fig. 60 were investigated to find their loadbearing capacity under snow load and their thermal insulation properties. A further six geometric forms were tested for maximum tension. Nine different membrane materials and six different filling materials were compared with each other in twenty technical properties. The prototypes were constructed out of nylon reinforced PVC membranes with Styropor filling.

65

66

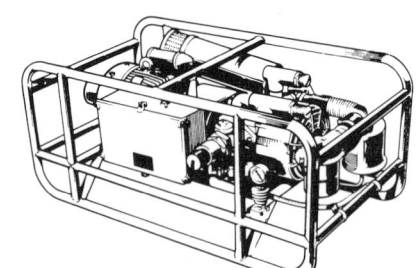

67

68

69

65–68. Inflatable shelters

Design and manufacture: M. L. Aviation Company, Ltd.

The series-produced pneumatic sandwich panels, of which an English example is shown, have excellent mechanical properties. The insulation capacity is also very high. The panels manufactured by the M. L. Aviation Company are 12 cm thick and consist of two membrane coatings each of which is built from several layers. The two coatings are held apart by nylon threads positioned very close to each other (further details see p. 145). The panels form segments of polygonal buildings of different sizes and shapes. The buildings are guyed on the outside so that the dome in Fig. 66, for example, can withstand wind velocity of 160 km/h without damage. Technical components such as air conditioning equipment, electrical switchgear and a generator to maintain the internal pressure of the panels (which can also act as a decompressor – Fig. 67) are also supplied with this highly developed system of construction. Fig. 68 shows a half cylindrical structure with spherical ends (as shown in Fig. 65, but with three central sections) when packed for transportation.

69. Inflatostair – emergency exit for aircraft

Design and manufacture: Goodyear Aerospace Corp.

Inflatable escape slides have long proved their value in aircraft crash landings on land and sea. Gangways, which can also be inflated in seconds, are more comfortable. The wall construction is similar to that of the pneumatic sandwich panels. The example illustrates the complicated individual forms that can be produced when using pneumatic elements. (Bibl. 44, p. 157 ff.)

70–73. Emergency first aid stations
Design and manufacture: Krupp Universalbau

This modular construction unit consists of a tent system, in which a light steel framework is spanned by "air mattresses". Only the covering for the access locks is made out of non-pneumatically stressed membranes. The additional metal support structure was chosen to prevent a collapse in the event of extensive damage to the membrane.

The air space in the "mattresses" provides heat insulation and can also be supplied with cooled air according to external conditions. The tent interior is under a light positive pressure of 10 to 15 mm of water pressure.

70

71

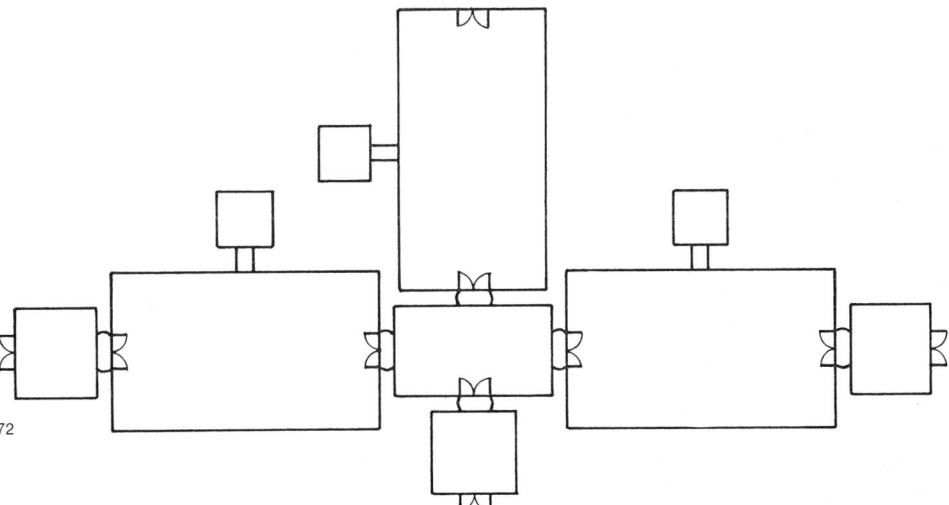

72

73

74

75

74–78. Dynamat
Design: Simon Conolly and Mark Fisher, 1971

The elements presented here also originate from the British Isles, as do so many innovations in the field of pneumatic structures. The Dynamat, whose prototype was exhibited at the DEUBAU 1971 in Essen, is one example of a whole series of similar attempts in which the user is given the opportunity of creating an environment for himself and of manipulating its form. The volume of the folded mat is over one hundred times smaller than the envelope when opened out.

The changes in shape are not achieved by mechanical means but by mere alteration of the air pressure in the individual cells, which are separated from each other by airtight divisions. These cells are connected to the central control by 3 mm tubing, and are held vertically by rigid alloy tubes.

By regulating the air input, individual or total movements of the structure are carried out centrally or decentrally.

The mat consists of individual square elements, which can be composed to form surfaces of any size by means of zip fasteners. The stage of development shown cannot, however, be viewed as final since the taking up of transverse forces, for example, has not yet been fully mastered structurally.

(Bibl. 38.)

77

78

76

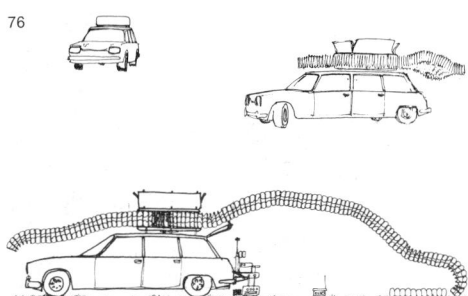

79–81. Air cushion roof over the stage of an open air theatre in Ratingen near Dusseldorf
Design and manufacture: Krupp Universalbau, 1962

The roof is only used during the season and is suspended by cables from a light steel mast. The lower membrane is additionally supported by a band of parallel cables, and the air cushions are connected to each other by tubes.

82–84. "Helium Lifted Canopy" for covering Wembley Stadium, London
Design: Arthur Quarmby
Structural advice: David Powell

The very flat helium cushion is anchored against wind loading by means of guy cables all round the Stadium. Whether the uplift of the gas is sufficient to prevent fluttering of the membrane which spans round the remaining area of the Stadium would still have to be investigated.

85, 86. Pavilion for a travelling exhibition of the U.S. Atomic Energy Commission
Design: Joseph Eldredge
Manufacture: Birdair Structures, Inc., 1964

The lower membrane was supported by cables which ran from the footings to the top of a mast standing in the middle of the building. Thus a strong profile was created from below, which considerably improved the acoustic quality of the room. On its outside the structure obtained the significant shape of a dome with guys at its perimeter between which the membrane forces were taken up by cables.

The travelling exhibition visited Central and South America.

79

80

81

82

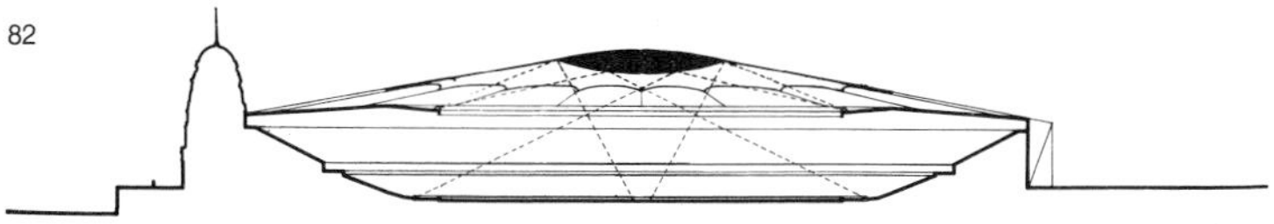

83

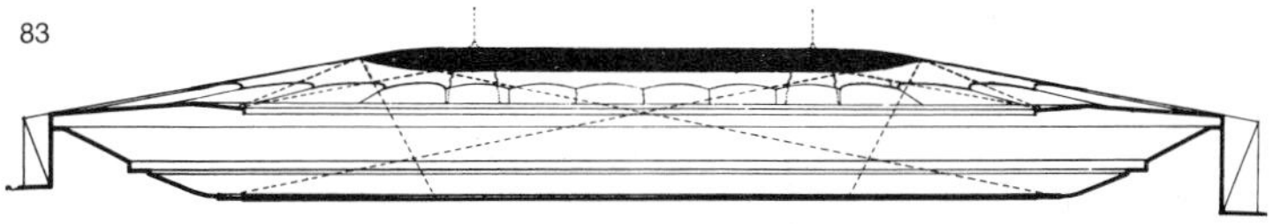

84

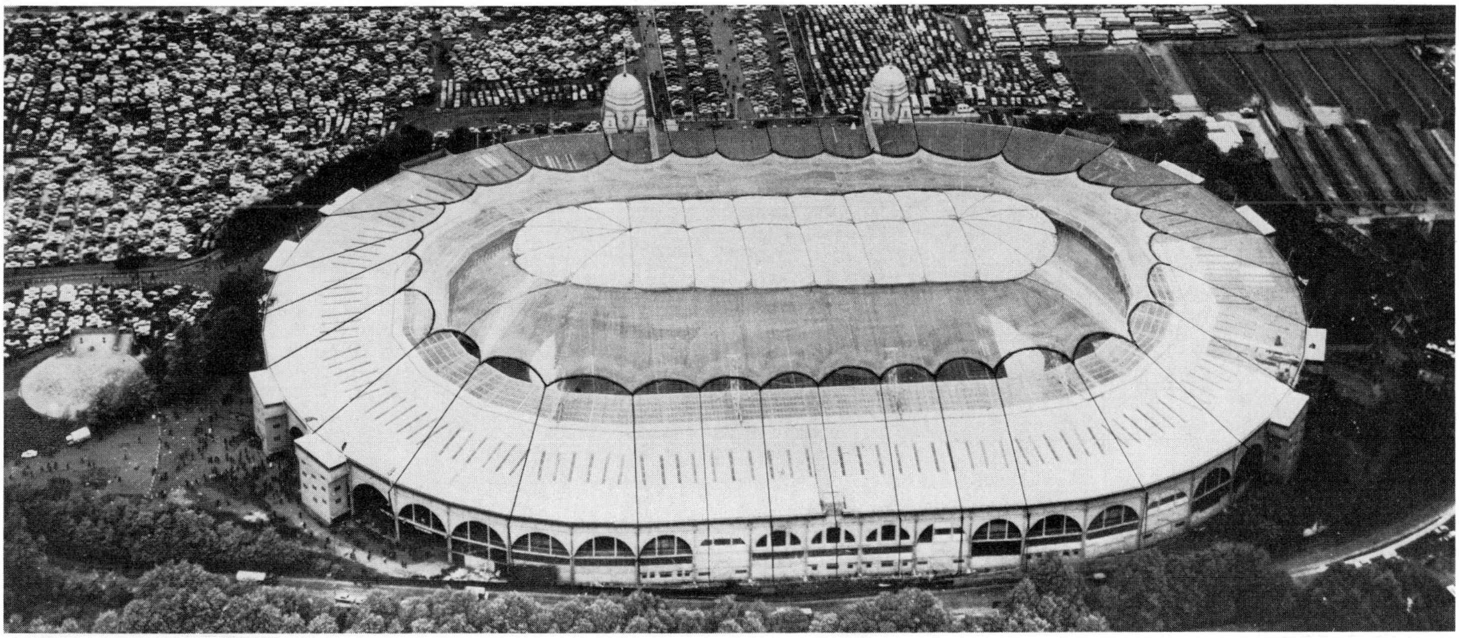

85

86

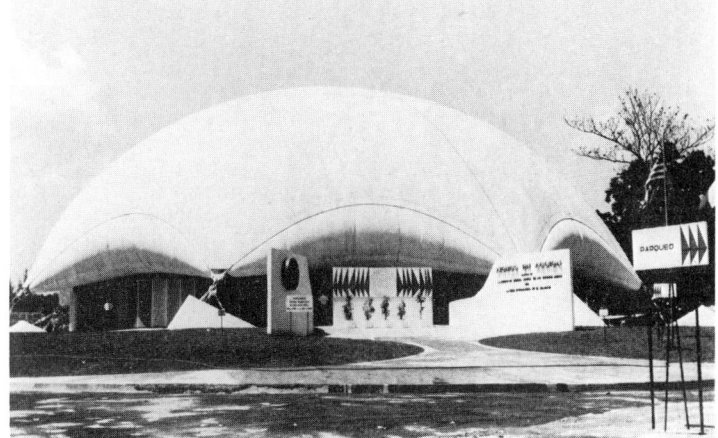

87

88

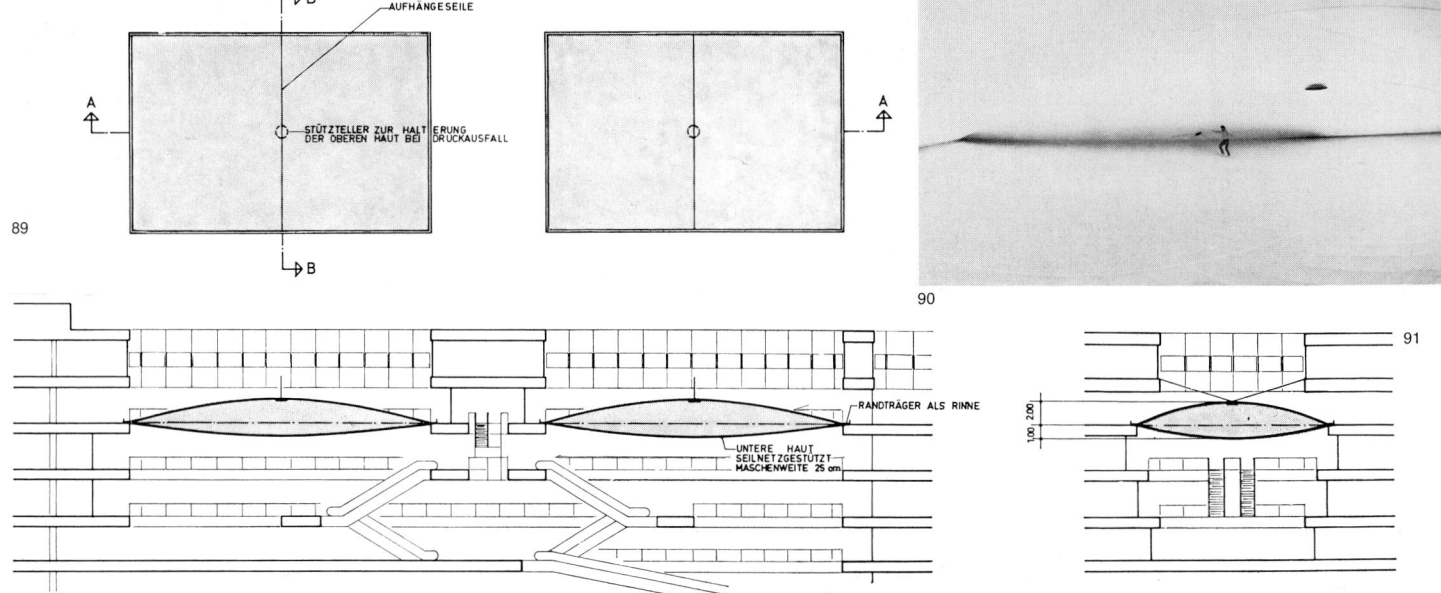

89

STÜTZTELLER ZUR HALTERUNG
DER OBEREN HAUT BEI DRUCKAUSFALL

AUFHÄNGESEILE

90

91

RANDTRÄGER ALS RINNE

UNTERE HAUT
SEILNETZGESTÜTZT
MASCHENWEITE 25 cm

87–91. Roof covering for the inner court of Forum Steglitz, Berlin
Design: Bernd-Friedrich Romberg
Manufacture: L. Stromeyer & Co. GmbH, 1970

There were somewhat less than four months available for the planning and execution of these pneumatics. The skeleton structure was already completed, and it permitted only low additional loadings from the added structure. On the narrow sides it was not possible to make a connection to the ceilings. Here IPB 800s are arranged as horizontal supports, able to receive a considerable load due to the low arching of the cushion in that area. As only a very slight sag was permitted in the lower skin of the cushion, additional support nets of 4 mm thick steel cables with a mesh size of 25 cm are stretched beneath it (see p. 156). The upper skin is arched with a rise of 2 m maximum over the top edge of the concrete ceiling. The internal pressure is 10 mm of water pressure. Both the upper skin and the lower net-supported skin are accessible for maintenance and cleaning purposes. Zip fasteners in the upper skins provide this access.

The interior of the cushion cannot be described within conventional categories. With almost shadowless illumination floor and ceiling meet along an equidistant horizon, and the floor yields at every step like a trampoline. This room – an unforeseen and unintended "by-product" of cushion design – seemed to the architect in retrospect to be the most important outcome of the work.

The air supply of the pneumatics is by means of continuous fans. As the low internal pressure is not sufficient for full snow loads and as, on the other hand, snow covering would seriously impede the passage of light, the air input is warmed to +45 °C in winter. The constant air flow produces a sufficiently high surface temperature on the outer skin for any snow falling on it to immediately melt and run off. The gutters are also heated.

In the tops of the cushions there are additional support discs which are carried by cables. When the air flow is interrupted, e.g. for maintenance and cleaning work, they prevent dropping and reverse sagging of the upper skin.

Inside lighting is installed which automatically switches on when daylight ends.

Each cushion weighs about 750 kg and receives 10,000 m³ air per hour. To avoid the formation of condensation inside, the air supply consists of 60% fresh air and 40% recirculated air.
(Bibl. 138.)

92

93

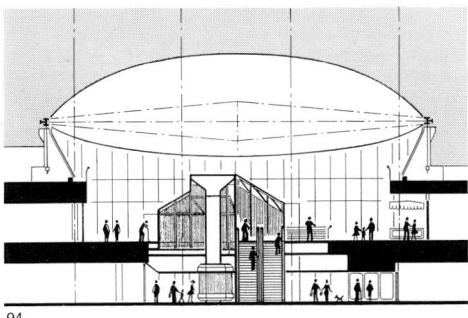

94

92–95. Roof covering of the shopping street in the "City" of Marl
Design: H. Kloss, P. Kolb + Partner – H. Drinhausen, 1970
Manufacture: Krupp Universalbau and Steffens & Nölle, 1974

As roof covering for the 185 m long shopping street, sliding roofs were first taken into consideration. However, the idea was renounced because of the free span being as much as 28 m wide. Steel and glass structures such as the 19th century passages had also to be ruled out – mainly for architectural reasons and on account of their architectural reasons and on account of their low suitability for mining subsidence areas.

The pneumatic structure eventually chosen consists of three not interdependent cushions supported by cables. The two outer cushions are 58.80 m long, the middle one being 67.20 m long, and their width throughout is 29.40 m. A steel framework with bending resistant edge girders and guyed compression struts, which take up the horizontal forces from the membrane cables, serves to stabilize the structure and to maintain its from. The membranes are made of PVC coated polyester fabric; the fibre materials is Diolen superfest produced by Enka Glanzstoff AG.

Each cushion has an inflation unit of its own. Each unit includes two radial fans one of which is enough to maintain the necessary pressurisation of 30 mm of water pressure in the interior of the cushions.
(Bibl. 169.)

95

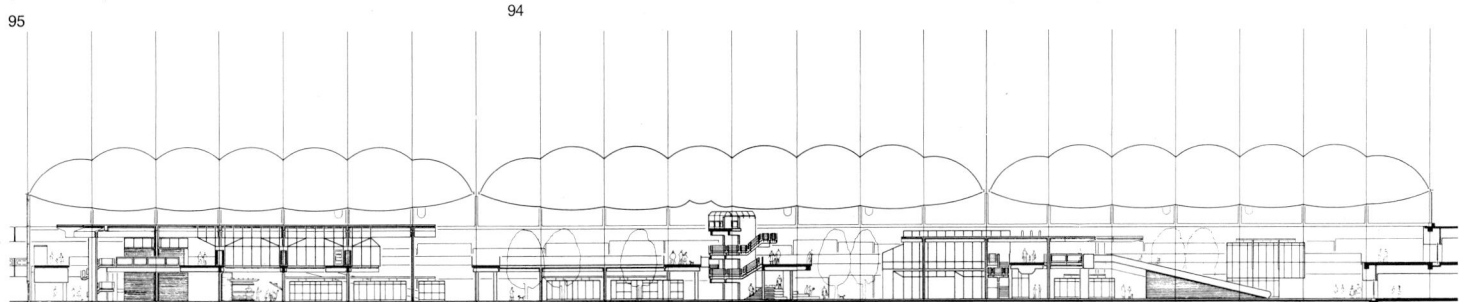

96, 97. Roof covering for radar and television aerials
Design: William Fischer and Sandy Hook
Manufacturer: Air Cruisers Company

In contrast to the usual radomes made of single membranes, this structure is so rigid that contact between the membrane and the antennae because of wind load is virtually impossible.
The membranes consist of neoprene coated Dacron with an outer Hypalon coating. They are divided up into eight curved sections, anchored to the ground by means of guyed flanges. In some cases tubes for water ballast are also built into the wall. The wall sections are connected to each other by stainless steel clamps through which runs a fastening cable (see p. 148). The dismantling of individual sections allows large items to be transported in or out. The internal positive pressure in each wall is 400 to 1,400 mm of water pressure according to wind loading, and is produced by electric compressors. The walls are separated into individual chambers by internal membrane ribs that give linear support to the external membrane.
The radome has a weight of 770 kg and a diameter of 13 m; the wall thickness is 2 m at the ground and 1.40 m at the top. Erection takes about 3 hours.
A total of about 20 such buildings have been erected throughout the world.
(Bibl. 131, p. 163 ff and p. 236; DBP 1 559 112.)

98–100. Transportable air supported bridge
Design and manufacture: Military Engineering Experimental Establishment (England), 1965

The 350 kg bridge under high pressure can carry a load of 1,000 kg at a span of 5.50 m. The central section as well as the two ramps are divided along their length into separate chambers by internal membrane ribs. In the tension zone there run wires similar to the reinforcement in reinforced concrete. The membrane consists of a three layered neoprene coated fabric.

101–103. Inflatable structures with variable shape and volume
Design: Michel Fourtané, 1969

These are flat structures which are put together from inter-connected chambers. Air is not only the support medium, for when its pressure changes air also causes the structures to change their shape and move in a predetermined manner. (DBP 1 961 523.)

96

97

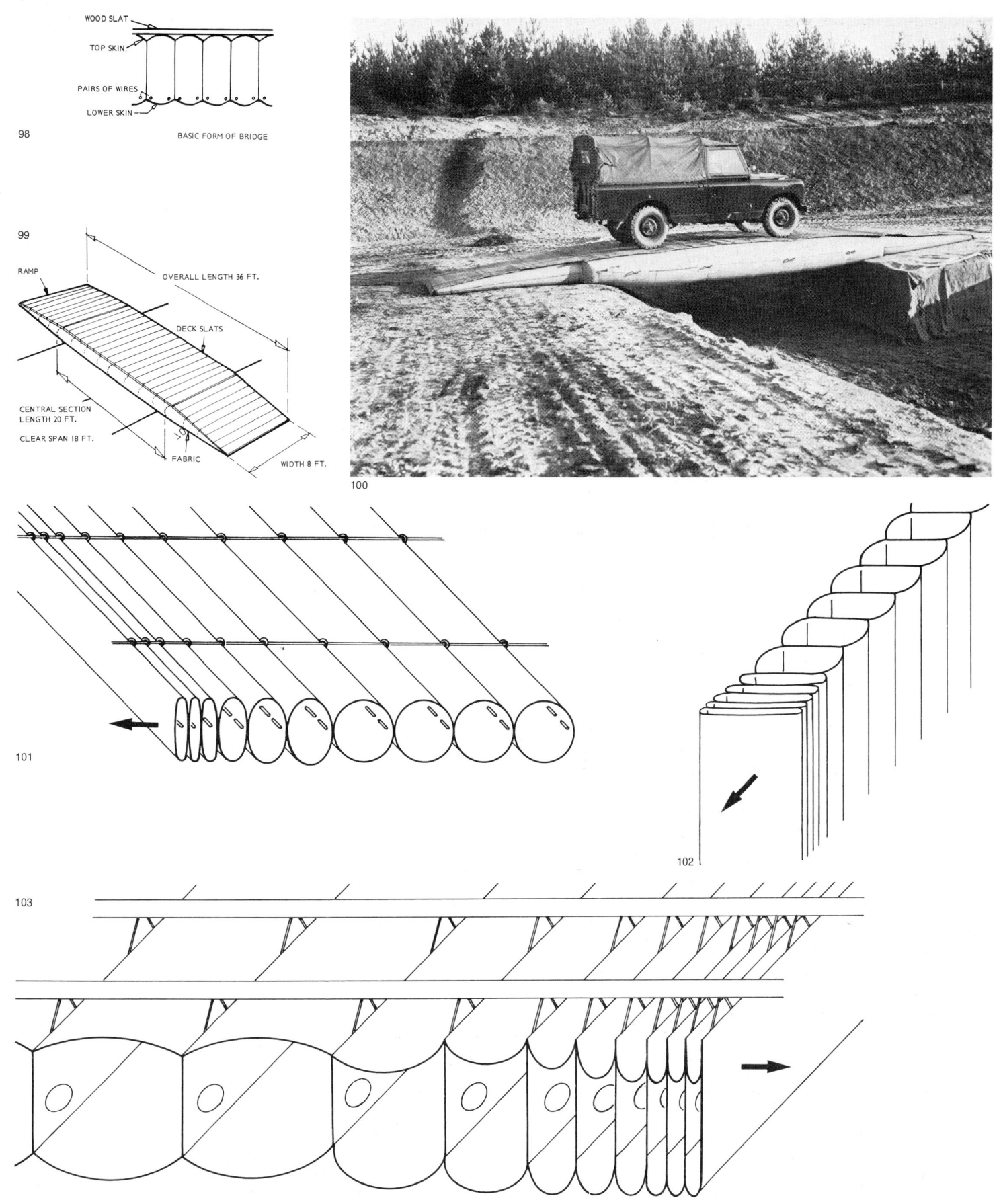

98 BASIC FORM OF BRIDGE

WOOD SLAT
TOP SKIN
PAIRS OF WIRES
LOWER SKIN

99

RAMP
OVERALL LENGTH 36 FT.
DECK SLATS
CENTRAL SECTION LENGTH 20 FT.
CLEAR SPAN 18 FT.
FABRIC
WIDTH 8 FT.

100

101

102

103

104. Pneumatic environment
Design: Quasar Khan
Manufacture: Kléber Renolit Plastiques, 1968

Khan, a Vietnamese living in Paris, who has become world famous through his blow-up furniture, has created this transparent environment out of PVC foil. The flat form of the room boundaries was produced by internal membrane ribs. The joints are welded.

104

105–108. Roof of the all weather swimming pool at Unterlüss, near Celle
Design: Bernd-Friedrich Romberg, 1968–72
Manufacture: Steffens & Nölle, L. Stromeyer & Co. GmbH, and C. Haushahn, 1972

A traversable envelope made of PVC coated polyester fabric and fixed to three of the steel arches spanning the swimming pool enables the installation to be used as an indoor pool when closed and stabilised with positive pressure of 30 min of water pressure, and as an outdoor pool when open.

The central arch has a span of 26.35 m and a rise at centre span of 13.58 m, while the two outer arches have spans of 20.70 m each and a rise at centre span of 10.75 m. In the longitudinal direction of the hall the arches are stabilised against tilting by means of two guy ropes, anchored to two terminal foundations far outside the hall. Six pulleys run along each arch, between which supporting wires span longitudinally. The envelope in the limp condition is folded up over these wires. On the firmly anchored side the envelope is fastened to a profile on the upper edge of the sloping ramp surface above the roof of the massive cloakroom block; the movable side is bounded by a square pipe. The envelope consists of two layers of polyester fabric with an intermediate air space of 3 cm which is additionally aerated for better insulation and in order to prevent condensation. The two layers are connected to each other at 50 cm centres by means of pointed plastic spacers.

The opening and closing of each takes five minutes only. The roof can be laid completely folded on the sloping ramp over the large block, so that when open only the three arches with the guy cables over the swimming pool can be seen. (Bibl. 154.)

105

106

107

108

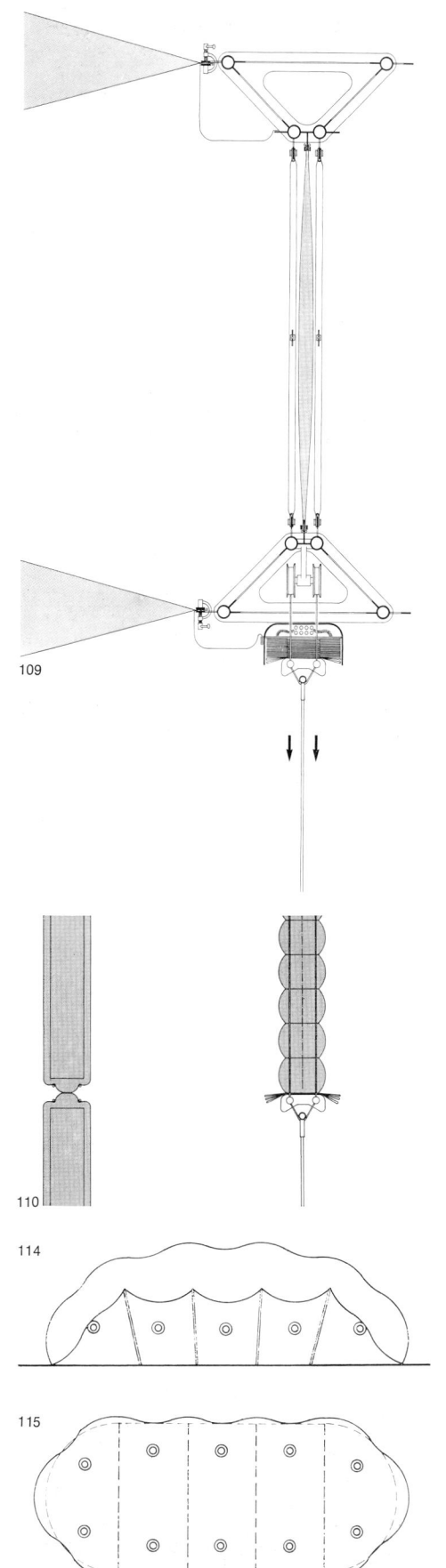

109

110

114

115

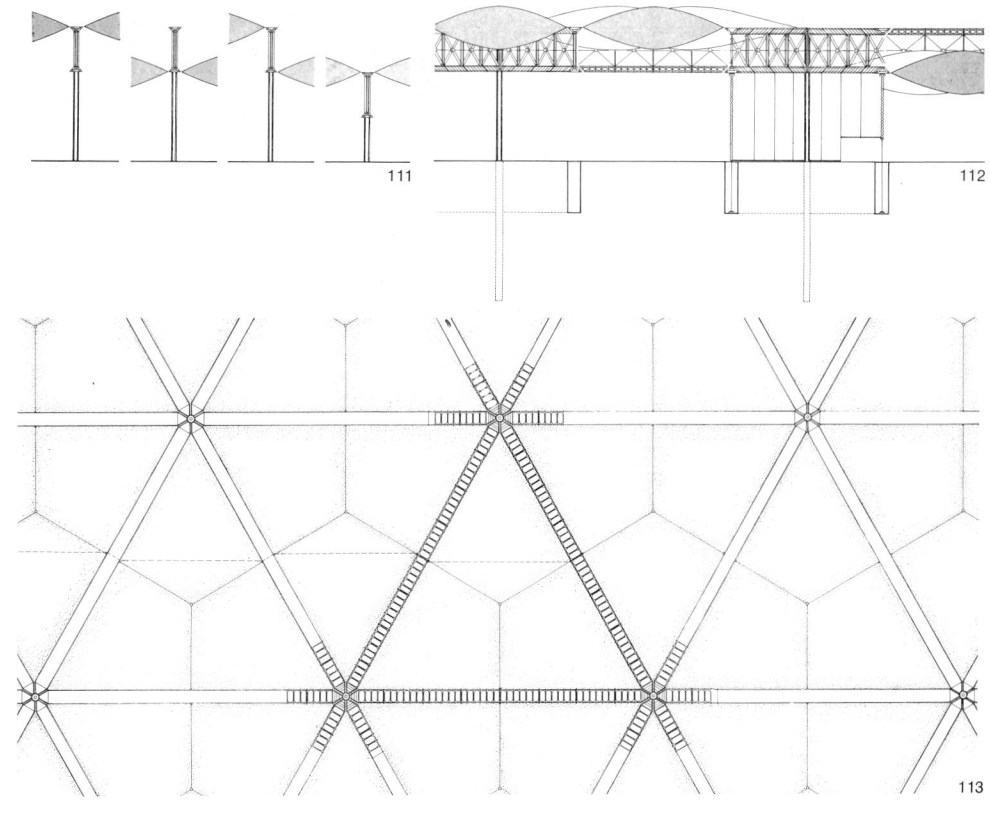

111

112

113

109–113. Alterable fairground halls
Design: Thomas Herzog, 1972

For the solution of the problem of spanning a fairground with an alterable roof a construction has been developed which is based on a 60° grid pattern with 28.8 m lateral sides supporting a steel frame structure with pneumatically stressed membranes as roof and façade elements. In consideration of the purpose to use the site either as an open air fairground or to cover the whole of it with a roof or to use only part of it to put up exhibition pavilions, a principle of construction appeared to be called for which would enable the structural members to change their volumes to an extreme degree. To obtain this objective pneumatic structures have so far proved to be much more suitable than any other principle of lightweight construction.
At all the crossing points of the grid there are

vertical shafts inside which steel pillars are sunk in the ground. These steel pillars can be extended up to a maximum length of 11 m by motors which are installed at the two ends of the 4 m high truss girders. If the drive assemblies are being reversed the girders "climb" up the pillars.
The façade elements are drawn up with the girders. They can be arrested at any point by a holding up device. The air not only serves to stabilise the façade parts but also to move them.

114–116. Hall roofs
Design: Ana Sklenar, 1972

The double membrane structure is point supported by means of rings on the outer membrane and linear supported by means of pressure arches on the inner membrane.

116

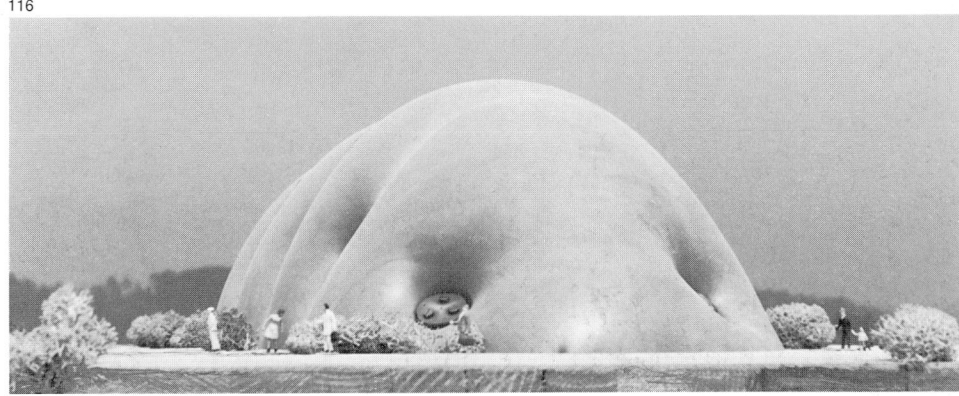

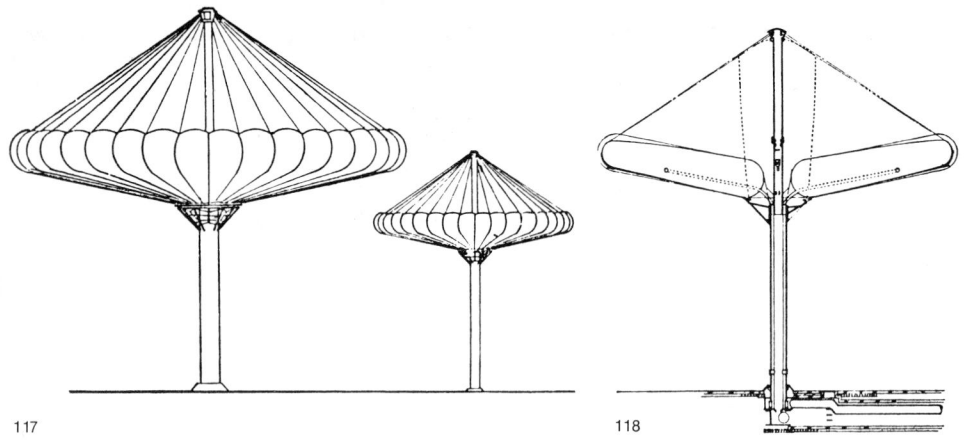

117   118   119

117–122. Mobile roofs at Expo '70, Osaka

Design: Tanero Oki & Ass.
Structural calculations: Shiger Aoki & Ass.
Manufacture: Taiyo Kogyo Co., Ltd., 1970

The mushroom-shaped red and yellow double membrane structure stood in "Expoland", the amusement park of the world exhibition. The diameter of the mushrooms when open varied between 15 and 35 m.

The pneumatically stabilised surfaces were guyed with radial cables to the centre masts, and if the cables were retracted the mushrooms closed.
When open the support pressure was 200 mm of water pressure; when closed it was 400 mm under wind speed loadings of over 15 m/sec. The membrane material was PVA and polyester fabric with PVC coating.

120

121

122

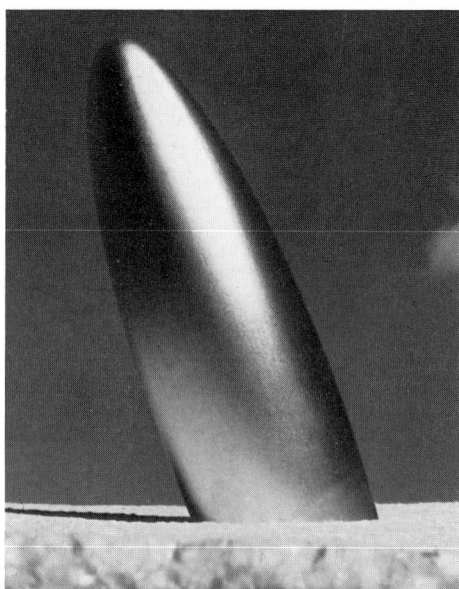

123

124

125

123–125. Envelope for oblong and oblique objects
Design: Frei Otto, circa 1960

The model shows a rotational body with harmonic stress distribution in different inclinations.

126. Pneumatic tower
Design and manufacture: RAVEN Industries, Inc.

The cone is some 30 m high. The membrane tensions are decreasing toward the apex, a fact which manifests itself clearly in slight wrinkles at the top.

127. Air supported tower
Design and manufacture: Birdair Structures, Inc., 1959

Here the tensions run more uniformly, but the tower is so slender that it has to be guyed. The height is 24 m.

128. Inflatable mast
Design and manufacture: Ministry of Technology, Research Development Establishment (England)

Here the cross-section is cylindrical right to the top. The height of this quickly erected mast is 30 m and the positive pressure is 1,400 mm of water pressure.

126

127

128

129, 130. Fabridam
Design: N.M. Imbertson and Associates
Manufacture: Firestone Coated Fabrics Company, 1957

The first dams constructed by Firestone were 1.50 m high and 40 m long. Today the height can be over 4 m and the length over 600 m.
The membrane tube made of heavy neoprene coated polyamid fabric is screwed on to a concrete base in the river and is filled with water or air. The overflow height can be raised or lowered simply through pumping or discharging.

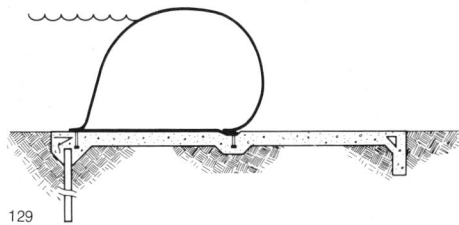
129

130

Within a few minutes the internal support medium can be completely emptied and the membrane will then lie flat on the concrete base; such a quick reaction to tidal waves is not possible with any conventional structure.
If the outside of the membrane is regularly given a new coat of Hypalon, then a life span of about 20 years can be anticipated.
Fabridams cost 75% less to manufacture than conventional dam constructions.
(Bibl. 10.)

131, 132. Dipole antenna
Design and manufacture: RAVEN Industries, Inc.

The membrane of this dipole antenna, which has been developed for use in outer space, is stabilised only by weak positive pressure.

133. Inflatable hoist for lifting persons or goods
Design and manufacture: Ministry of Technology, Research Development Establishment (England), 1970

A cradle made of a pair of rollers and a strap device for carrying goods is fixed to a fire hose. The rollers are closely fixed so that they clamp the hose airtight. If this is inflated from one end, then the rollers and the goods travel to the other end. To raise a man of average weight vertically, a positive pressure of 2,000 mm of water pressure is required (the maximum load capacity is 280,000 mm of water pressure).
The British Central Electricity Authority uses the inflatable hoist for the inspection and maintenance of overhead lines. Small rockets shoot a retaining line over a suitable attachment point. The compressor is carried in the car.
(Bibl. 44, p. 152.)

131

132

133

134

135

136

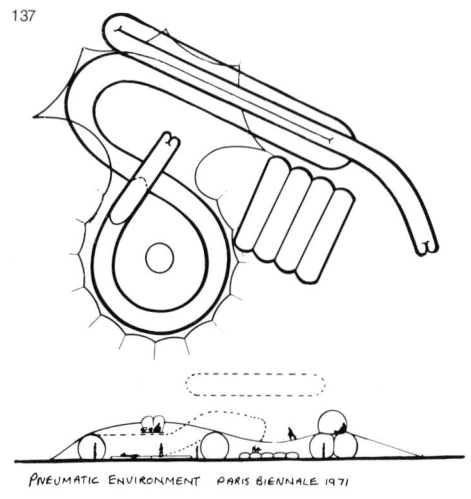

137

PNEUMATIC ENVIRONMENT   PARIS BIENNALE 1971

138

139

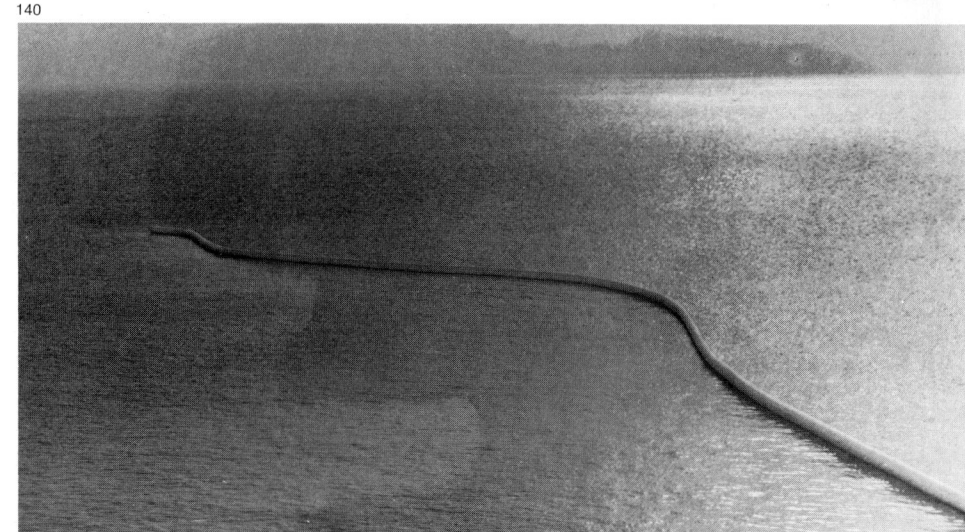

140

141

139, 140. Wavetube
Design: Graham A. Stevens, 1967

Concerned are ways and means of transporting persons and cargo through long tubes stabilised by air pressure.

In its simplest form the Wavetube is a locked-up tube floating on water, in which a kind of hovercraft is "surfing" shorewards with the waves. To translate this into a system usable over ground and independent of waves, a smaller, open ended tube with air passing through it is laid along the bottom. A hoverpallet serves as vehicle. Its weight creates a partial seal and the pressure build-up creates a wavefront which pushes the pallet forward.

The system is simple and if adequately elaborated can be used in a wide range of conditions and dimensions.
(Bibl. 155.)

141, 142. Passage connecting two buildings in Milan
Design: Studio d'Architettura e Industrial Design, 1968
Manufacture: Plasteco Milano, 1968

This PVC tunnel, which connected the Palazzo Esposizioni with the Padiglione della Produzione Italiana at the 14th Triennale, was 60 m long. Four positive pressure tubes formed the loadbearing part and the walls were single membranes not pneumatically stabilised, which were stressed by a cable running above the tunnel. The window frames were small pneumatically stressed ring tubes.

142

134, 135. Bridge over the Maschsee, Hanover
Design and manufacture: Eventstructure Research Group, 1970

This tube bridge was installed during the street art programme in August 1970. It was 250 m long and had a diameter of 4 m. In order to prevent rotation of the structure a small waterfilled hose was installed underneath which acted as ballast as soon as it came out of the water at one side.

The upper part of the tube consisted of 0.4 mm thick PVC sheet, the lower of coated fabric. The bridge was fixed to the floor of the lake at 4 m intervals.

136–138. Pneumatic Environment
Design: Graham A. Stevens, 1971

This arrangement was designed by Stevens on the occasion of the Septième Biennale de Paris. It was intended to demonstrate possible means of expression in form rather than technical functions.

143–149. Inflatable Kindergarten

Design and manufacture: Seminar Pneumatische Konstruktionen, Institut für Umweltplanung, Ulm, under the direction of Gernot Minke, 1972

The 60 m² trial structure consists of 120 m² PVC sheet with a thickness of 0.5 mm and 60 m² PVC coated polyester fabric.

Access is by "lip doors"; the membrane pouches, which open inwards, are pressed together by positive pressure so that they are self-sealing.

The fan has a power input of 0.16 KW. With 15 changes of air per hour it produces a positive pressure of 10 mm of water pressure.

143

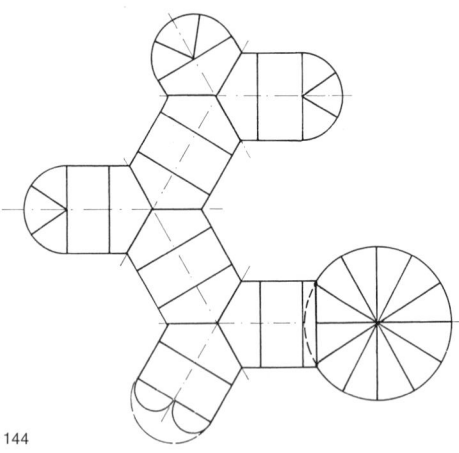

144

145

146

147

148

149

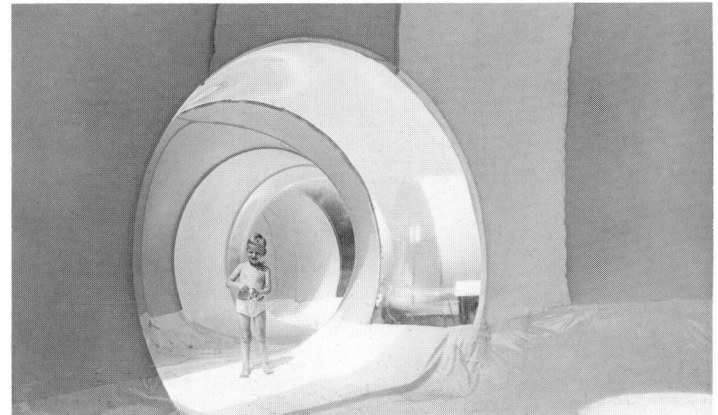

150

151

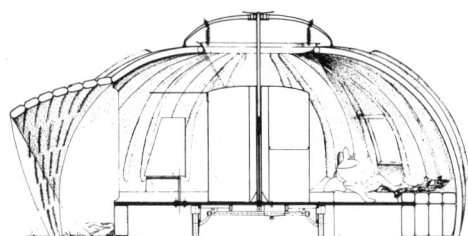

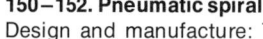

153

150–152. Pneumatic spiral
Design and manufacture: Tom Colborn, Paddy Acheson, William A. Hanna, and Robin McKelvey under the direction of Graham Hardy, 1968

Because of its cutting pattern the spiral has a tendency to draw together so that the coils are pressed hard on one another. Thus in addition to the normal membrane stresses, there are also torsional and bending stresses, which act like "prestress" on the structure and give it a certain stiffness against lateral forces. It would be even better if the individual lays of the spiral could be prevented, by means of additional connections, from displacing each other.

153, 154. Caravan with variable volume
Design: Jean Louis Lotiron and Pernette Martin-Perriand, 1967

During the journey the volume of the caravan is reduced to the utility unit and takes up a mere 6 m³. On arriving at its base, the small compressor attached to the car motor can inflate the dome in less than half an hour. The dome consists of individual tubes and covers an area of almost 25 m².
The floor of the dome is formed by the folding-down casing of the utility unit; its top part is a pneumatically raised light and ventilation dome. Six pneumatic beds. and six swivelling wardrobes are supplied as furniture.
(Bibl. 50, p. 16.)

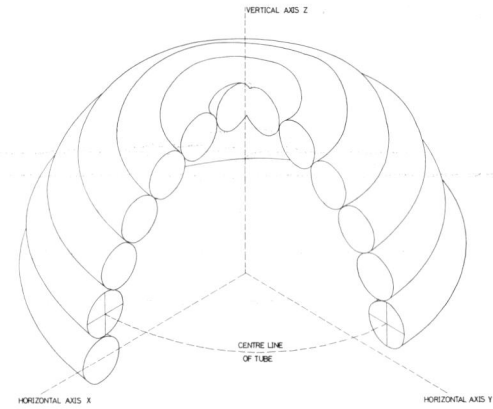

152

157, 158. RFD rescue islands
Design: RFD-GQ, Ltd.
Manufacture: RFD-GQ, Ltd.; Autoflug GmbH

These "floating shelters" are supported by two tube-like pontoons that inflate automatically and with great reliability in seconds. They are carried on ships and aeroplanes.

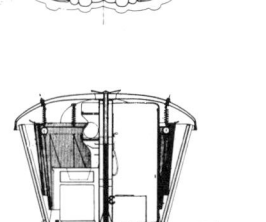

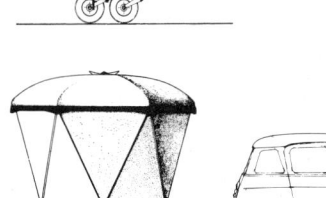

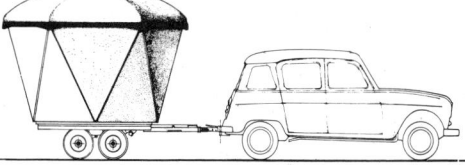

155

156

155. MUST (Medical Unit, Self-contained, Transportable)
Design and manufacture: Air Cruisers Company

MUST consists of sections of 12 tubes each with a cross-section of 30 × 50 cm, which when placed one behind the other form barrel-shaped rooms of any length. The air spaces in the tube are separated from each other but are inflated simultaneously. The membranes are made of Dacron fabric with neoprene coating and an outer Hypalon coat. For safety reasons, inside every tube there is a second tube made of nylon with a neoprene coating. The positive pressure is 1,000 mm of water pressure.

A building made of four sections can be erected by eight people in 30 minutes. It is subsequently secured by guys and can thus withstand wind speeds of up to 130 km/h.

About 400 buildings of this kind have been used since 1968 as field hospitals in Vietnam.
(Bibl. 131, p. 161 ff.)

156. Numax house
Design and manufacture: RFD-GQ, Ltd.

The Numax house is mainly used in the military field and as a shelter in disaster areas, where the small volume of the pack and quick erection (some eight minutes) are particularly important features.

The tubular framework supporting a weather proof membrane is called a low pressure structure (1400 mm of water pressure) by the manufacturer and in the technical literature (Bibl. 44, p. 138; Bibl. 131, p. 211). Air expansion bags are attached to the tube structure in order to compensate for the increasing pressure due to heating by solar radiation.

Inflation of the structure is achieved by use of a fan that can be connected to car batteries, or by means of bellows. The valves are at the front ends.

157

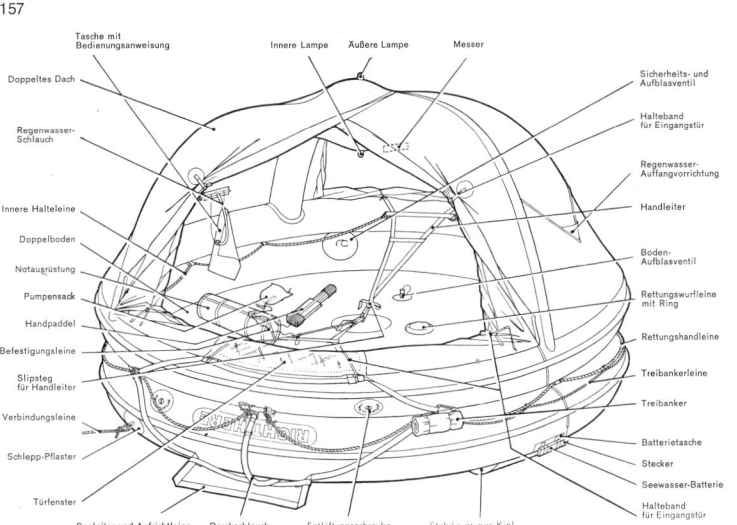

158

159

160

159, 160. Beach pavilion
Design: Frei Otto, before 1962

The double cones are connected at the points of contact and can therefore be inflated from one point. The compressor is situated under one of the cones.

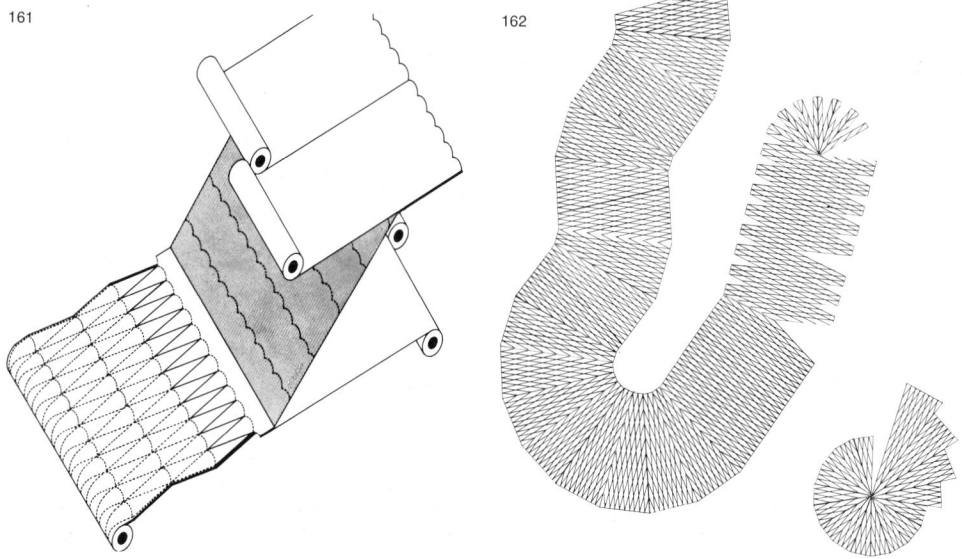

161

162

161–171. Construction system for pneumatic shells
Design: Winfried Wurm, 1968

In contrast to the double cones of Frei Otto's beach pavilion, these elements are not buckled on the inside, thus allowing also curved cross-sections of the building to be produced.

The factors to be considered in the manufacture of these structures from two membrane layers are of special interest (Fig. 161). As the seams run through there are no connection problems between the elements of a section. The inner membrane is made up by the coated fabric running from the roll without any cutting pattern or any processing. The outer membrane is folded by the roll into lengths corresponding to half the length of the double cone and sewn as folded into the shape of an arch. Then the inner and outer membranes are matched and sewn together in a straight line.

Sc, Mc / 1 di / 0

If high thermal insulation and resistance to snow loads are required, rigidising foam is used instead of pressurised air in the chamber. This foam is a mixture of resin, an expanding agent and a hardening agent that rises up of its own accord and produces an adjustable positive pressure before it hardens after about three minutes. In this construction system the membranes are, in fact, being stressed by a pneumatic process; however, this process serves to create a sandwich-like structure densified towards the upper surface which no longer represents an air supported structure.

The number of breaks in the curve of a structure can be arbitrarily laid down. On it depends the ratio of slenderness of the double cone. There are mathematical formulae that on the one hand represent the form relationships between the number of breaks, the length of the cone and the largest cone diameter, and on the other hand permit the statically correct dimensioning of the structure.

By combining differently constructed individual sections many different plan forms can be produced (Fig. 162).

In collaboration with the DLW Aktiengesellschaft and the Kunststoffbüro München GmbH & Co., a prototype was made with a span of 9.5 m and an average diameter of the cone of 0.8 m. PVC coated polyester fabric with a tensile strength of 300 kp/5 cm was used for the membrane. Polyurethane foam with a density of 60 kg/m³ and a compression strength of 4 kp/cm² served as a foaming agent. The structure corresponded to expectations in terms of manufacture and costs. However, an early structural failure occurred under snow and wind load. A span of only 4.5 m could be achieved with semicylinders and of 9.8 m with domes.

By using thicker foam and firmer membrane material domes with spans of 50 m are supposed to be feasible.

(DBP 1 930 563.)

164

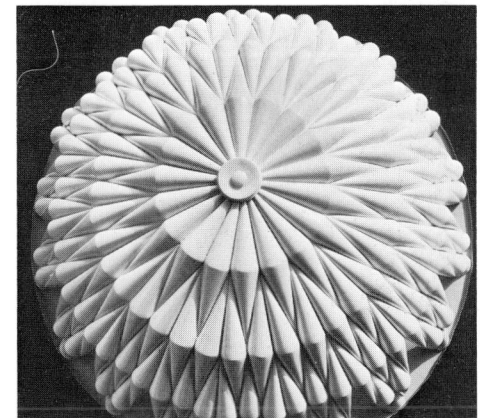

165

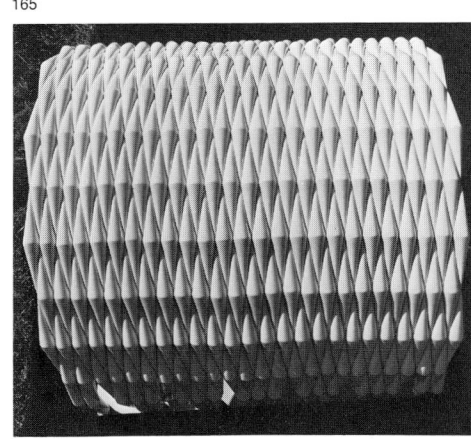

166

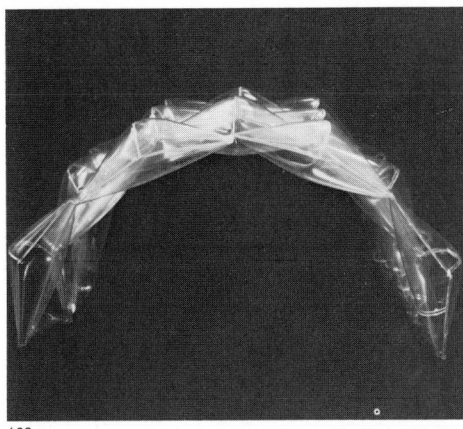

168

169

170

167

171

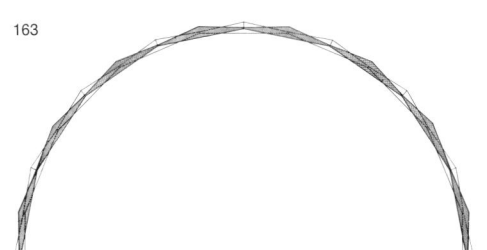

163

172–183. Fuji Pavilion, Expo '70, Osaka
Design: Yutaka Murata
Structural calculations: Mamoru Kawaguchi
Manufacture: Taiyo Kogyo Co., Ltd., and Ogawa
Tents Co., Ltd., 1970

This construction, which is the largest multi-
membrane structure so far built, has become
world famous for its magnificent organic form. It
consisted of 16 arched tubes with a diameter of
4 m and a length of 78 m, whose bases defined a
circle with a diameter of 50 m. This arrangement
caused the two ends of the building apex to jut
forward by 7 m. The tubes were held together at
4 m intervals by an encircling horizontal band of
50 cm width. The centre arch was semicircular,
the others arched higher and higher as their
bases came closer together. The openings of
10 m width at both fronts served as entrances.

The tubes consisted of PVA with a tensile
strength of 200 kp/cm and a weight of
3.5 kg/m². The exterior was coated with Hypalon, the interior with PVC. The lower ends were
anchored in steel cylinders.

The internal pressure was normally 1000 mm
of water pressure and 2500 mm in storm conditions. All tubes were connected to a central
turbocompressor by means of a peripheral system of steel pipes. This turbocompressor was
very efficient and could react even in a strong
wind.

A slowly evolving turntable was installed inside
from which a fascinating 20 minute show could
be watched. Multi-projectors specially developed for this purpose threw their light on to the
white interior of the tubular wall (the tubes were
only coloured on the outside). Under the turntable was a pneumatic single membrane structure which sealed off a control room from dust,
as well as small double membrane domes hous-
ing a bar and toilets.

When the World Exhibition was over the pavilion
was destroyed (Fig. 182). In the brief time available for planning it had not been possible, in
view of the many basically new technical developments, to make the building completely dismountable.

(Details see pp. 147, 148, 154.)

172

173

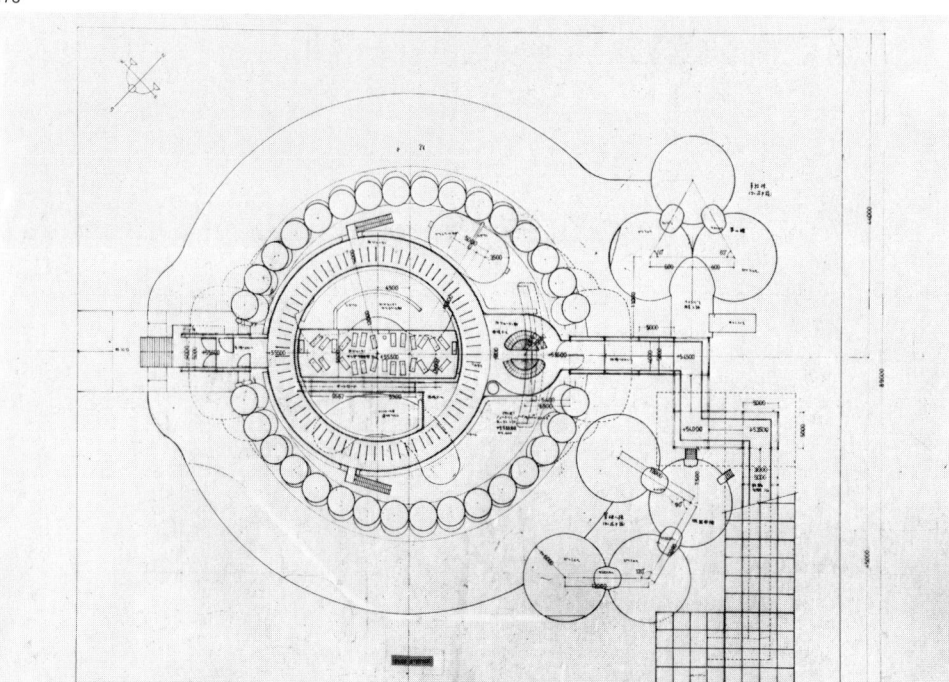

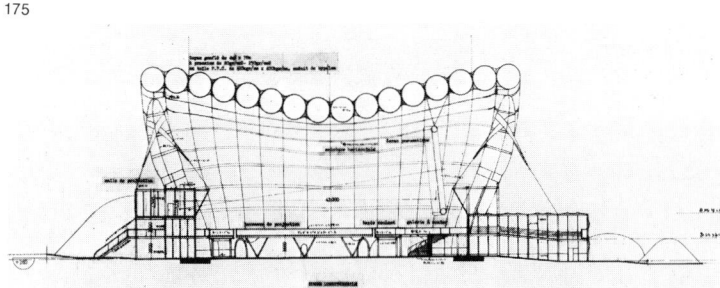

174

175

176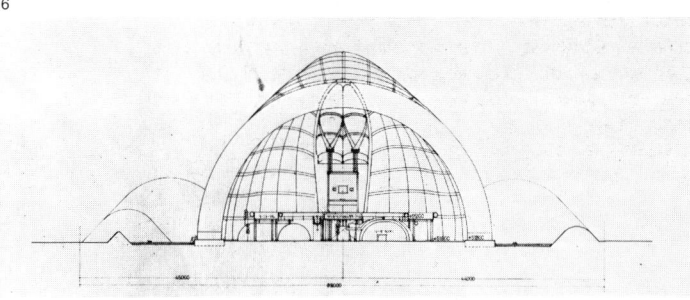

177

178

179

180

181

182

183

184–191. Federal German Pavilion for Expo'70, Osaka. Competition design
Design: Wolfgang Rathke and Eike Wiehe, 1968

This structure could certainly have held its own against its Japanese competitors. However, the jury gave preference to another design.
From the architects' explanatory report:
"The proposal includes the integration of building and exhibition, as desired by the competition committee, by means of the many possibilities in the programmed play of structure, space, light and shade. The cloud-like roof formation of PVC balloons can be made of transparent, translucent and coloured material that is changed in appearance by the introduction of light in the programmed play. The possibility of free agglomerations of the cellular roof formations offers every capacity for adaptation to special demands for different exhibitions or operating cycles . . . External and internal wall joints are effected by prismatically connected glass or plastic lamellae which are introduced into the balloon structure of the 'cloud'".
In order to test the basic theories a model, scale 1:2, was erected out of 100 "pneumatic rods" under the supervision of Professor Polyoni at the Technical University in Berlin. The recorded results formed the basis of the expertise given by him, which states:
"The structure described below uses a new kind of loadbearing system that is produced by the joining together of individual air filled, vertical cylindrical balloons by means of several horizontal belt nets. This loadbearing system, which seen as a whole resembles a plane loadbearing structure in the usual sense, is able to resist bending tension and because of its low deadweight is suitable for covering large spaces with any kind of ground plan".
The structure consists of any number of cylindrical balloons and at least two horizontal belt nets. The balloon should have a diameter of 1.25 m and a height varying between 5 and 15 m. The material used is PVC coated Diolen (Polyester). At the front of each balloon there is an air inlet with a nonreturn valve. The belt nets are also made of Diolen; alternatively other fabrics or wire ropes can be used. The fabric forms a hexagonal net whose nodal points are formed by metal connection elements.
The balloon membranes are put into the prefabricated belt nets in a non-inflated state. All the balloons are connected by air tubes and all receive the required calculated pressure. Thus initial tensile stress is present in the belt net. The balloons press one against the other and take on a hexagonal form, similar to a honeycomb. Thus the two main loadbearing elements of the structure, balloon and belt net, are under tensile stress while in an externally unstressed state. The lower belt layer takes on further tensile forces when used as roof covering, while in the upper belt the tensile forces are reduced. The internal pressure of the balloon remains largely constant throughout. Thus tensile stresses occur in the structure even under load.

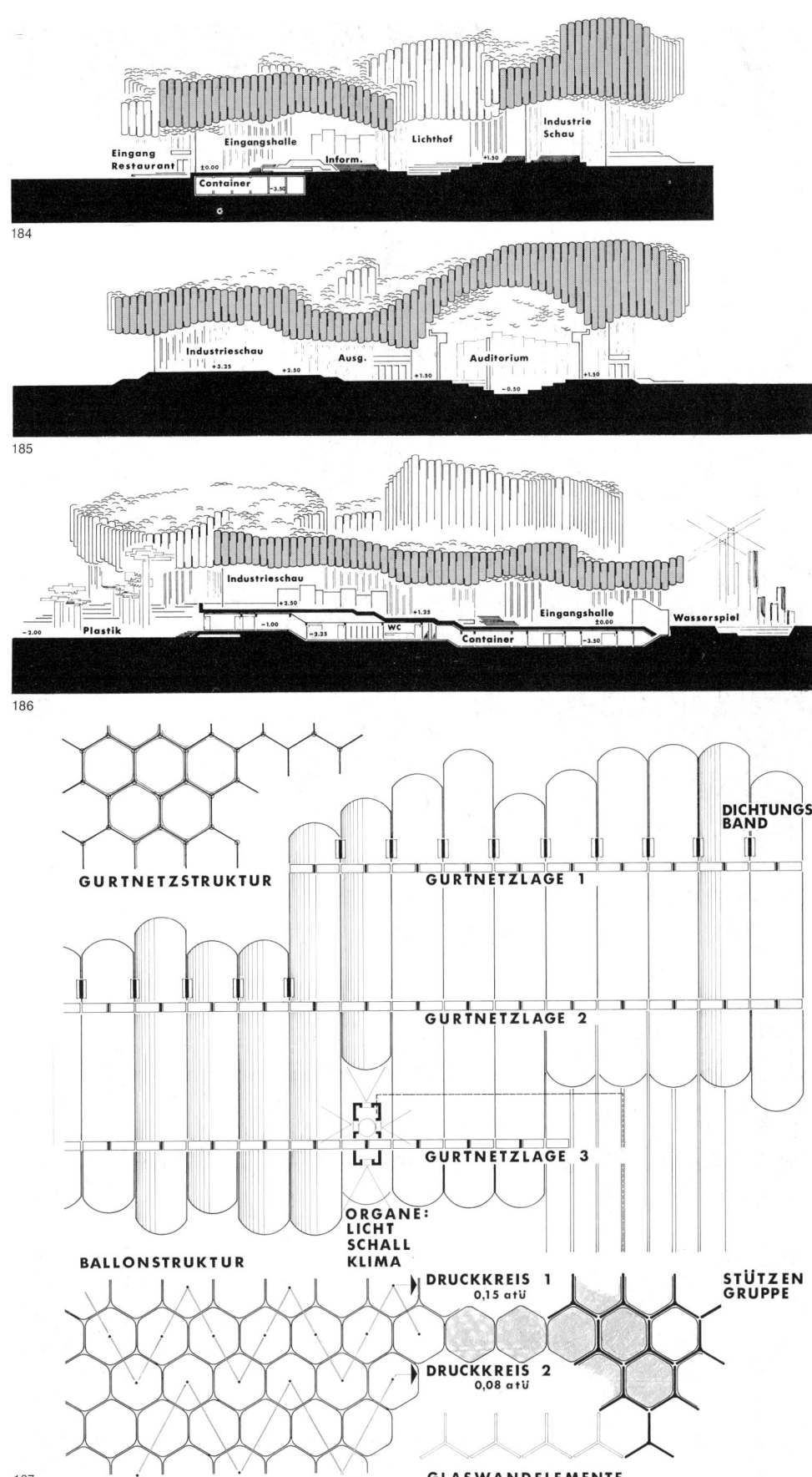

188

189

190

191

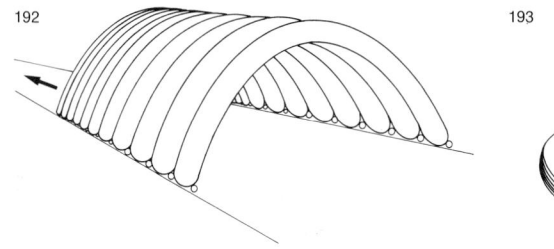

192

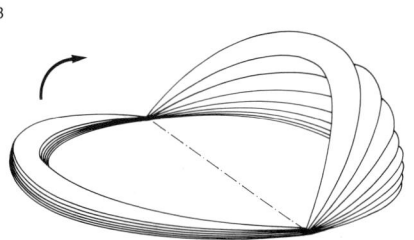

193

192, 193. Inflatable structures with variable form and volume
Design: Michel Fourtané, 1969

The similarity of these two mobile structures, both termed roofs, to Ramstein's theatre (Figs. 194–197), is striking. Certainly in the next few years buildings will be erected according to these principles.
(DBP 1961 523.)

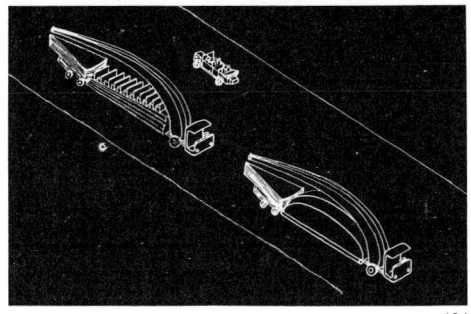

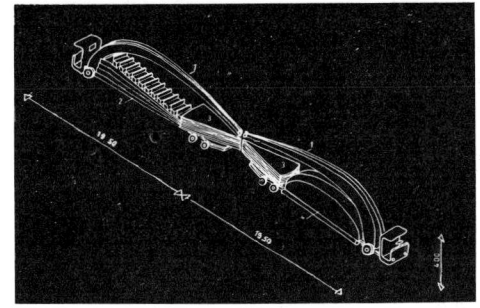

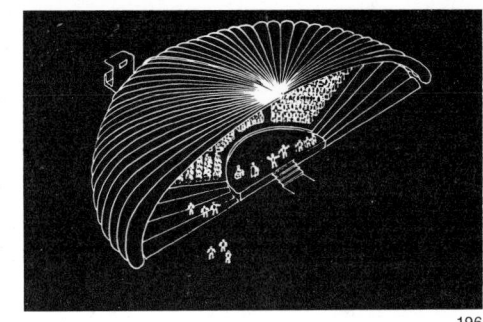

194   195   196

194—197. Travelling theatre
Design: Willi Ramstein, 1961

The significance of this project lies in the fact that as long as 15 years ago the designer recognised the opportunities offered by compressed air as a stabilising medium and as a means to move building components.

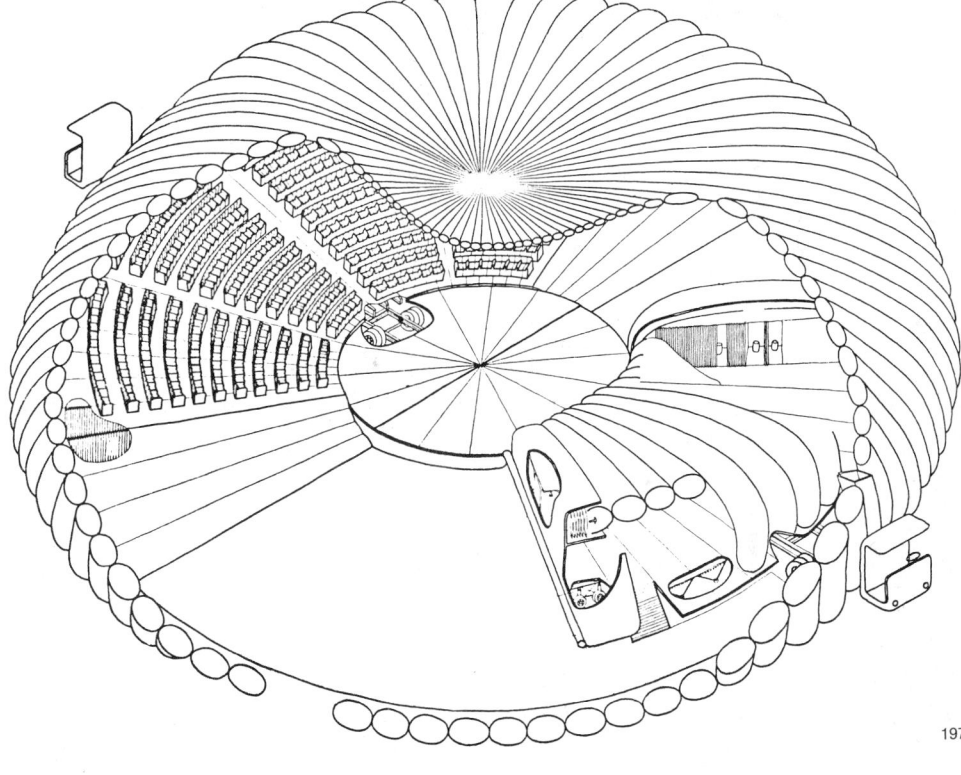

198, 199. Mobile roof covering
Design: Robert Laporte, Pierre Malachard des Reyssiers, 1970

This is a kinetic pneumatic structure in which the technical details have been carefully worked out.

The two halves of the building are individually inflatable and form two self-supporting structural units. When deflated each half lies in a box shaped cross-section. The uppermost tube is connected to the lid of the box which is elevated during inflation. On the lid is a stiff element made of transparent material which serves to illuminate the inside of the building and to which is attached an apparatus for connecting the two halves of the building.
(DBP 2 053 702.)

197

199

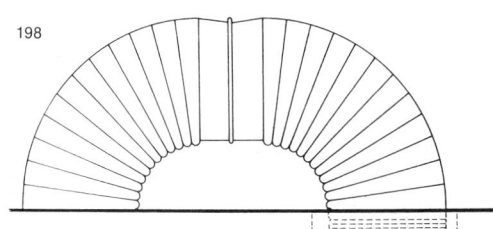

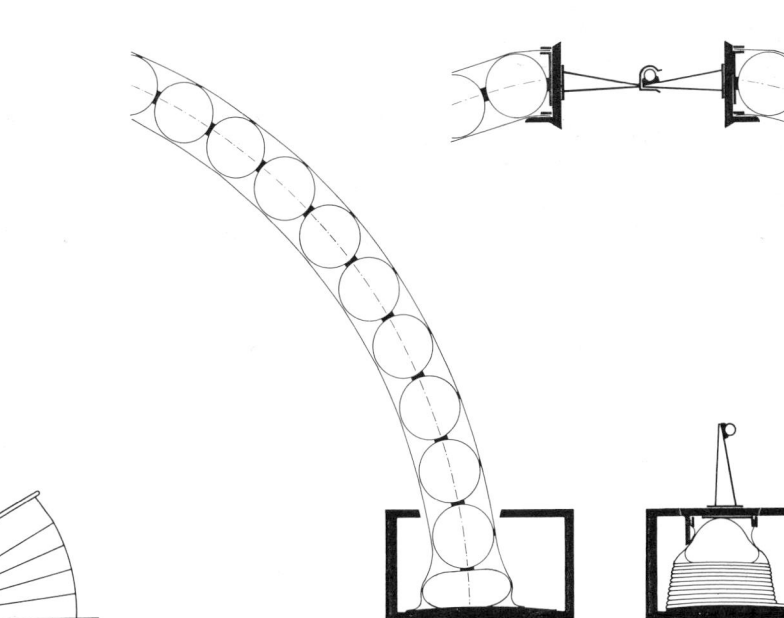

198

age and steel cables as guys against wind loading.

Christo said afterwards that the most important experience he gained from the project was to be made aware of the enormous difference between project and realisation (Bibl. 50, p. 50) – a recognition of great significance for pneumatic forms in general.

The installation shows clearly how linear support of the membrane by cables leads to a change in the radius of curvature.

201, 202. Air cushion roof for the Waldstadion, Frankfurt am Main
Design: Bernd-Friedrich Romberg, 1971

For the World Cup in 1974 FIFA required a great number of covered stands. The Frankfurt stadium, one of the grounds selected, would, according to Romberg's proposal, have a stand

roof made of open membrane sections and supporting steel trusses. The architect wrote of his project (Bibl. 139):

"The envelopes forming the surfaces are made of durable, very strong, coated and flame resistant material and are mounted between the steel trusses above and below as well as on the inner and outer edge of the roof. The junction of the envelopes and the T-shaped upper and lower chords of the trusses is achieved by means of clamping bands which, together with the chords, hold the edges of the envelopes. An almost airtight connection is achieved by means of tightly screwing the components. The envelopes thus mounted form a continuous accessible inner space which stretches over the whole stadium. This space is maintained at a specific pressure by means of compressors. These stabilise and support the whole envelope without external prestress having to be applied. The roof structure is transparent, and this is of par-

200. 5600 Cubic Meter Package, 4. documenta, Cassel
Design: Christo
Engineering consultants: Dimiter S. Zagoroff and R. Trostel
Manufacture: Wülfing und Hauck, 1968

On 25th June 1968 a first attempt was made to realise this sensational project. Not until 3rd August – at the fourth attempt – was the plan successful. The total costs (borne by Christo by selling everything he had), including the expensive helium gas for inflation which was imported from the USA, amounted to 60,000 dollars.
The object was 85 m high, had a diameter of approximately 10 m and was made of 2,000 m² of PVC coated polyester fabric with a thickness of 2.5 mm. Ropes served as "string" for the pack-

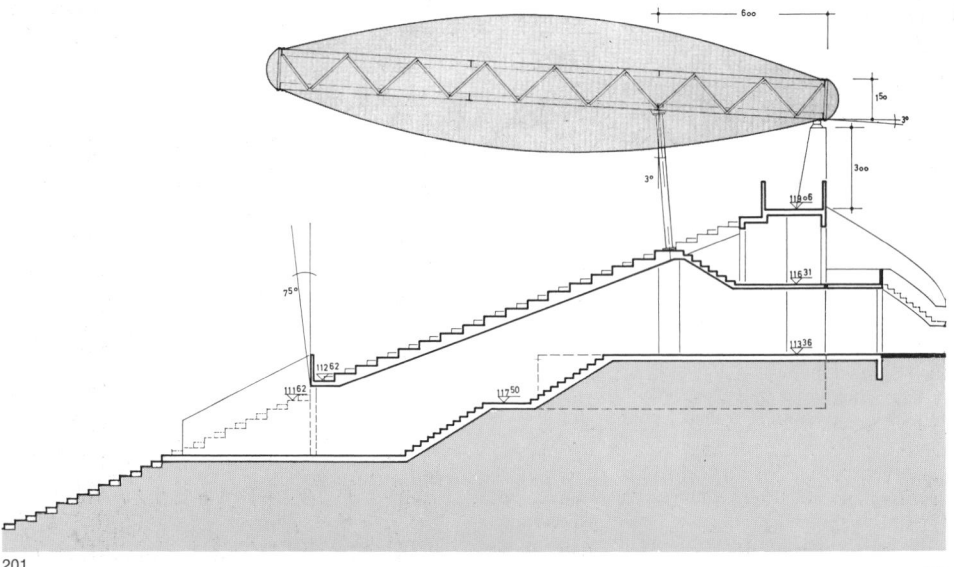

201

202

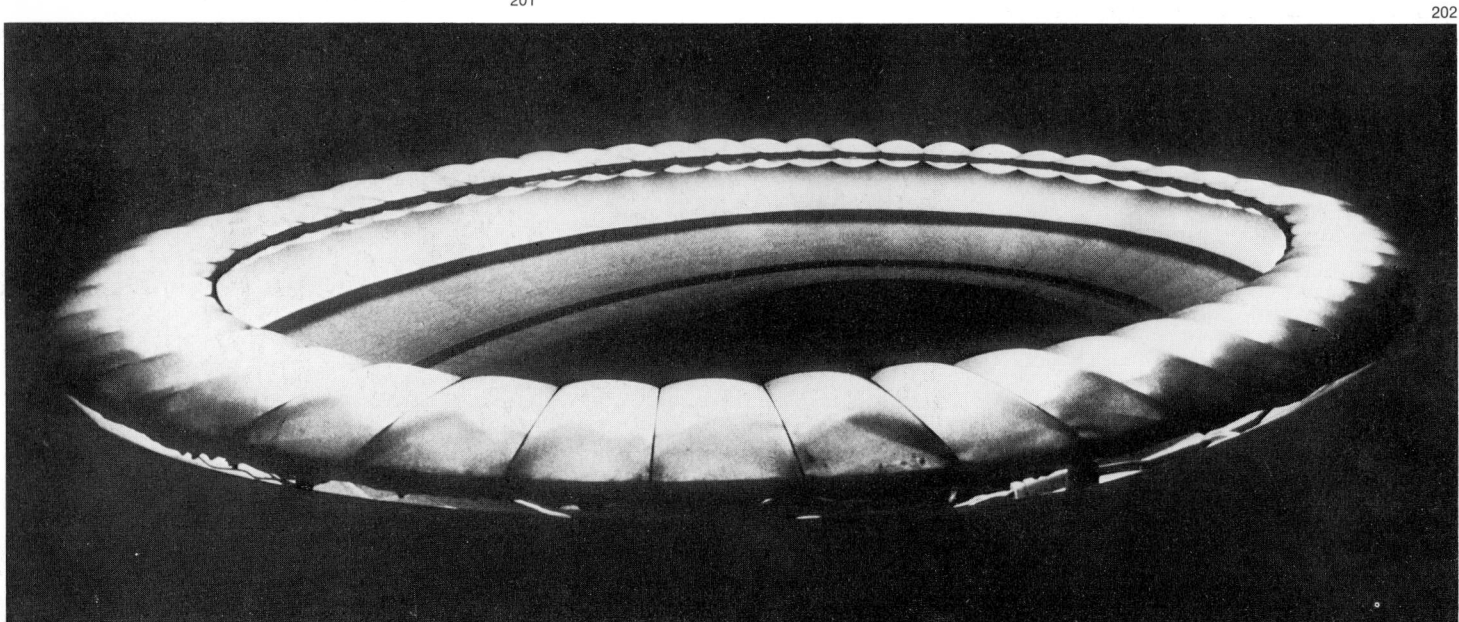

ticular importance for television transmission. Floodlights installed in the protected interior of the roof produce a very good, non-dazzling illumination of the stands. The internal pressure is safeguarded by four compressors. These compressors are adjusted so that for normal operation one of them is sufficient. The internal pressure can be adapted to external weather conditions. A key-operated device raises the internal pressure by increasing the speed of the compressor when there is increased loading on the membrane. This system would also function if there was a leak . . .''

To be precise, this circular tube structure should be classified as a series of open membranes, because the membrane sections between the steel girders are independent of one another. However, because the interruption of the membrane is so insignificant this point is disregarded.

203–205. Diolen Compositum, Hanover Fair 1970

Design: Wolfgang Rathke and Wilfried Lubitz
Manufacture: Conrad Scholtz AG, 1970

Eight pneumatic tubes made of PVC coated Diolen with a length of 10 m and a diameter of 45 cm are held together at each end by two light metal nodes. The two pairs of nodes are connected by a guy rope running in the centre of the horizontal axis. The interior expands when the cable is tightened. By tying the tubes together it is possible to "control" exactly the shape of the space inside.

The space cell was fixed on a cushion base filled with granules. Entry to the inner space was obtained through three smaller cushions.

This modification of the system of a pneumatic cell structure (see pp. 79, 80) proposed by Rathke and Wiehe for Expo '70 was meant to demonstrate the material properties and packaging possibilities of PVC coated Diolen fabric. The sponsor for the study was Enka-Glanzstoff GmbH.

204

205

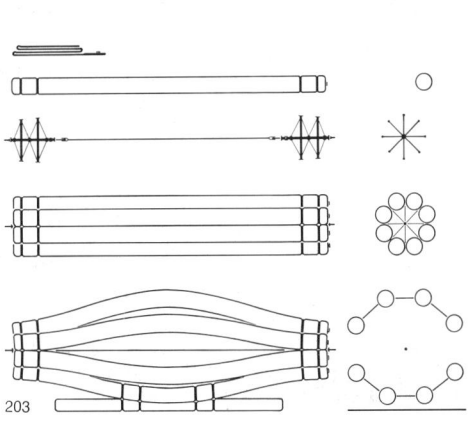

203

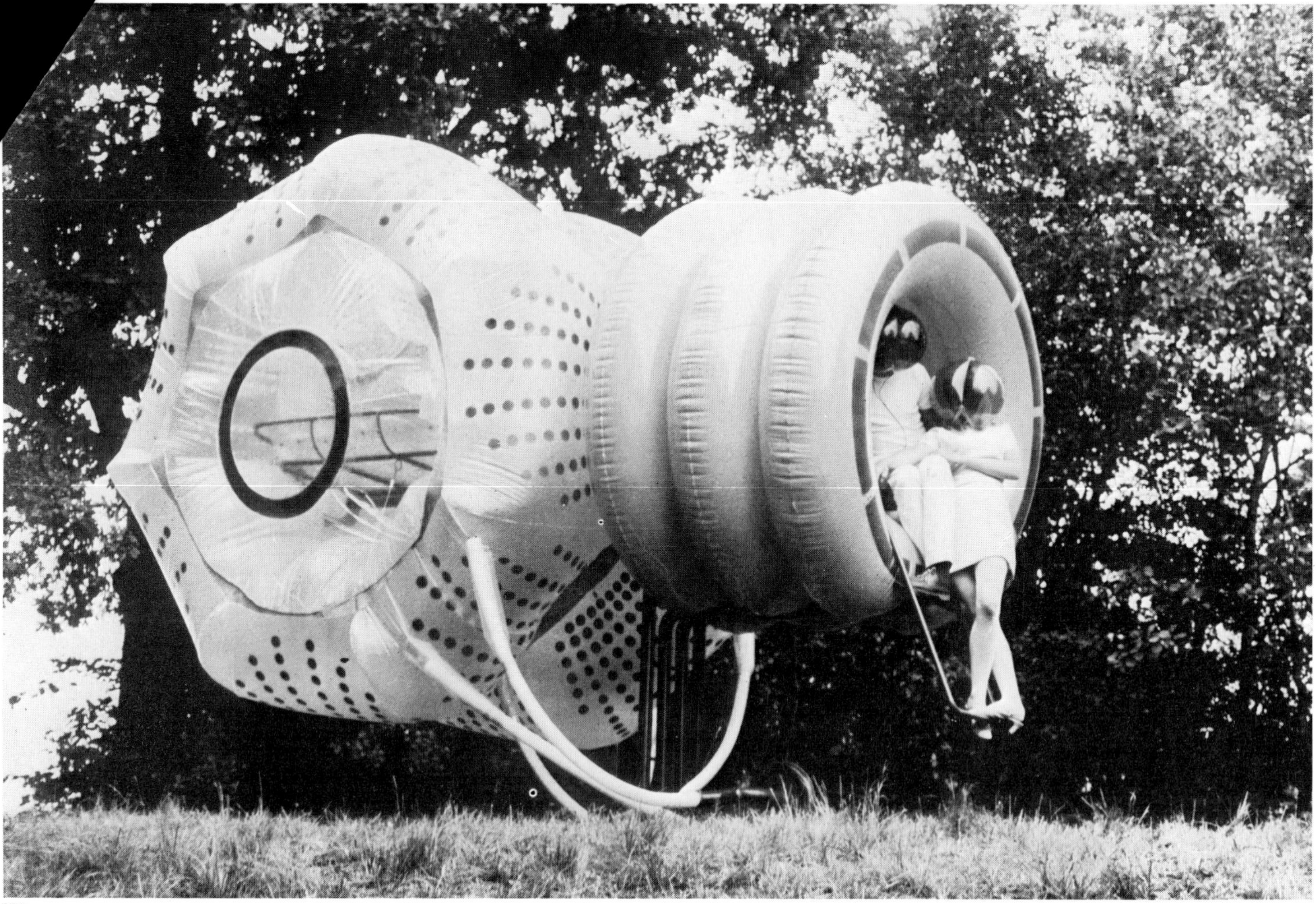

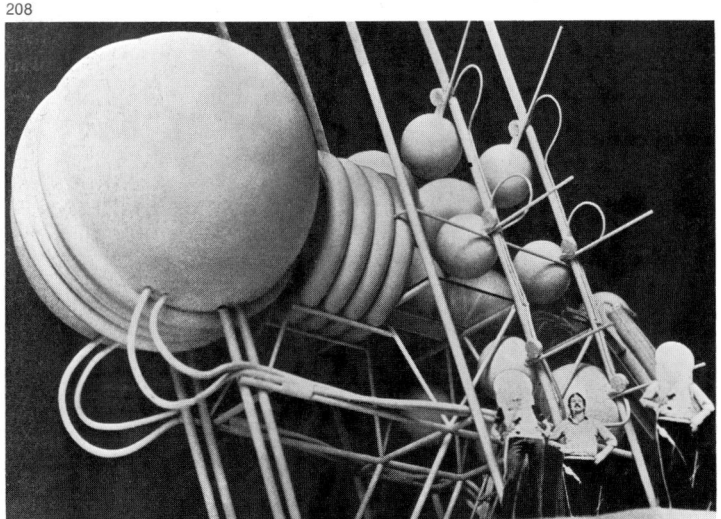

206

207

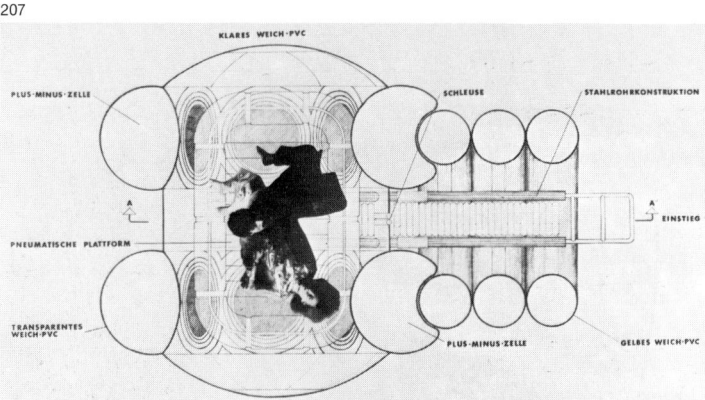

KLARES WEICH-PVC

PLUS-MINUS-ZELLE   SCHLEUSE   STAHLROHRKONSTRUKTION

A   A   EINSTIEG

PNEUMATISCHE PLATTFORM

TRANSPARENTES
WEICH-PVC

PLUS-MINUS-ZELLE   GELBES WEICH-PVC

208

209

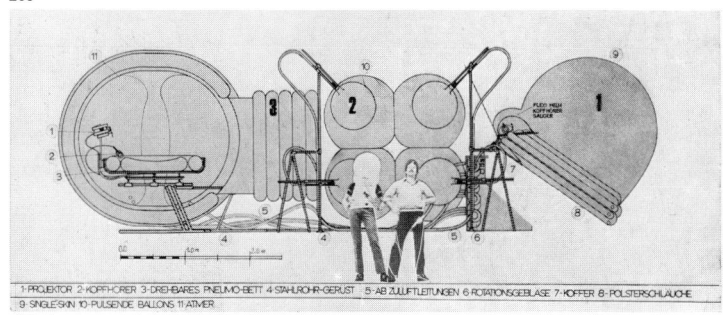

1-PROJEKTOR 2-KOPFHÖRER 3-DREHBARES PNEUMO-BETT 4-STAHLROHR-GERÜST · 5-AB ZULUFTLEITUNGEN 6-ROTATIONSGEBLÄSE 7-KOFFER 8-POLSTERSCHLÄUCHE
9-SINGLE-SKIN 10-PULSENDE BALLONS 11-ATMER

Sc, Mc / 3 di + 2 di + 1 di / 0

206, 207. Gelbes Herz (Yellow heart)
Design and manufacture: Haus-Rucker-Co., 1968

The "heart", made of PVC foil, expands and contracts rhythmically by means of a corresponding control of the air aggregate. As the foil is marked by points, the alteration in the spatial dimension can also easily be seen on the surface.

208, 209. Villa Rosa
Design: Coop Himmelblau, 1969

A pneumatic recreation room which is variable, pulsating and has a swivelling couch, is fixed in a framework of steel tubes. Light, shade and smell can be controlled inside.

210–212. Pneumacosm
Design: Haus-Rucker-Co., 1967

The design is of interest as a translation of the idea of a pneumatic living cell into the dimension of the town. The great number of problems existing there, will, however, hardly be solved this way. "Oasis No. 7", exhibited at documenta 5 in Cassel, was a kind of prototype. Transparent PVC and very strong Trevira were used for the membrane. The diameter was 7 m.

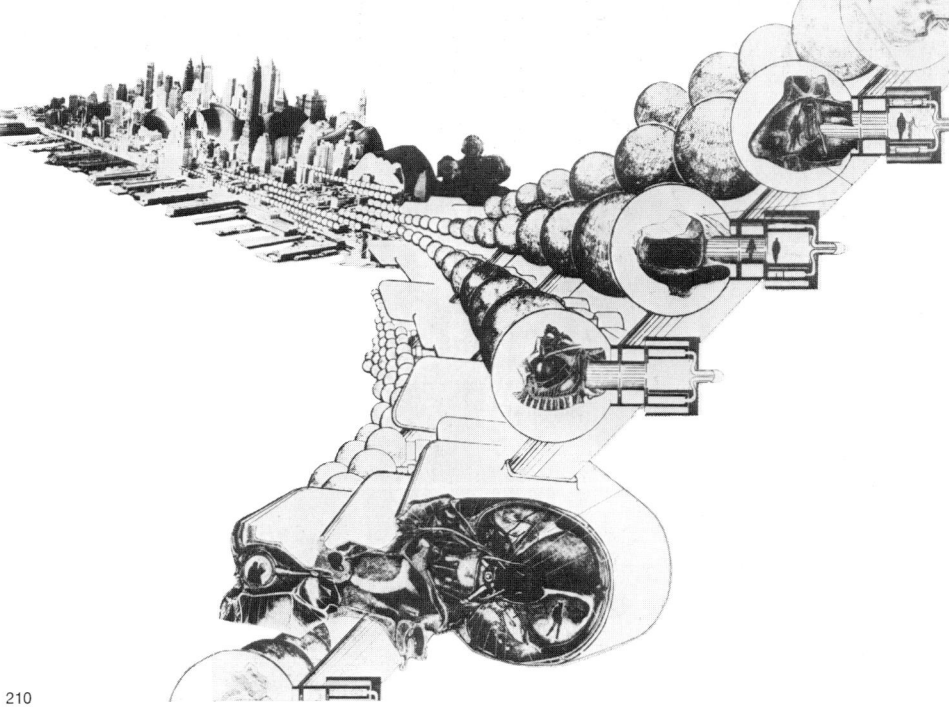

210

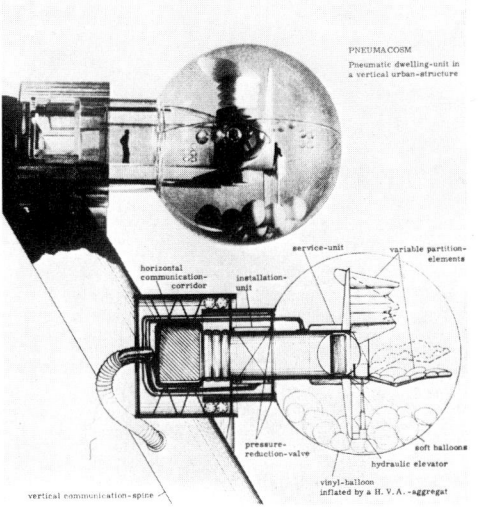

211

212

213–221. Dyodon

Design: Jean Paul Jungmann, 1967

Jungmann seems to have carefully studied the laws of form for pneumatically stabilised structural elements made of closed membranes. However, the expenses required for his design are well beyond the amount firms are at present prepared to spend for trial structures; it was clearly less important for him to find an economical solution by using identical parts, than to design an object which illustrated a great spectrum of pneumatic forms.

A framework of tubes in the form of a polyhedron is infilled by means of rigid and flexible filling elements. The building is stabilised against wind loads by guy ropes in several places.

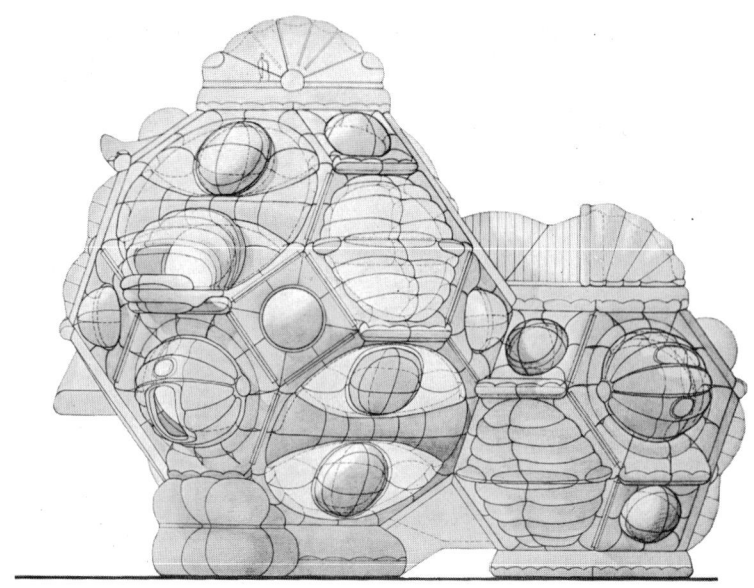

213

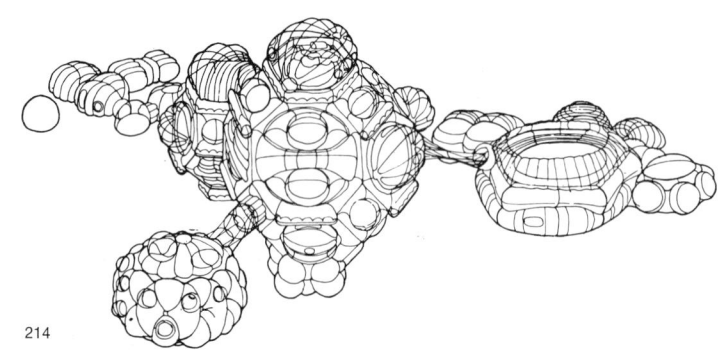

214

215

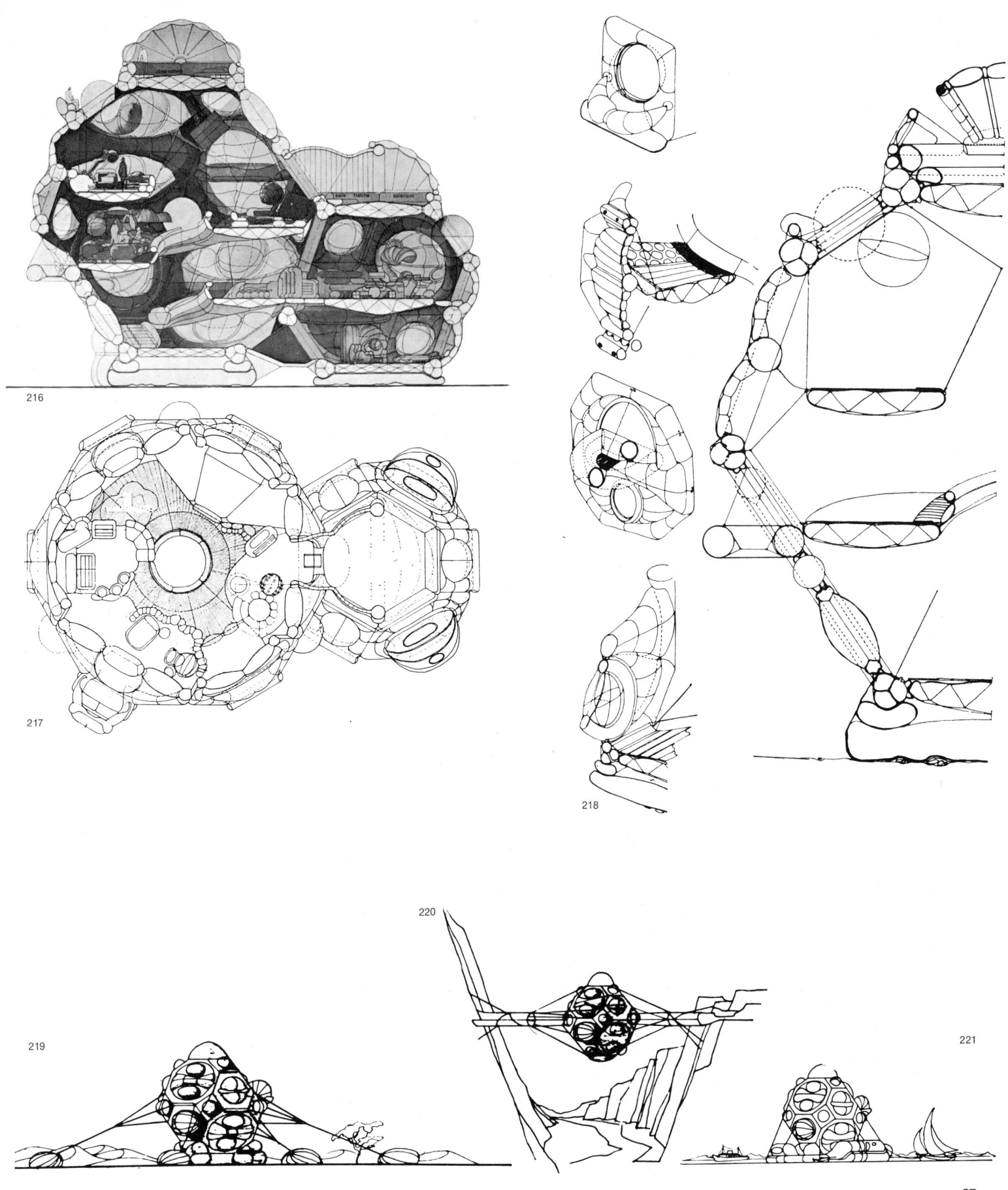

216

217

218

219

220

221

Sc, Mc / 3 di + 2 di + 1 di / L

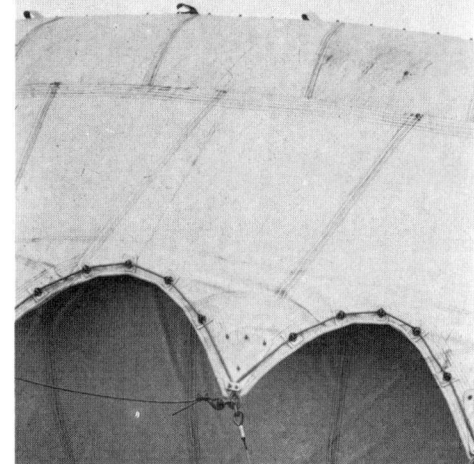

222

223

224

222–228. Floating theatre, Expo 70, Osaka
Design: Yutaka Murata
Structural calculations: Mamoru Kawaguchi
Manufacture: Ogawa Tent Co., Ltd., 1970

The upper roof membrane spanned three tubes made of two layers of PVC coated polyester fabric, and the lower membrane was held by five steel cables. The space in between was under a negative pressure of 10 mm of water pressure below atmospheric, which was raised to 20 mm of water pressure below atmospheric in storm conditions in order to prevent membrane flutter. The tubes had a positive pressure of 1,500 mm of water pressure in normal wind conditions and 3,000 mm in storm conditions.
Behind the theatre seats a steel pipe ran in the hem of the membrane, and the two ends of this pipe were fixed to the floor by hinges. The steel arch was held in place by a cable pull which counteracted the negative pressure. When it was released it elevated and made a wide entrance opening for visitors. (Detail see p. 154.)

The building stood in a small artificial lake on a circular steel structure which in turn floated on 48 PVC balloons. An automatic control system adjusted the air pressure and thereby the buoyancy of the individual floating bodies so that the theatre always stayed horizontal, regardless of the changing loading caused by visitors.

229–231. Truckin' University
Design: Ant Farm, 1970

The Truckin' University, which takes all its equipment with it, is formed by quickly inflatable pneumatic structures corresponding to the concept of great mobility.
A circular tube acts as an opening and for air distribution, and the net over it as wind and lightning protection.

225

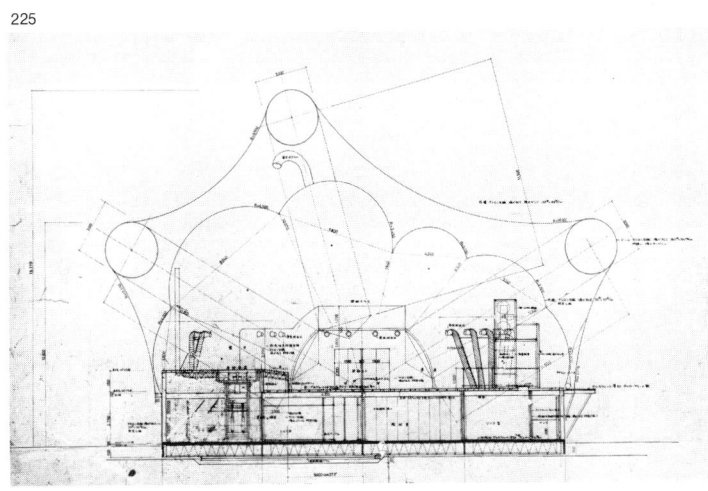

226

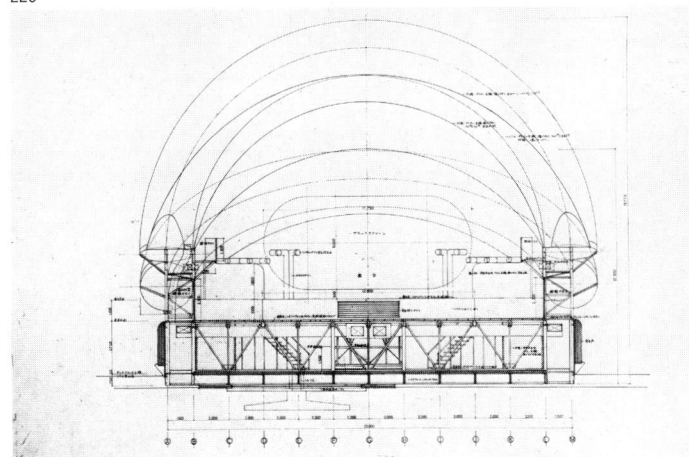

227

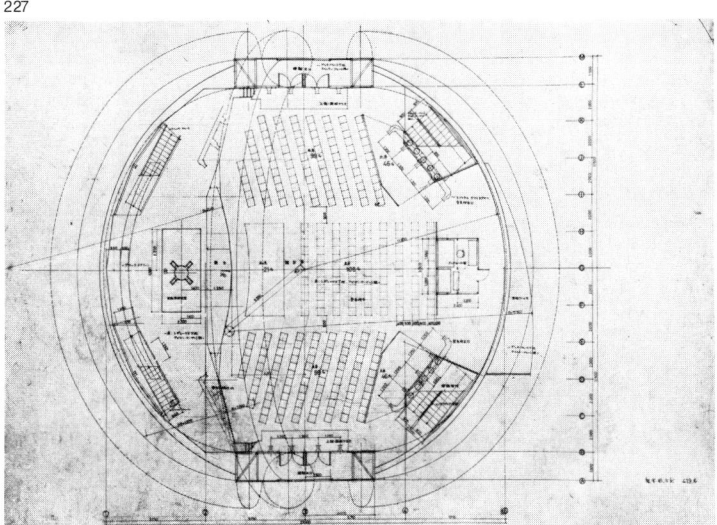

228

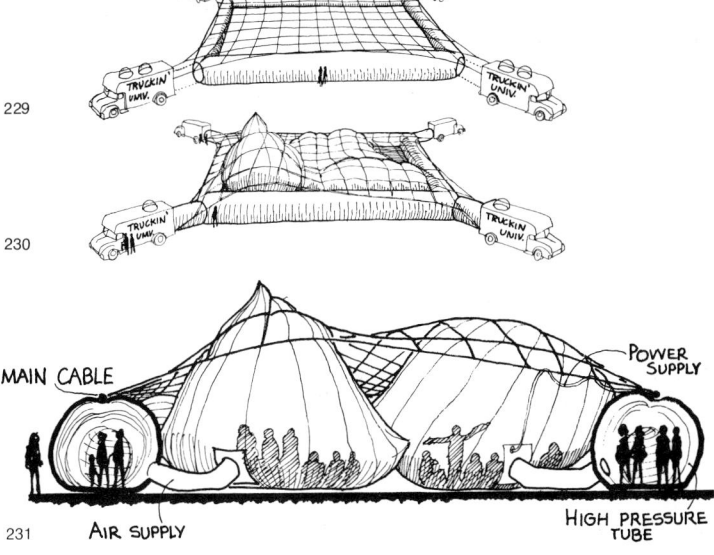

229

230

231   AIR SUPPLY   MAIN CABLE   POWER SUPPLY   HIGH PRESSURE TUBE

232–236. Trial structure in Delft

Design and manufacture: Gernot Minke with students of the Technische Hogeschool Delft, 1971

An arch-shaped high pressure tube with a height of 3.70 m, a span of 7.05 m and a diameter of 1.20 m, together with two elastic high pressure globes, supports a saddle-shaped cotton membrane. The tube is made of PVC coated polyester fabric; pressure is 2,000 mm of water pressure.

234

232

233

235

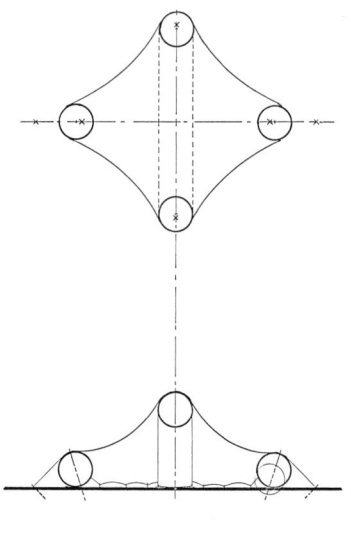

236

Sc, Mo / 3 di + 1 di / 0

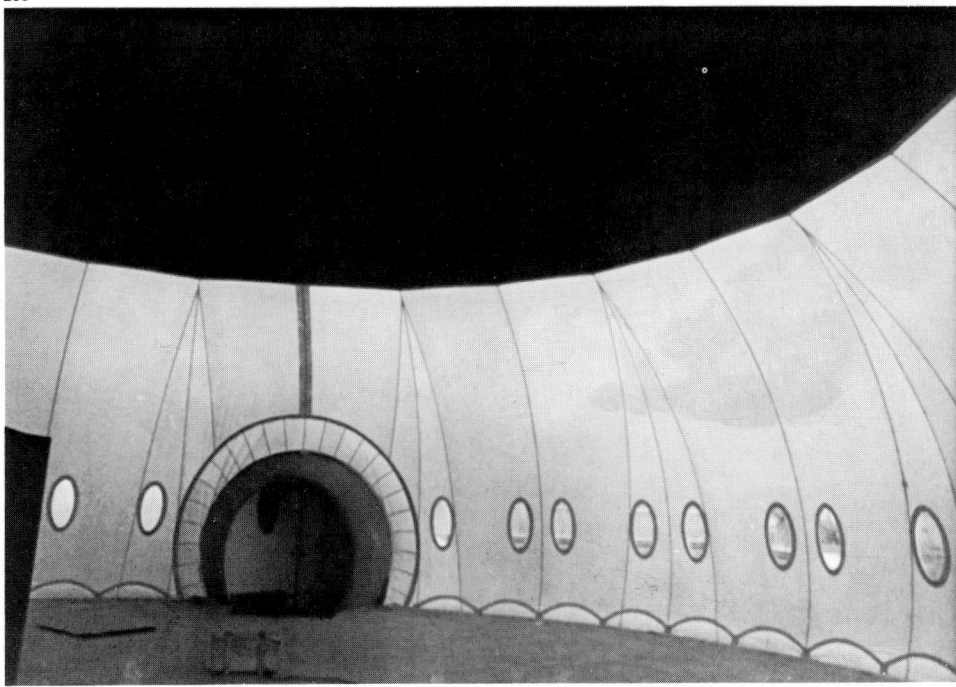

237

238

237, 238. Dome
Design: José Miguel de Prada Poole
Manufacture: Tolder S.A., 1971

This small hall has had many predecessors throughout the world. However, what makes it worthy of mention is the detail design which is usually completely disregarded in such simple buildings, for example the anchor hem, the window, the door, etc. As the formation of wrinkles shows, however, the cutting pattern for the connection collar to the door was not quite successful.

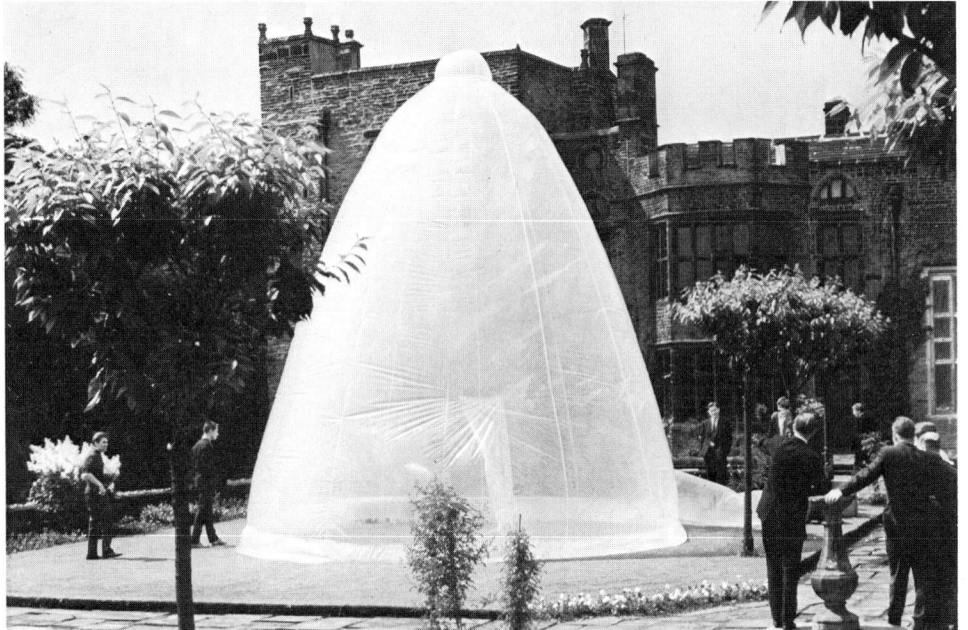

239

239. Pneumatic paraboloid

Design and manufacture: Arthur Quarmby with students of the Bradford School of Architecture, 1963

The experiment was meant to show where the limit for low pressure systems lies, but it was clear that it was not reached with this trial structure. The "wart" on the top was developed for reasons of production engineering. The individual lengths made of polyethelene foil could be more easily welded together with this special cutting pattern. The structure was anchored by means of a water filled circular tube with a diameter of 38 cm. The dome was 7.5 m high and could be inflated in two minutes.

240   241

240, 241. Pneu-dome

Design and manufacture: Chrysalis, Los Angeles

Not only the pneumatic envelopes of these elegant domes, but also their foundations are made of membrane material. The latter is a water filled circular tube lying on the ground. Its diameter is over 8 m.
The night photograph shows the cutting of the membrane and the slipping doors especially well.

244–247. Radome at the Sternwarte Bochum ▷

Design and manufacture: Krupp Universalbau, 1964

The PVC coated membrane fabric has a thickness of 1.2 mm. The sheets are welded. The total weight of the envelope, which has a diameter of approximately 40 m, is 3,500 kg. The erection of the radome over the already installed antennae was tested by the manufacturer in its individual phases on a model (scale 1:10). At the bottom the radome is encircled by a 4 m high wall. The internal pressure of 40 mm of water pressure is increased in wind to a maximum of 100 mm. The normal supporting pressure can be produced by just one of the four compressors. (Bibl. 8, p. 12 ff.)

242, 243. Radome
Design: Société Industrielle L'Angevinière et
Joué-les-Tours, 1963

This French invention (DBP 1255908) is primarily concerned with the method of anchoring the membrane which is here meant to be connected to a steel cylinder (see also p. 153).

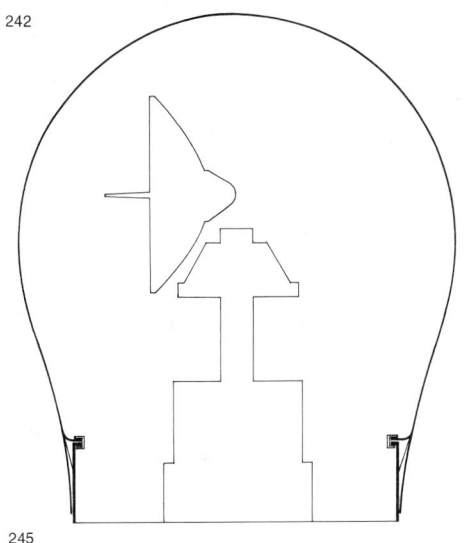

242

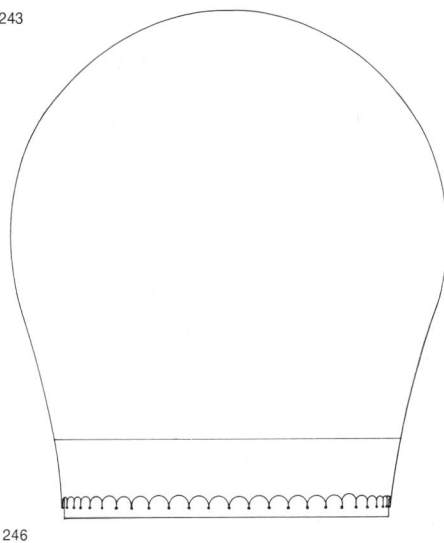

243

244

245

246

247

249

248

248—251. Pavilion for travelling exhibitions

Design and manufacture: Krupp Universalbau, 1958

As underground anchoring is often not possible because of destruction to road surfaces, for this construction a transportable ground and anchorage structure was developed with individual, easily demountable circle sections arranged round a central steel ring.

250

251

252, 253. Pentadome

Design and manufacture: Birdair Structures, Inc., 1958

The five large domes were exhibition areas for the US Army. The centre dome had a diameter of 49 m, while the diameter of the smaller domes was 33 m. The long cylindrical structure served as a lock unit for large exhibits, e.g. rockets. Twelve compressors in all were installed, producing a positive pressure of 20 mm of water pressure. The building covered an area of 4,650 m² and was in its time the largest air supported building used for military purposes. (Bibl. 119, p. 35.)

253

252

254

255

257

258

256

254. Exhibition hall
Design: Seminar Pneumatische Konstruktionen, Institut für Umweltplanung, Ulm, under the direction of Gernot Minke, 1971

In this example individual supports guyed inwards are expanded to bulged surfaces. The author does not know of any pneumatic structures of this kind that have been put into practice.

255, 256. Studies in form
Design: Frei Otto, circa 1960

Frei Otto's model tests show very clearly the influence exerted on a very elastic rubber membrane by guying with additional linear elements. The pressure inside the membrane is constant. It can be seen that such support of the membrane does not only bring the advantage of reducing the radii of curvature and thereby the membrane tensions (part of the tensile stress is absorbed by the cables) but also considerably affects the total appearance – a point that is of great significance for the design of pneumatic objects.

257, 258. Test structure in Eindhoven, the Netherlands
Design and manufacture: Gernot Minke with students of the Technische Hogeschool Eindhoven, 1972

The test building is the first example of a webbing stabilised air supported structure which is composed of prefabricated standard elements (regular hexagon, regular pentagon, half of a regular hexagon).
The webbing straps which are connected to each other to form a net take up the main forces and transfer them from the skin to the anchorage. In case of rupture the tear cannot extend farther than a strap, since the straps, being connected with the membrane, act as tear stoppers.
The straps consist of stretched polyester fibres. In contrast to to steel cables they can be calculated so as to have about the same expansion as the membrane. This is a special advantage in the event of extreme wind loads. Due to the webbing stabilisation the internal pressure necessary is only 10 mm of water pressure.

The focussing effect of the reflected sound waves usually occurring with cylindrical and spherical shapes is avoided by the mode of curvature of the elements.

With this structure a new manufacturing technique was successfully tried: the elements were joined together by stainless steel staples shot by means of a pneumatic tool; the staples had been specially treated so that they automatically bent. Apart from cementing this is the only available technique which allows the jointing of elements in two axes without any limitation in size since the material need not be rolled up in the process. The whole membrane of the structure could thus be prefabricated in one piece.

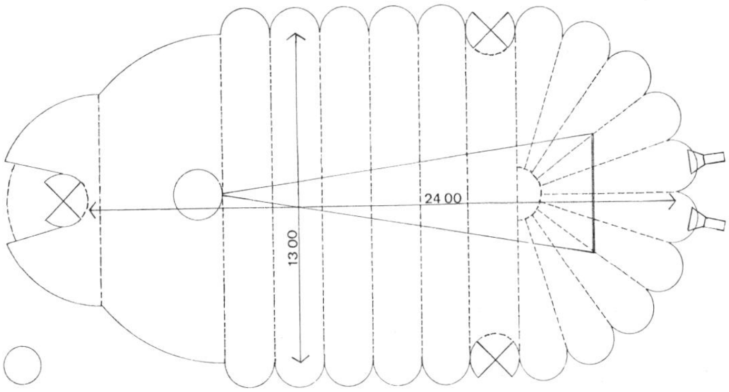

259

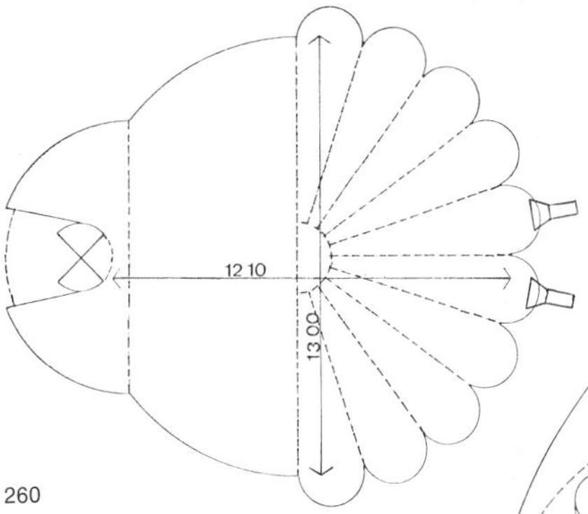

260

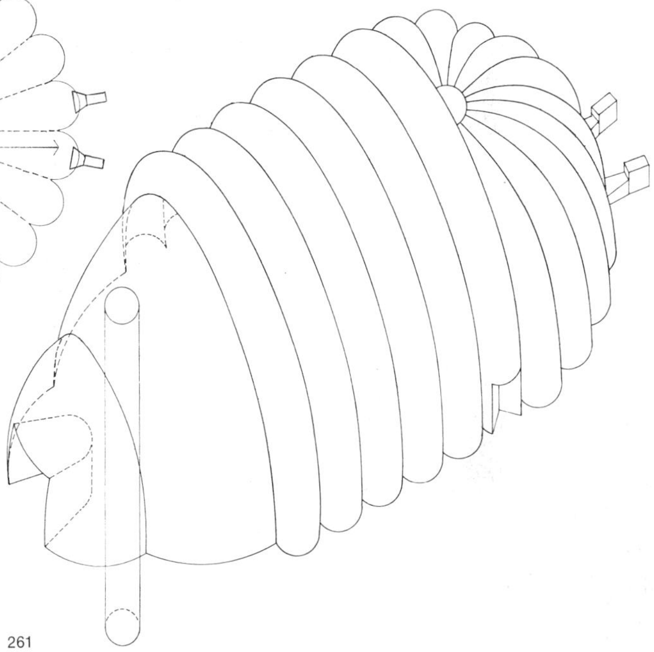

261

262

259–261. Exhibition building for Hidronor S.A. for the 1st Exhibition of South Argentina, Patagonia and Comahue
Design: Airestructures SRL, 1971

In this project the cable supports have the main function of plastically refining the shape of the structure. The axially symmetrical design of the ground plan and the arrangement of the entrances and compressors illustrate the strong aesthetic desire that is the basis of this project. In some places, however, clear deviations from the planned ideal form would probably occur if the project was carried out.

262. Dome
Design and manufacture: Birdair Structures. Inc.

A clear reduction of the tension in the transverse direction results from the arrangement of radial cables as in this American dome.

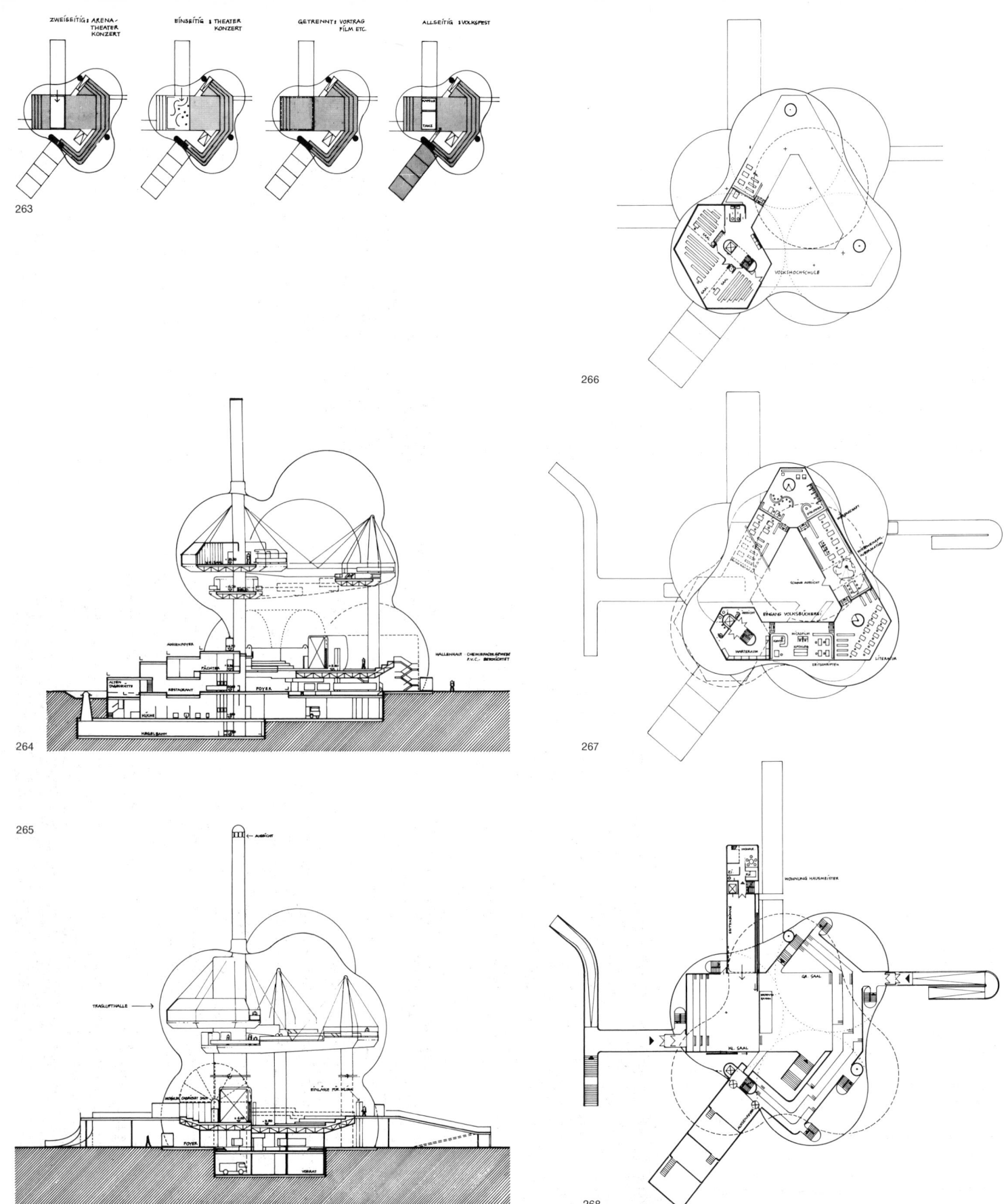

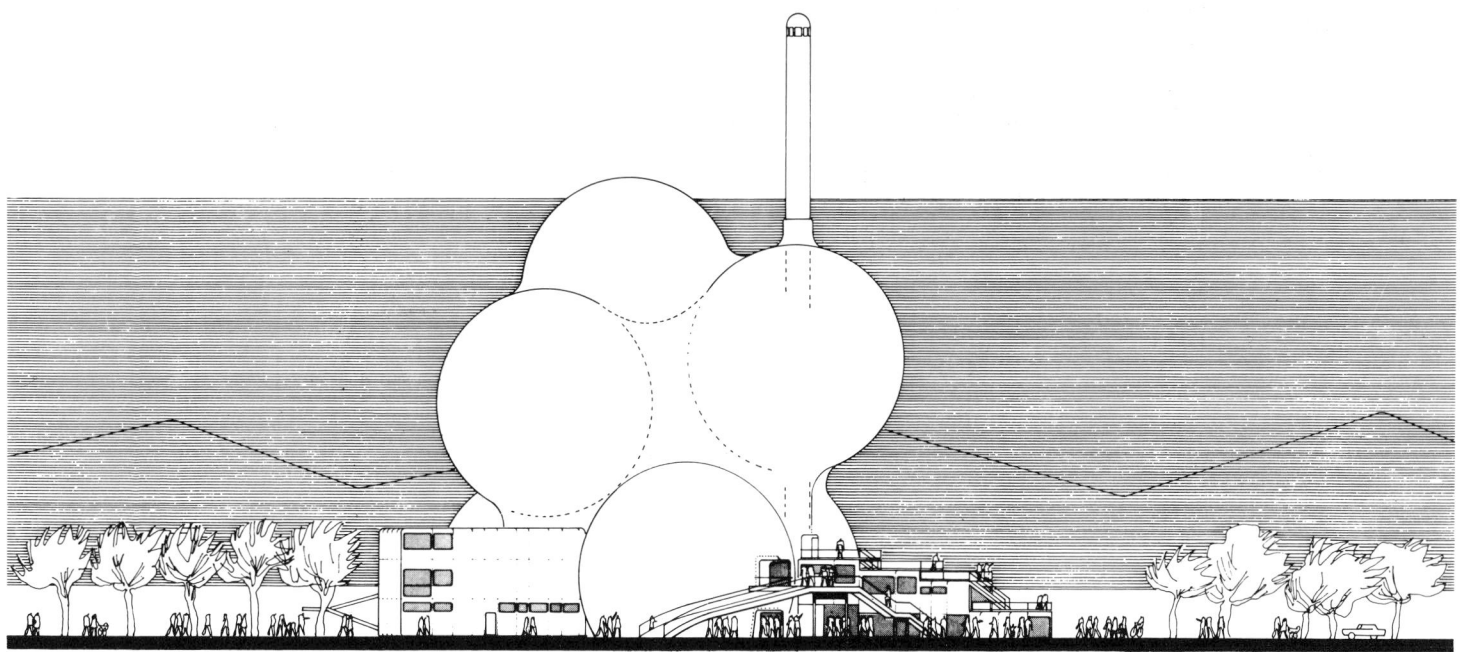

269

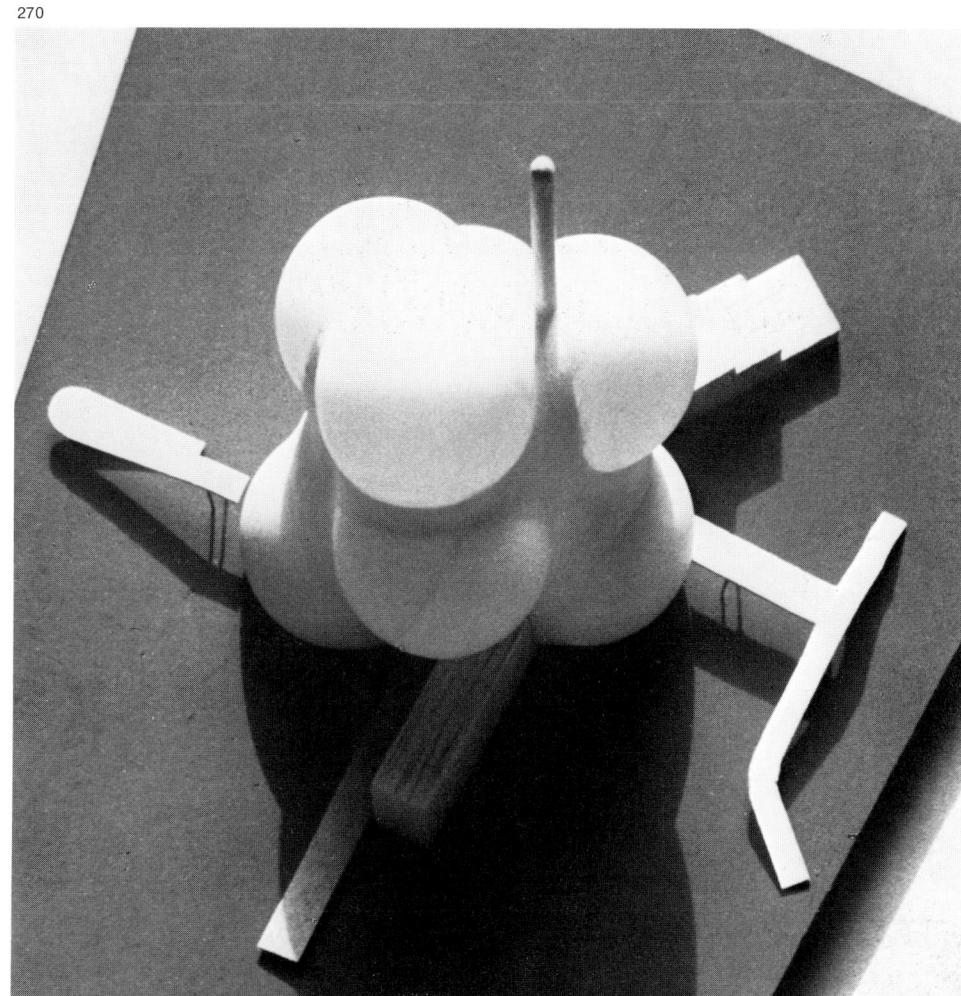

270

263—270. Civic centre in Sprendlingen, near Frankfurt am Main
Design: Manfred Schiedhelm, 1967

The envelope was to consist of transparent and opaque sections. As in the German pavilion in Montreal and the project for the French pavilion in Osaka (see p. 118) this membrane is independent of the steel platforms inside. Additional support of the membrane was to be achieved by means of localised reinforcement of the skin. However, structural problems would in any case arise on account of wind loading, and stabilisation of the skin with guy cables would certainly be necessary in many places, both inside and outside.

271

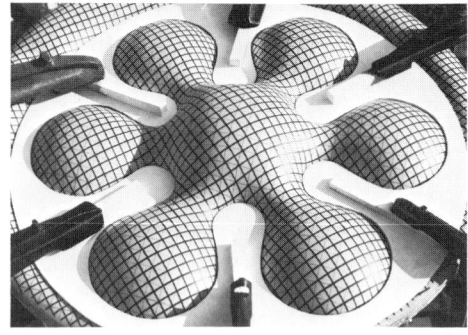

272

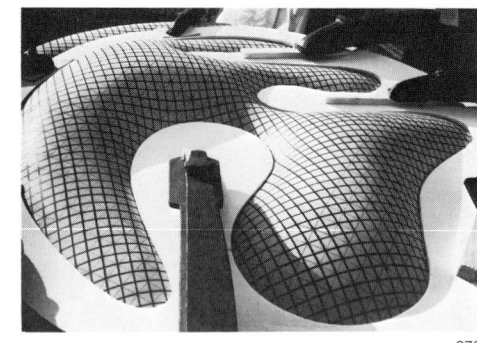

273

271–273. Test model
Design: Frei Otto, before 1962

Otto's model tests show something of the multiplicity of possible design forms for plane structures. Constrictions and undercuts present no problems. As one can see, anticlastic areas occur on the surface.

274

274, 275. Information pavilion in Sonsbeek, Holland
Design: Eventstructure Research Group
Manufacture: Hoogerwerff B. V., 1971

The peculiarity of this structure was its membrane surface. It was coated with synthetic grass which hardly differed from the adjacent lawn and made the building look as though it was a swelling in the park ground.
Also of interest are the entrances, now the subject of a patent application, in which two lip-shaped membranes are pressed against each other by internal positive pressure.

275

276

277

276-279. Swimming pool roofs
Design and manufacture: Birdair Structures, Inc.

The problem of transitions from the membrane to openings, such as doors and windows, is solved here by a particularly interesting method. Arched cables take up the membrane tensions and transfer them to the anchorages. Within these areas different skin sections which are more strongly curved and thus have less tension can be inserted.

278

279

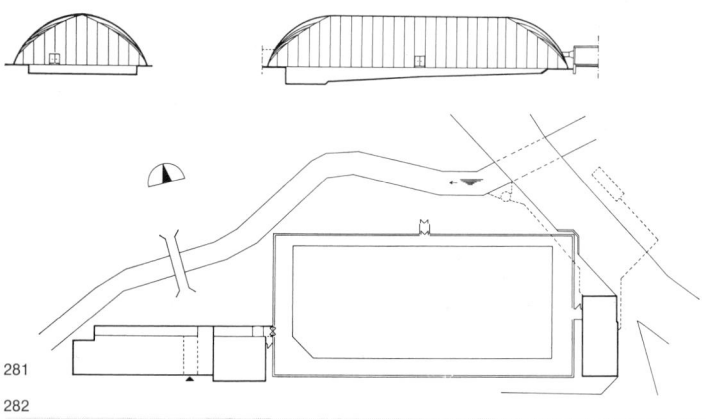

280

281

282

280–282. Roof covering for a former open air swimming pool in Kaufungen, near Cassel
Design: H. Volmar
Manufacture: Wülfing und Hauck, 1970

The hall is 57.5 m long, 26 m wide and is 10 m high at the highest point; the volume is 12,500 m³. The construction was manufactured in four parts and was assembled and made airtight by using approximately 2,000 aluminium clamping plates and as many stainless steel bolts.

The membrane is translucent and consists of high strength synthetic fabric coated on both sides.

Two electric hot air compressors with controllable fresh and recirculated air operation each have an air output of 36,000 m³/h and a heat output of 450,000 kcal/h. Heating is oilfired. One of the compressors is in reserve. Water heating is by means of a thermal pump. The total apparatus is automatically controlled and during the pool's hours of use gives out an air temperature of +28 °C, a water temperature of +24 °C and a relative air humidity of 40%.
(Bibl. 150.)

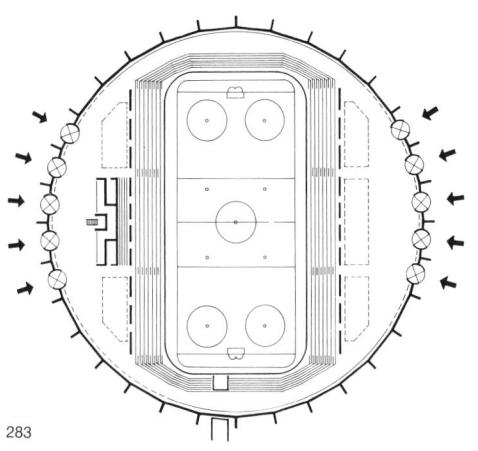

283

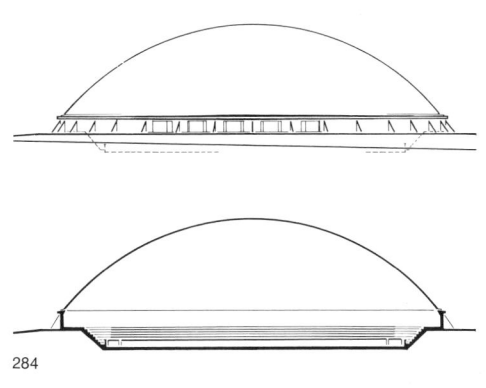

284

285

283–285. Roof of the ice rink palace in Forssa, Finland
Design: Antti O. Bengts
Structural calculations: Timo Suällström
Manufacture: Oy Urheilli-Vaatetus-Ruka, 1971

The hall has a ground area of 4,000 m² and at present is the largest air supported sports hall in the world. The diameter of approximately 71 m and the shallow curvature was possible because of the use of high strength PVC coated synthetic fabric and extensive preliminary investigations on a model hall on which the exact cutting pattern was also determined. As the temperature in the top of the building is always a minimum +12 °C, snow loading can be disregarded in the calculations as the snow melts immediately.
The material was supplied by Enka Glanzstoff GmbH.

286, 287. Cover – roof for Haus Lange, Krefeld
Design: Haus-Rucker-Co.
Manufacture: Wülfing und Hauck, 1971

In the foreword to the catalogue for the exhibition "Cover – Überleben in verschmutzter Umwelt" (Cover – survival in a polluted environment) Haus-Rucker wrote: "Cities are buried under coverings of smog. The dust that is swallowed by the inhabitants of these cities can be measured in lorryloads. The streets have changed into gas chambers, the rivers into viscous poison brews. The sun has become a 40 watt bulb; wandering rubbish dumps eat grass and trees . . .
Despite massive campaigns against environmental pollution the continuity of the process and constant adaptation to the ever worsening conditions prevent an instinctive grasp of the extent of the danger. Death will not come as quickly through the environment as through a H bomb, but it will be just as final.
"Cover" makes a jump in time and shows the situation that will arise if increasing contamination of the environment continues: life in artificial reservations.
As an example, the Lange house, a one family house in its structural concept, is enveloped in a pneumatic protective covering (air supported structure). A climatic "island" is created which, equipped with the necessary technical apparatus, becomes a self-sufficient life cell. Quartz lamps are suns. Audio-visual sceneries simulate changing weather and landscapes. The illusion becomes a substitute for the real experiences that are lacking. A small synthetic cosmos surrounds the house from which it is impossible to break out.
Noah's ark is launched again.
"Cover" tries to make people conscious of this situation. Consciousness through total simulation of future conditions . . ."
Inside the house further pneumatic "climate zones" were installed to satisfy basic human needs – true objects of horror of future existence.

286

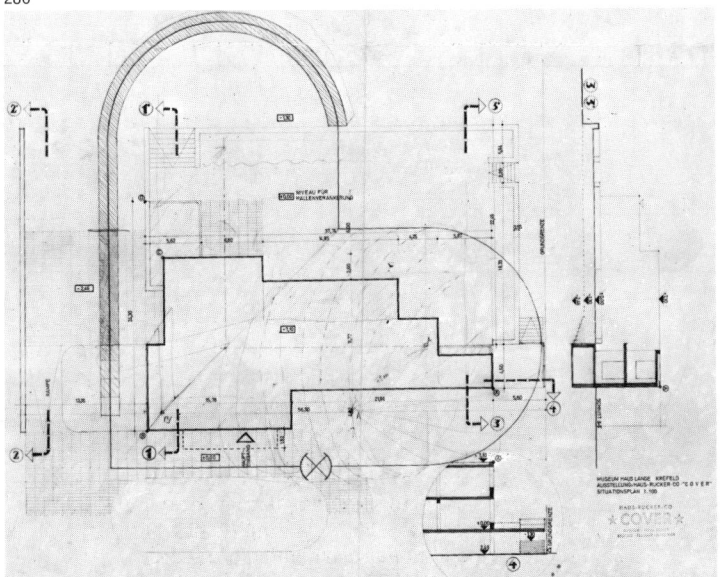

287

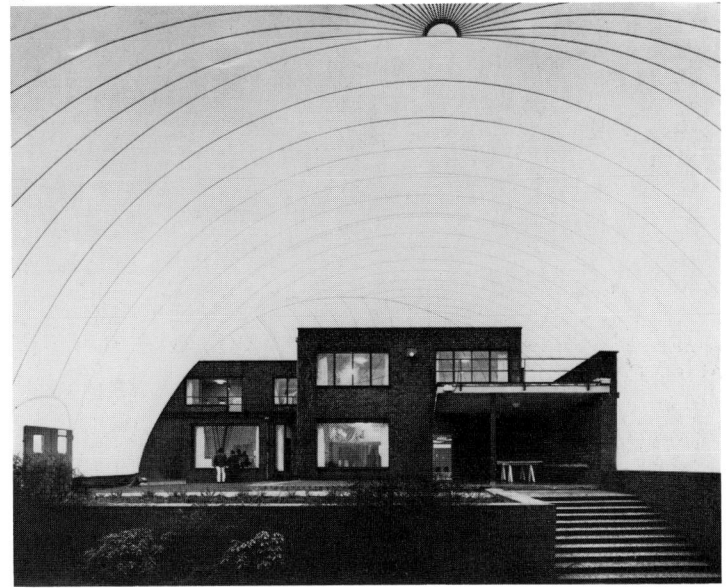

288

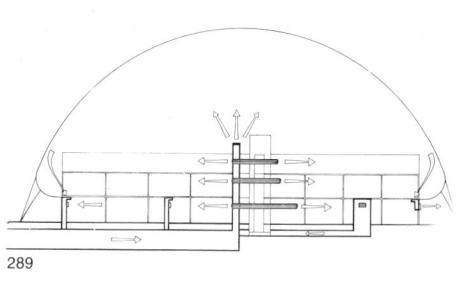

289

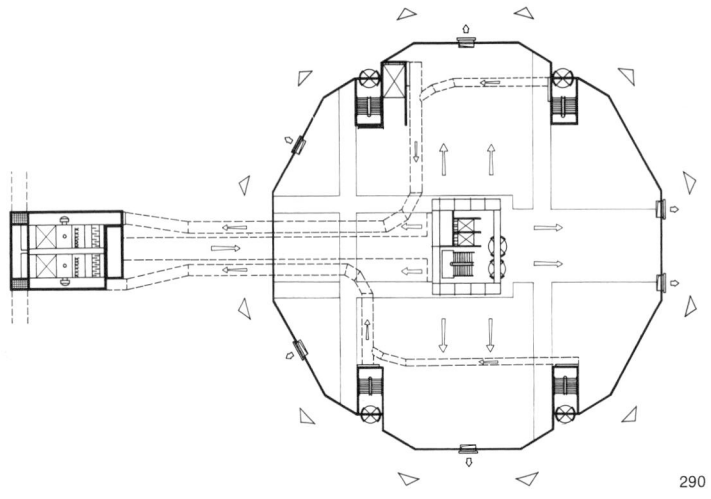

290

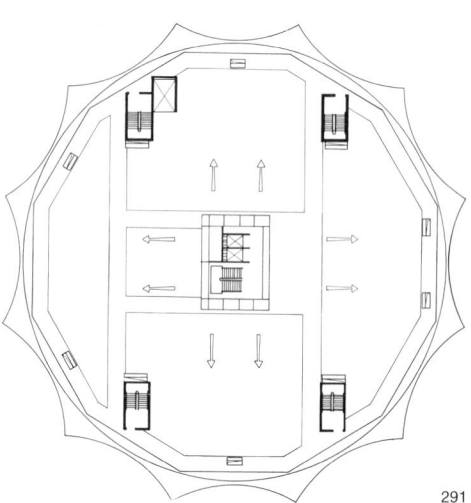

291

292

293

288–291. Air supported pavilion with arcade

Design: Friedrich Krupp GmbH, Zentralinstitut für Forschung und Entwicklung, 1970

The study is part of a research work on pneumatically stabilised membrane structures (Bibl. 91). A spherical segment with a diameter of 65 m is placed on a frustum of a cone cut by 12 parabolic arches. Above the arches the membrane skin is divided into two. The inner part of the skin seals the hall while the outer part conducts the membrane forces into the ground anchors. Through this profiling an arcade is created around the pavilion. On the inside are shop windows, revolving doors and lorry ramps. By dividing the interior into storeys the ground floor can remain pressure-free. Thus expensive lorry and personnel air locks are avoided.

292, 293. Roof covering for the skating rink in Anegasaki, Japan

Design: Japan Engineering Consultant Co.
Manufacture: Taiyo Kogyo Co., Ltd., and Taisei Construction Co., Ltd., 1971

The building has a certain similarity to the preceding project as far as the transfer of membrane forces to the ground and the formation of the membrane apron is concerned. The bottom edge is divided into two parts, one of which serves as an anchorage and the other as a seal. A clear difference in colour turns this structural measure into a distinguishing motif.
The structure covers an area of close to 3,300 m². It has a length of 79 m, a width of 53 m and a height of 16 m.
(For details of the fixing of the membrane to the anchorage see p. 153, Fig. 70.)

294

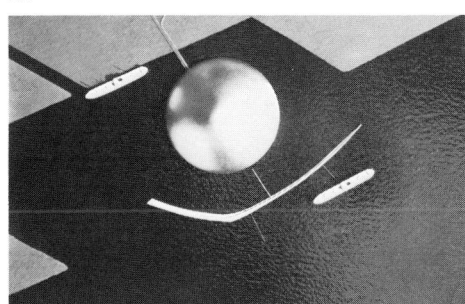

295

294, 295. Pneumatically stressed bulk goods container

Design: Frei Otto, before 1962

Conical surfaces, like cylindrical surfaces, are only curved in one direction and therefore offer the advantage of a favourable material cutting pattern. As the longitudinal tensions decrease considerably towards the peak, the difference in the skin tension of the cone must be counteracted by a rounding-off of the tip.
The use of this form suggested by Otto as a bulk goods container has the advantage of optimum use of space. The fuller the container becomes, the less supporting air is required to keep the internal pressure constant.
(Bibl. 119, p. 78 ff.)

296

296. Project for covering Wembley Stadium, London

Design: Arthur Quarmby, 1967
Structural consultant: David Powell

In this alternative proposal to the "Helium Lifted Canopy" (see p. 57) a single membrane structure is to cover the football pitch. The whole stadium would be under positive pressure, which would entail considerable sealing measures in the present buildings.

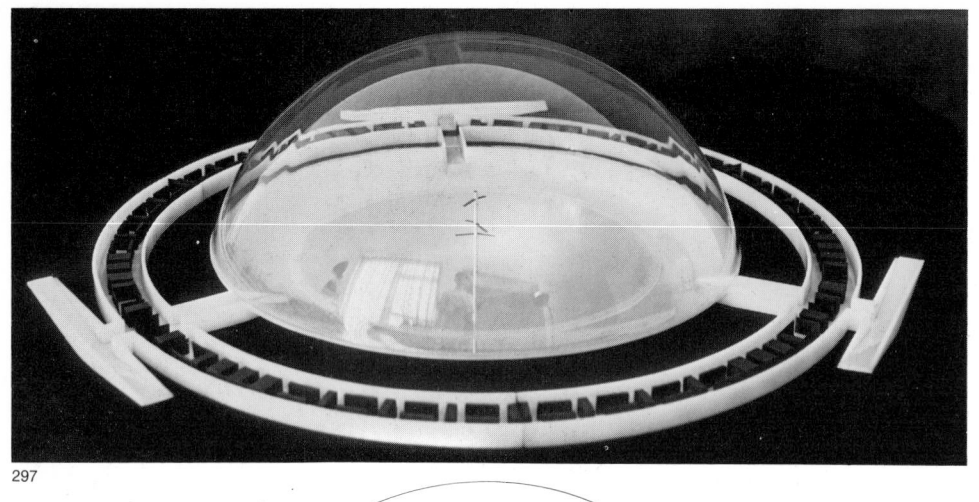

297

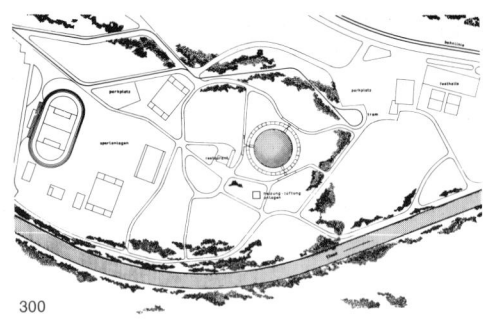

300

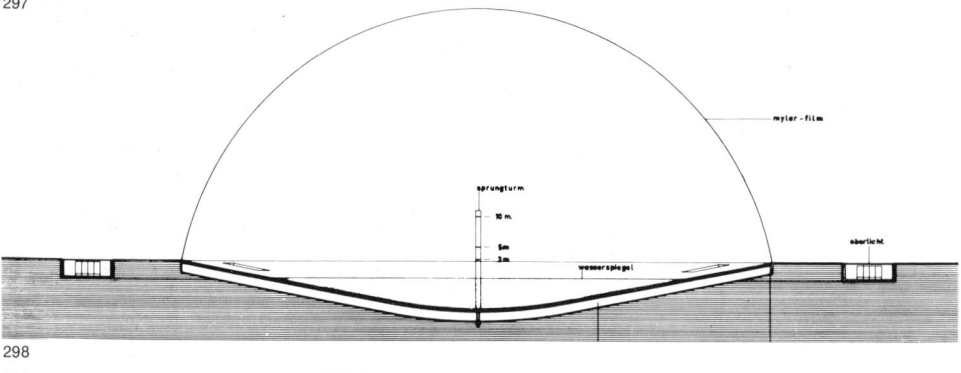

298

299

297–300. Indoor swimming pool
Design: Gil Hirt and Willi Ramstein, 1959

The authors wrote at the time about their project, which is fascinating by the clarity of its concept:

"The projection of this object was based on the following ideas:

The passage from land to water should not be by a direct separation (e.g. steps) but by a gentle slope of about 17%. In order to produce this transition from beach to water on all sides, the pool is constructed as a flat spherical segment with a diameter of 100 m. The water surface itself has a diameter of about 76 m and is surrounded by a 12 m wide beach.

Permanent use can be guaranteed by covering the pool. An unhindered entry of light, strong visual connection with the green of the surrounding parkland and an easily produced structure led to a solution in the form of a pneumatic roof covering. The envelope of transparent Mylar film is supported by a light, scarcely noticeable positive pressure.

On the beach, seats can be created for almost 5,000 onlookers. This centre will be used for cultural as well as sporting purposes. For example, theatre, variety, shows, fashion shows, etc. can be performed on floating walkways and stages.

In conjunction with dwelling units the swimming pool can be placed in a spacious park. No superstructure disturbs the landscape and only the light envelope, in which sun and water are reflected, rises into the air".

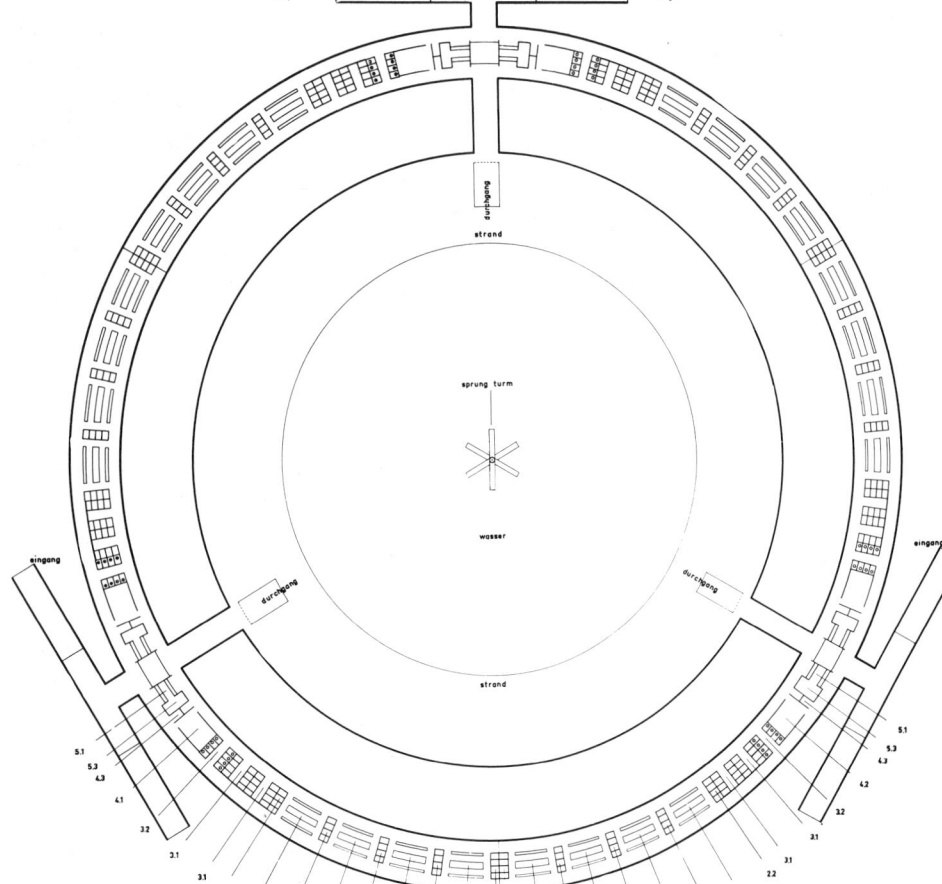

301

301. Dome for covering part of New York City
Design: Richard Buckminster Fuller, 1962

Fuller's dome for New York City is one of the best known Utopian pneumatic structures in architecture. Even if it were possible to realise such a project – apart from the fact that our knowledge of the climate (heating, cloud formation) created inside is greatly lacking – the attendant city structure is also lacking; it is one without exhaust gases, without dust, without fire, more or less sterilised, with buildings whose exteriors no longer have the function of protection against the weather . . .
This is, therefore, a Utopia whose realisation must begin with the changing of the environment that the dome encloses.

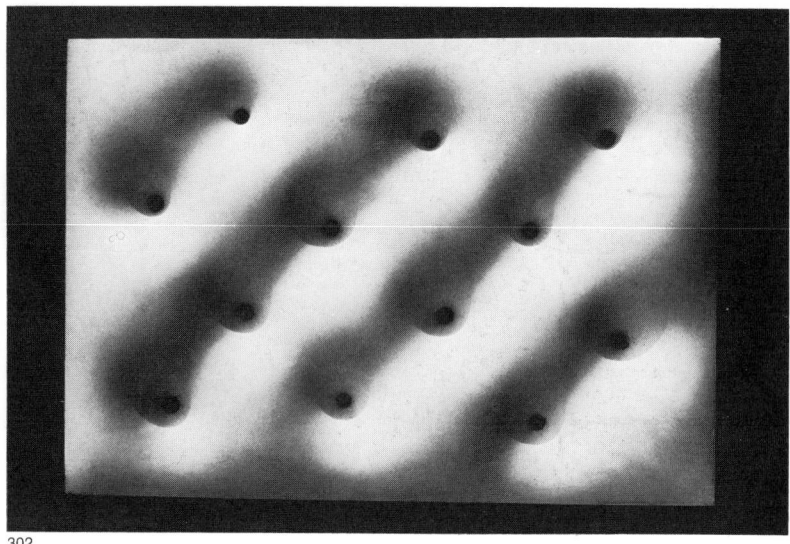

302

303

304

305

302, 303. Form study of a membrane with internal drainage
Design: Frei Otto

The plaster model shows a rectangular membrane with 12 guys.

304, 305. Exhibition pavilion
Design: Gernot Minke with students of the Technische Hogeschool Delft, 1971

The ground plan of the pavilion is circular; the membranes are anchored at three low points inside.

306

306. Multi-purpose structure
Design: Seminar Pneumatische Konstruktionen, Institut für Umweltplanung, Ulm, under the direction of Gernot Minke, 1971

By means of high pressure balloons used as support elements a rather uniform stress distribution is achieved in the membrane which is stabilised by negative pressure.

307–309. Demountable exhibition pavilion
Design: Seminar Pneumatische Konstruktionen, Institut für Umweltplanung, Ulm, under the direction of Gernot Minke, 1971

This pavilion too is a negative pressure structure. By means of circular individual supports the outer membrane is held apart from the inner membrane, which is only fixed at the base.

309

307

308

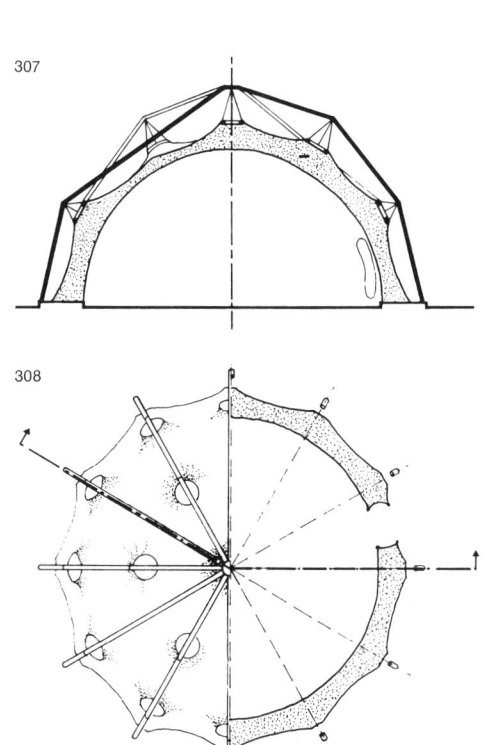

310–312. Form studies
Design: Frei Otto, before 1962

Frei Otto's model tests show the influence of cables and nets on the design. The total radii of curvature are increased by nets; within the mesh the membrane forms surface areas with smaller radii of curvature (Fig. 312).

In the use of cables, care has to be taken that the tensile forces of the membrane on both sides are applied at equal tangential angles if possible.

313. Exhibition pavilion
Design: Gernot Minke with students of the Technische Hogeschool Delft, 1971

The membrane is arched over a circular ground plan. It is additionally supported by means of two cables forming grooves.

314. Cable reinforced pneumatic for exhibition purposes
Design: Friedrich Krupp GmbH, 1973

The exhibition hall is put together from spherical sections in whose seams cables are laid. This formation causes small radii of curvature and correspondingly low membrane forces. Air supply and recirculation take place on both sides of the longitudinal axis of the hall, which is 228 m long, 72 m wide and 25 m high.

310
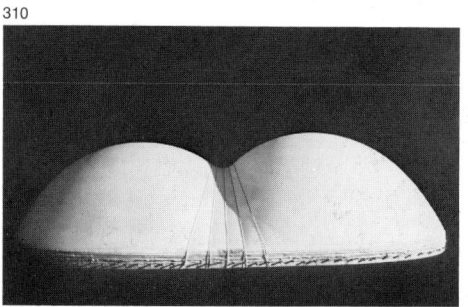

311

312
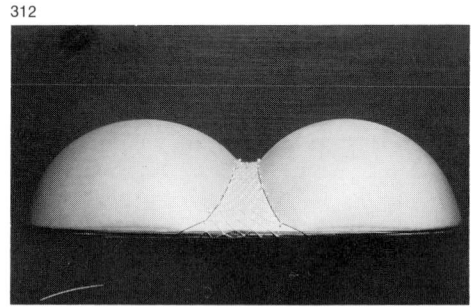

313

314
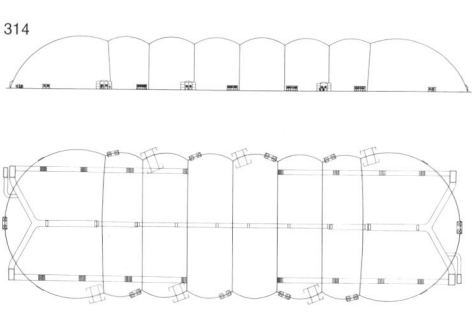

315

315. Multi-purpose hall on the English South Coast
Design: Gernot Minke with Croucher and Salt, 1969

A pneumatically stressed positive pressure membrane reinforced by intersecting cables was to span an area of 2,000 m².

The model demonstrates how, by means of diagonally running cable supports, the appearance of a building can be changed as compared to the previous examples.

316–318. Large span roof for a greenhouse
Design: Frei Otto, before 1962

Support of the membrane is achieved in the interior of the structure by restraining membrane ribs. The cross-section is slightly arched toward the outside to facilitate drainage. Additional ties to the ground reduce the tensions in the cables. (Bibl. 119, p. 104.)

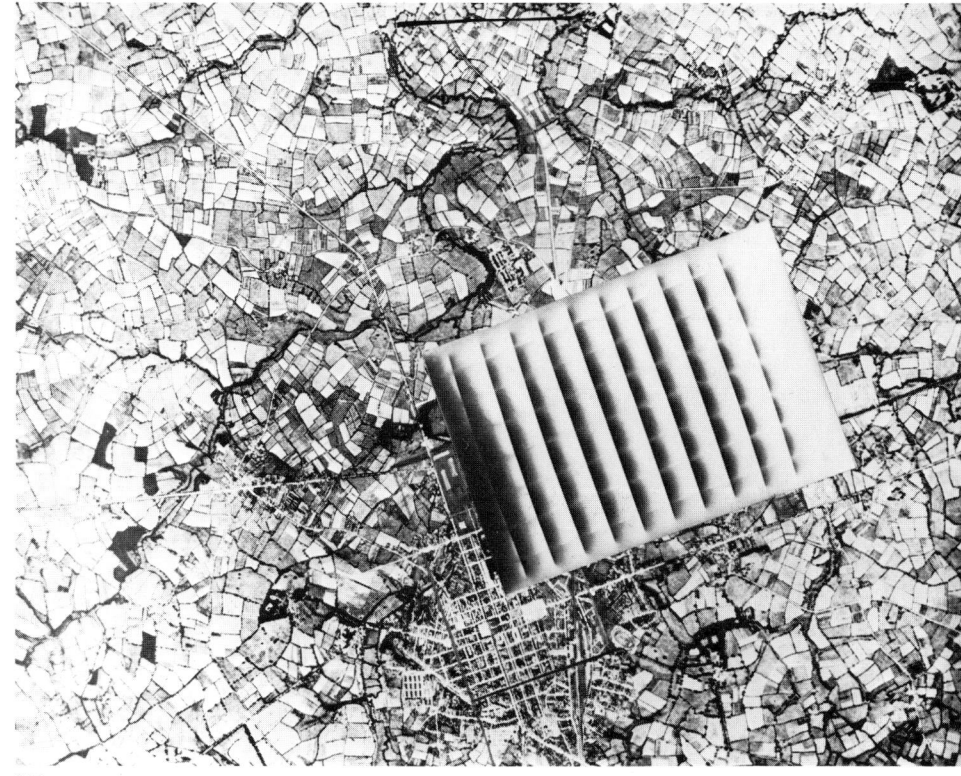

317

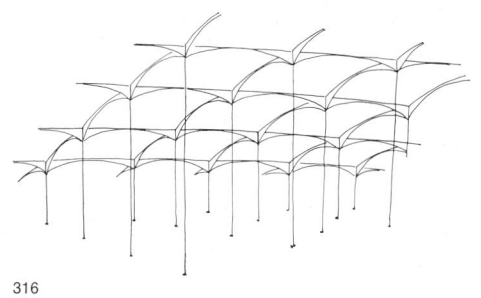

316

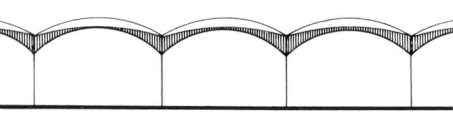

318

319, 320. Plastic foil structure stabilised by positive pressure
Design: Otto Walter Neumark, 1962

The design is based on the idea of covering a cable net, which is anchored at its edges only, with overlapping, laterally fixed foil lengths. The overlapping edges are to be tightly pressed against each other by the internal pressure.
In the (somewhat modified) drawings omitting a uniform scale the possible total dimension and the overlapping covering can be recognised. (DBP 1434630.)

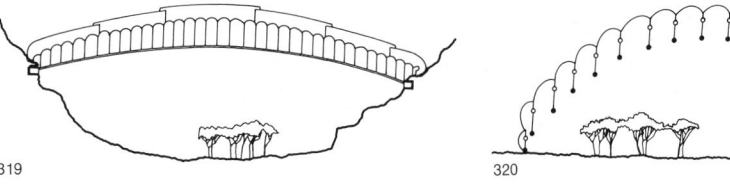

319

320

321. Structure supported by positive pressure with cable net
Design: Paul Desmarteau

As in Frei Otto's greenhouse the cable tensions here are greatly reduced by means of intermediate guying. (DBP 1219656.)

321

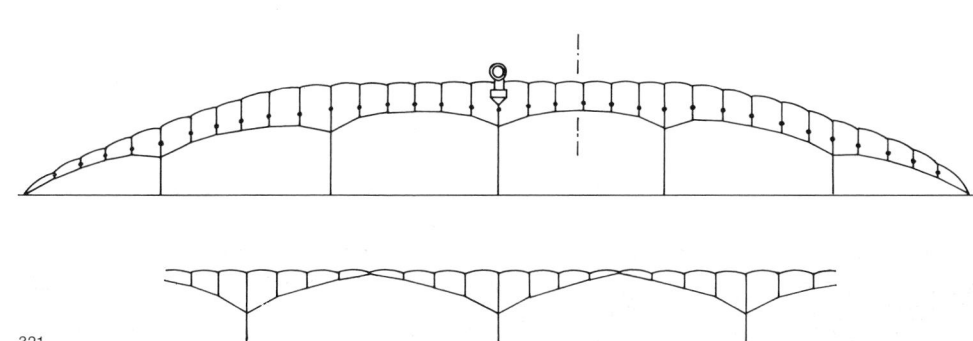

322

322–324. Padiglione Abitare, Milan
Design: Studio d'Architettura e Industrial Design
Manufacture: Plasteco Milano, 1969

The cable reinforcements crossing at 90° divided the building into four equal sections. By using a transparent strip in the lower area a spatial relationship was created between interior and exterior. The membrane consisted of high frequency welded PVC lengths 1.2 mm thick.

323

324

325

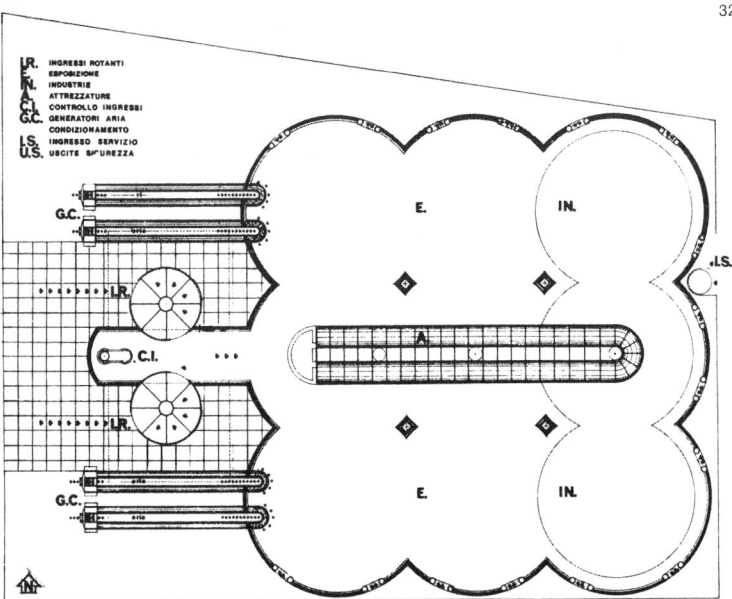

326

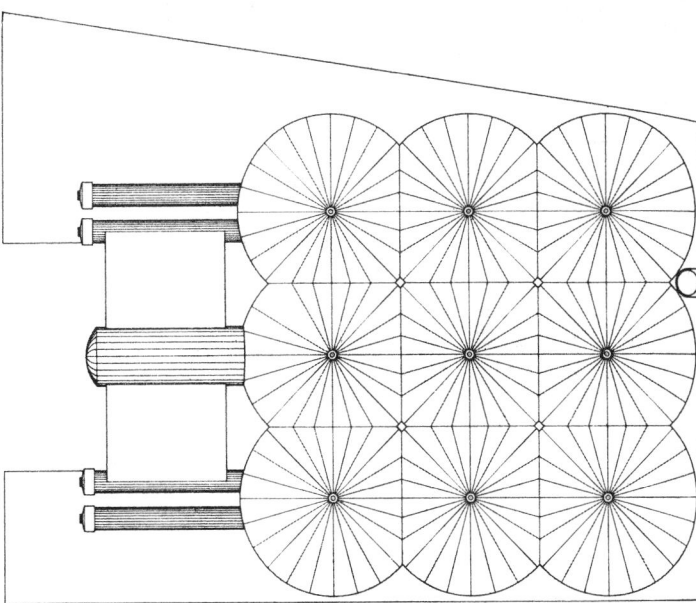

327

325—328. Italian pavilion for Expo '70, Osaka. Competition design

Design: Studio d'Architettura e Industrial Design, 1968

The design represented a further development of the structural system already used by the designers in the Padiglione "Abitare". They proposed a large envelope of dome-shaped membrane sections, which were to be assembled in the manner of modular construction. The 17 m high spherical segments were to be made of 24 lengths of high frequency welded PVC and cover an area of 3,850 m². The internal positive pressure was to be 20 mm of water pressure.

Two months were estimated for the total production of the structure and one week for erection.

328

329

329–332. Pneumatic roofs for recreation centres

Design: Yutaka Murata, 1972

The projects hardly differ from the two preceding Italian examples. The basic idea aims at offering a modular pneumatic construction system with interchangeable elements. For part of the fixtures and fittings Murata also proposes pneumatic elements which are particularly light, cheap and easily produced.

330

331

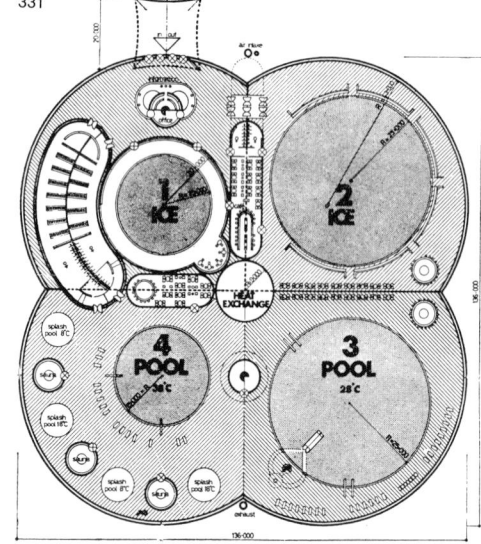

332

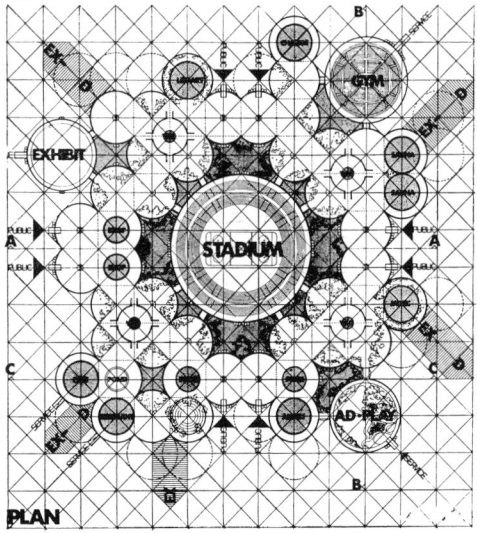

333-335. City in the Arctic

Design: Atelier Warmbronn (Frei Otto and Ewald Bubner), 1970–71
City planning advice: Kenzo Tange & URTEC
Structural advice: Ove Arup & Partners
Material tests: Farbwerke Hoechst AG

The study project commissioned by Farbwerke Hoechst AG provides for a pneumatically stabilised, climate-regulating shell with a diameter of 2,000 m, a height of 240 m and a dome radius of 2,200 m for a city with 15,000 to 45,000 inhabitants.

The transparent membrane is supported by a net of intersecting cables. The researches revealed that a cable net made of polyester fibres is superior to a steel cable net. Because of the low E modulus and the resulting high flexibility of the polyester material, considerable changes in volume (in the region of 200 to 300%) must oc-cur before the cables fracture or slacken, and deviations of this order of magnitude can be disregarded. The project provides for cables with a diameter of 270 mm at 10 m centres. Under normal load conditions this means a safety factor against breaking of 12. Because of the low loading there is no fear of creep. Anchoring is by means of a ring foundation.

Fig. 334 shows the stages of the construction process for the large shell. Before the cables are laid out on the ground, erection balloons are distributed and anchored. Then the cable bands are attached to each other, lightly prestressed and the membrane is fastened. Finally the foundations are closed, the erection balloons filled and the shell is inflated. The internal pressure varies between 25 and 40 mm of water pressure. The designers predict that in the early eighties the first projects of this kind will be realised. (Bibl. 77.)

333

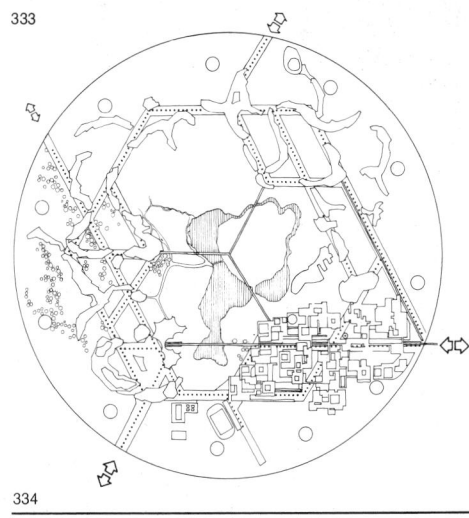

334

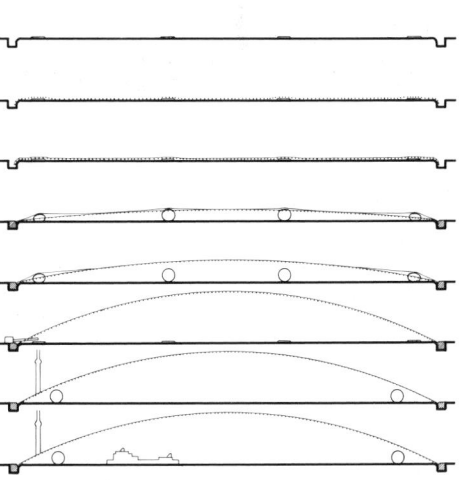

335

337

336

338

339

336–342. USA Pavilion, Expo '70, Osaka
Design: Davis, Brody, Chermayeff, Geismar, De Harak Ass.
Structural calculation: David H. Geiger
Manufacture: Ohbayashi-Gumi, Ltd., and Taiyo Kogyo Co., Ltd., 1970

The oval covered by the membrane was 142 m long and 83 m wide. The membrane had a rising height of only 6.10 m. The shallow curvature was made possible because the envelope was subtended by a net made of 32 cables, each having a diameter of 48 mm. The weight of the cable net was about 45,000 kg; that of the membrane,

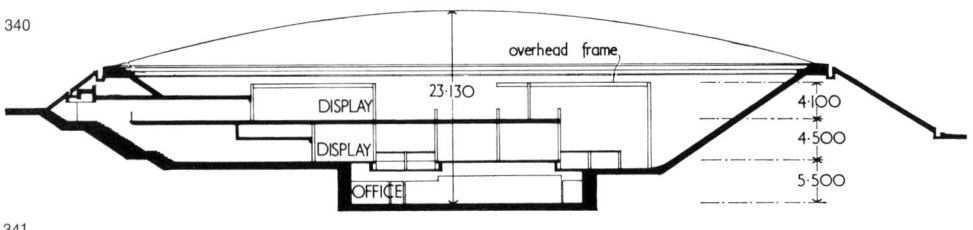

340

341

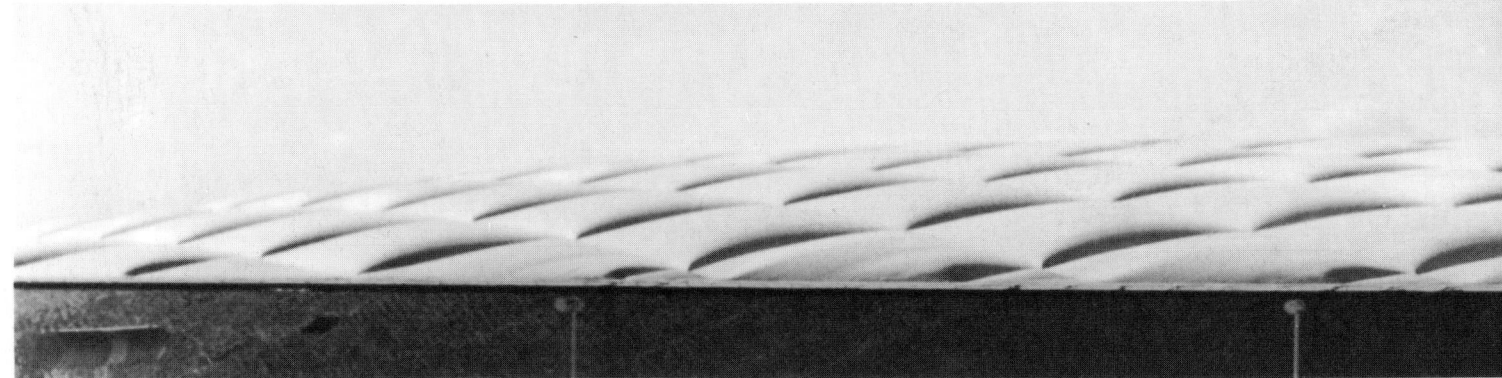

which was made of PVC coated high frequency welded glass fibre fabric, was about 15,000 kg. The internal pressure was 27 mm of water pressure, even in a strong wind; under snow loading it was raised to 63 mm of water pressure. (Details see pp. 153, 155.)

The Pavilion was one of the largest buildings at Expo 70. Its effect lay not in its size, however, but in the mixture of understatement and sophisticated design.

As the section shows, the structure was partly sunk into into the site. The excavated soil was piled up as a sloping wall around the periphery. Inside were steel platforms which were used for the display of the exhibits, and as traffic areas for visitors.

The restraint exercised with regard to the outer appearance was interpreted in Japan as a reflection of American political morale in South East Asia. However the great curtailment in the budget seems to have been at least as decisive a factor for the design.

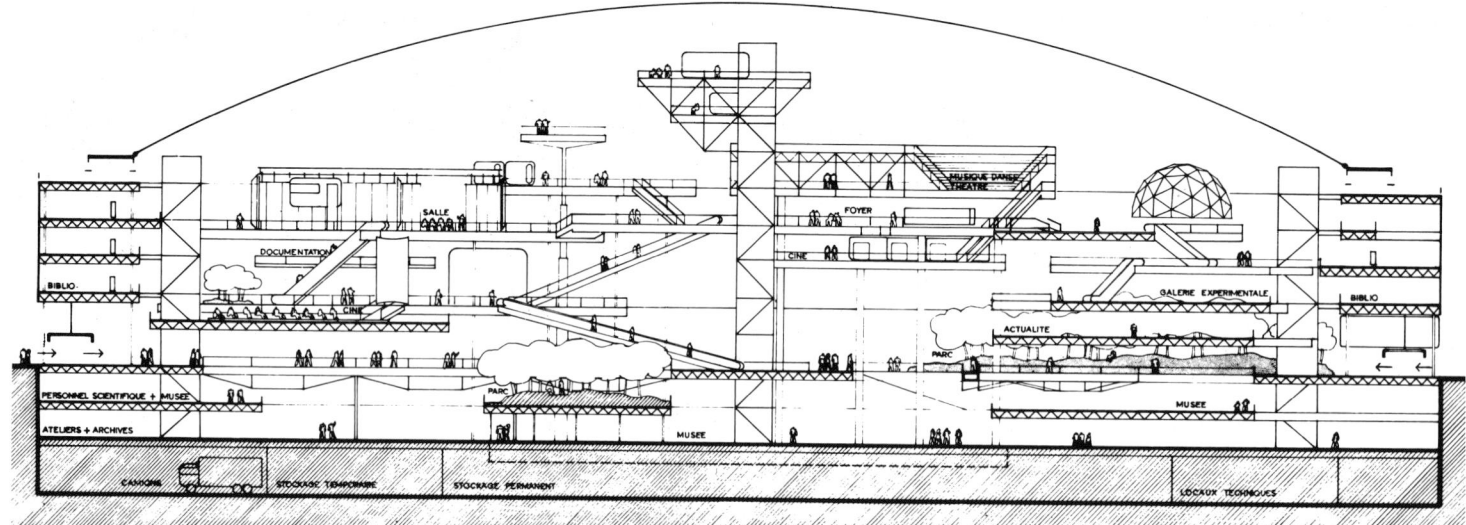

343

**343–345. Centre Beaubourg, Paris.
Competition design**

Design: Manfred Schiedhelm with Atilla Berker and Myra Wahrhaftig, 1971

The structure, a mixture of building and quarter, is covered by a pneumatic dome which is supported by a steel cable net.

The interior is a large spatial garden with small wooded areas on different levels. Supported by the latest communications techniques, cultural activities of all kinds were to take place here. The areas for library, exhibitions, theatres, museum, etc. are solidly installed yet demountable.

Despite the large total expanse of the structure the scale in its interior corresponds to that of the surrounding built-up area.

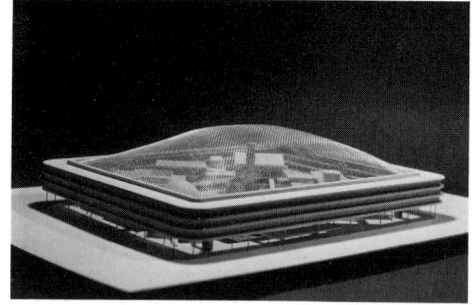

344

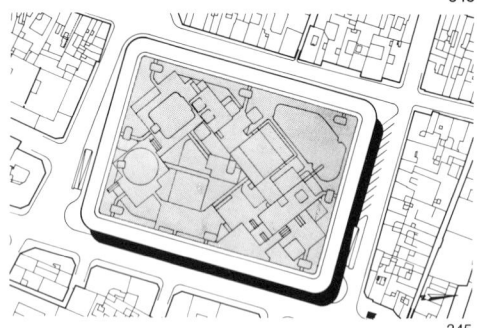

345

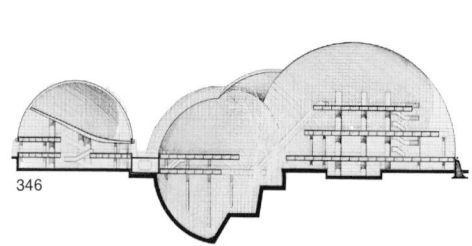

346

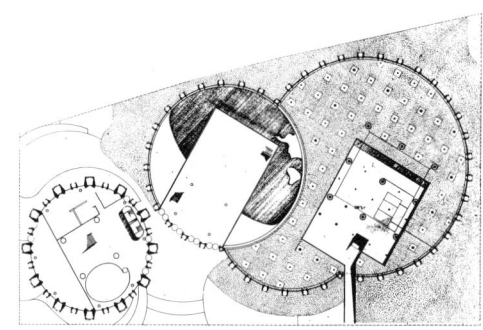

347

**346, 347. French Pavilion for Expo '70, Osaka.
Competition design**

Design: Jean le Couteur and Denis Sloan, 1967

Concrete shells in the ground were to be supplemented above ground by pneumatic membranes to form complete spheres. Inside multi-storeyed steel platforms were planned, similar to those in the German tent in Montreal. These platforms were not to touch the envelope at any point.

The project was not executed primarily because there was a feeling of uncertainty about the technical problems.

348–350. Investigation into the covering of university buildings with pneumatic structures

Design: Rurik Ekstrom, Charles Tilford and Blair Hamilton (Antioch College, Columbia, Maryland), 1970
Manufacture of prototype: Students of Antioch College, 1972

The investigation resulted in a prototype (Fig. 350) which spans an area of 60×60 m. Technical support for its execution was given by the Goodyear Corporation whose Research Division carried out the relevant experiments. The envelope is made of two PVC sheets with a 36 to 60 cm air gap between them to reduce the build-up of heat; this is also restricted by air conditioning. The two main cables have a diameter of 27 mm; they are each guyed down twice. The diameter of the secondary cables is 7 mm. The entrances are inset into a surrounding embankment.
The building contains art studios and theatre facilities, as well as lecture, seminar and faculty rooms for about 100 students.
In May 1973 the first Conference on Pneumatic Structures in the Educational System took place in which architects, engineers and educationalists from all over the world took part.

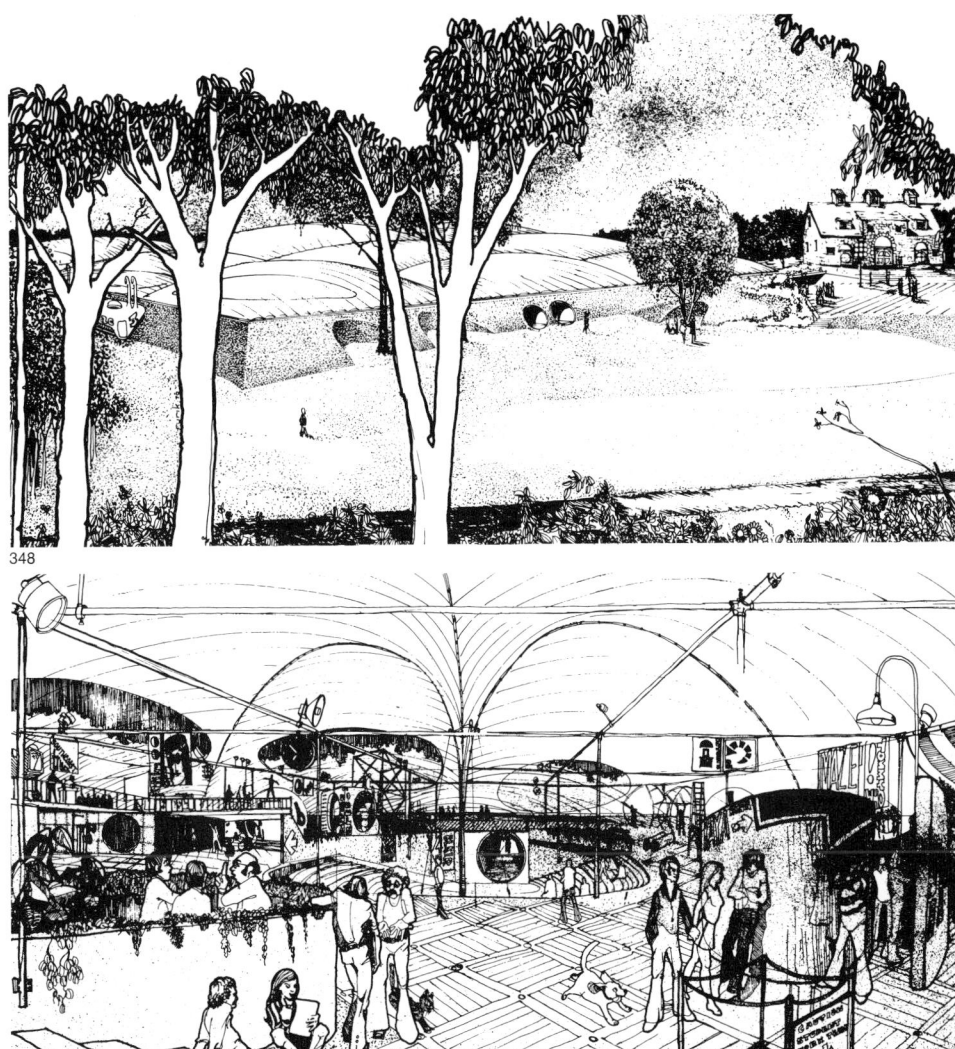

348

349

350

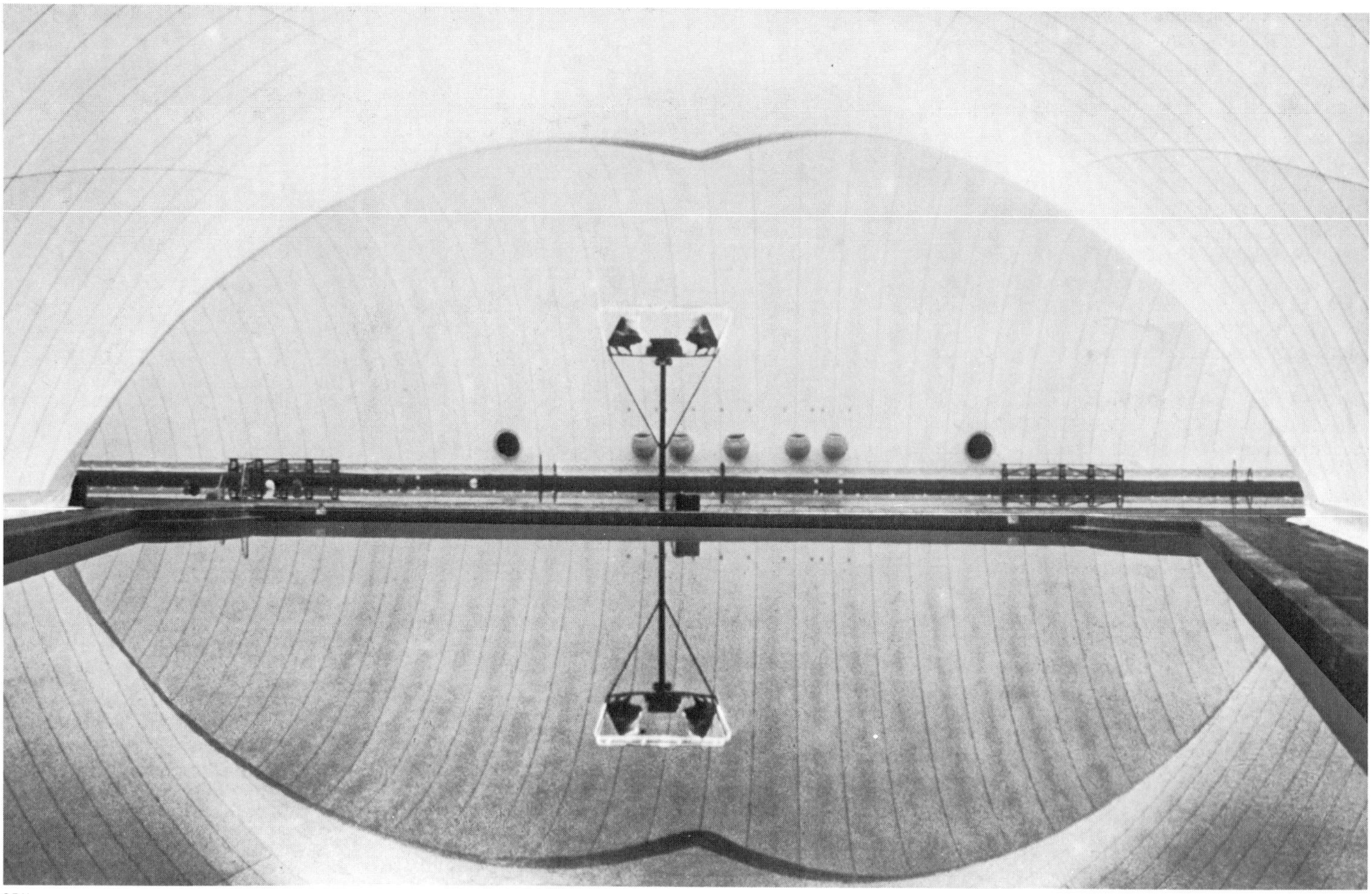

351

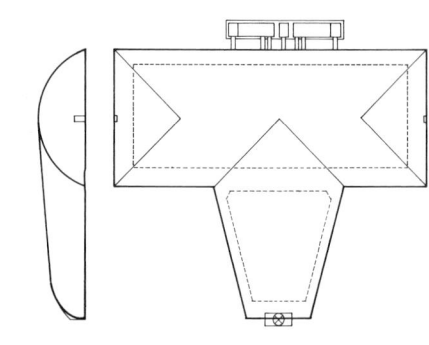

352

353

354

351, 352. Indoor swimming pool
Design and manufacture: Krupp Universalbau, 1971

As well as the swimming pool a non-swimmers pool with a trapezium-shaped ground plan had to be covered. The problem of manufacture lay in the cutting of the three dimensional elements, a cylinder section and a frustum of a cone.
In order to take up the linear forces which increased from the peak of the hall to its base, the membrane was reinforced along the cutting lines with several layers of the material. All the calculated cutting patterns were tested on a model, scale 1:20, and corrected where necessary. In the arc of the frustum the sheets had to be lapped, as the sloping silhouette made strongly curved cutting patterns necessary. For visual reasons the lapped edges were arranged in continuous lines.
(Bibl. 28.)

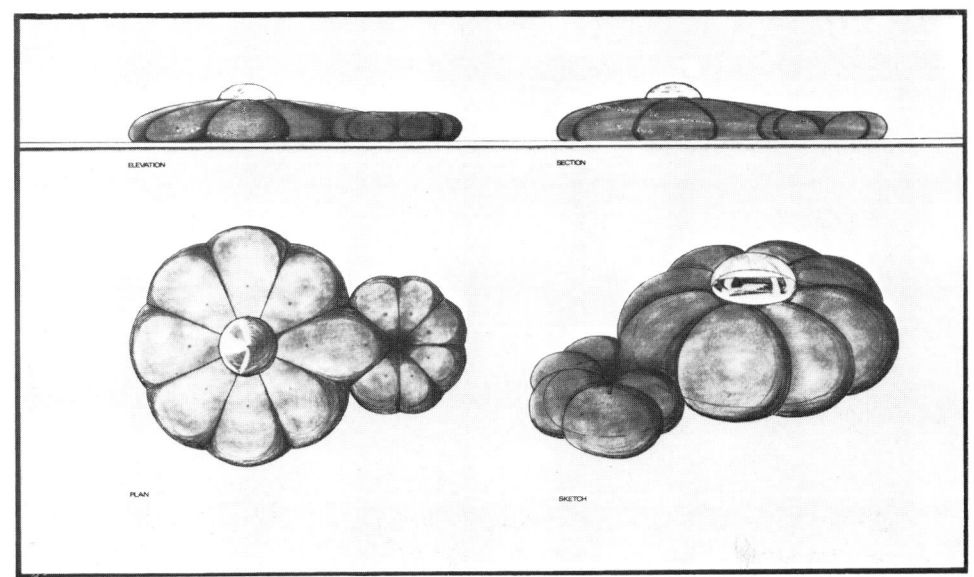

355

353, 354. Roof of the Alpamare wave bath, Bad Tölz
Design: Gernot Minke
Manufacture: L. Stromeyer & Co. GmbH, 1971

The demountable membrane of PVC coated polyester covers a ground area of 1,400 m². On account of the six radially arranged cables restraining the structure a favourable organisation of the interior could be achieved and the volume of air for heating and circulation could be reduced.
Two electric hot air fans with automatically controlled fresh air/recirculated air operation bring a heat output of 400,000 kcal/hour and an air output of 32,000 m³/h. Temperature and humidity are automatically controlled. In the event of power failure a reserve fan will switch on.
The shadow outline of the surrounding trees on the skin acts as a graphic element inside. Thus even when the pool is covered a relationship to nature is created.

355. Project for Brighton Marina
Design: Arthur Quarmby

The two inter-connected roofs were to have diameters of 93 and 150 m. This study illustrates particularly well the design possibilities provided by the arrangement of radial cables.

356. Exhibition hall on the English South Coast
Design: Gernot Minke with Coppin and Galloway, 1969

The membrane, which is to be fixed to a 3 m high surrounding concrete base, is divided by restraining cables into five sections.

357, 358. Cabledome
Design and manufacture: Birdair Structures, Inc., 1971

Spans of over 200 m can be achieved with this structure. The membrane is overspanned by a net made of three bands of cables which intersect each other at the same angle.
The uniform distribution of the net over the surface permits the formation of an almost exact spherical segment – in contrast to the flatter domes with varying curvatures which are restrained by radial cables. The cables do not need to be spliced at the points of intersection but require only positioning clamps; this greatly simplifies manufacture and erection. The fixing of the membrane to the cables is achieved by membrane flanges which are joined in situ over the cables and sewn to a closed loop profile.
(Details see p. 155.)

356

357

358

359

360

361

362

366

363

367

359–369. Krupp standard halls
Design and manufacture: Krupp Universalbau

Fig. 359 shows the two most frequently used types of standard hall: one is constructed on a rectangular ground plan; in the other semicircular sections are attached to the narrow side of the rectangle. Figs. 360 and 361 give examples of the range of these two types; the structure built on pure rectangles covers an area of 12,600 m², the other an area of 6,500 m². Figs. 362–365 show various access and connecting structures. Figs. 366–369 give an idea of the erection of a standard hall from transport of the packaged membrane to inflation.

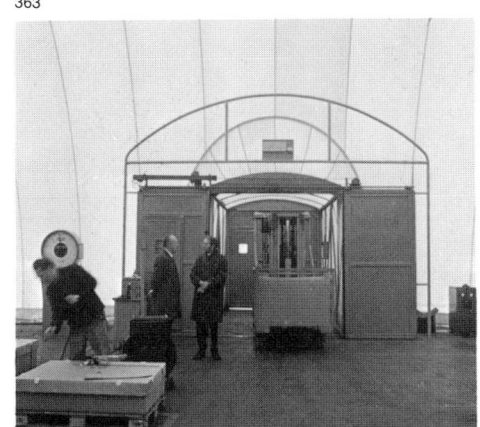

364

365

368

369

370

371

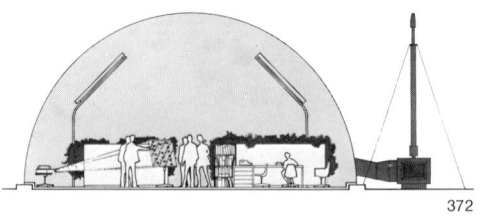

372

370–375. Temporary office building for Computer Technology, Ltd., Hemel Hempstead, England

Design: Foster Associates with Loren Butt
Manufacture: Polydrom, 1970

In the quickly growing computer industry it is often necessary to create new buildings at very short notice. After a number of alternatives had been' investigated, this air supported structure proved to be by far the cheapest solution. Eight weeks elapsed between commissioning and going into operation and erection took only 55 minutes. After some two years use the 60 m long and 12 m wide building that was erected on the asphalt of a car park was dismantled again.

In the event of an emergency a double row of lamp standards would have supported the membrane and kept the escape ways free. The lights were fitted with neon tubes and indirect light sources were projected on to the white membrane which in turn acted as a reflector.

A combined heating and ventilation system with an output of 150 KW produced a maximum input temperature of +50 °C in winter. The ventilation supply was about 10,000 m³/h. The room temperature was found to be quite acceptable in winter, but in summer the air heated up as high as +32 °C. In the case of a longer working life of the building a cooling system would have had to be connected to the ventilators.
(Bibl. 44, p. 192 ff.)

373

374

375

376. Air supported structure for a winter building site in Anzère, Switzerland
Design and manufacture: Sarna-Hallen AG, 1967

In order to be able to work on a hotel site at 1500 m above sea level even in winter, a construction firm commissioned the erection of the illustrated air supported hangar which is 67 m long and 27 m wide. The membrane was manufactured from soft PVC sheeting reinforced with a polyester lattice web. It weighed 3,500 kg and was produced, transported to the site and erected within four weeks.

Two compressors with an output of 10,000 m³/h kept the pressure inside at 10 to 15 mm of water pressure. In strong wind forces the internal pressure was increased to approximately 30 mm of water pressure.
(Bibl. 11.)

377, 378 Krupp exhibition hall, Hanover
Design and manufacture: Krupp Universalbau, 1966

The air supported hall has a length of 106 m, a width of 35 m, a height of 17.5 m and covers an area of 3,300 m². Four cold air compressors provide a frequent change of air during the period of the fair. The air is ducted to the hall in concrete channels lying under the anchorage structure and is dispersed vertically upwards. At each deflection of the air flow, air deflectors keep the unavoidable loss of pressure to a minimum. The

376

378

377

surplus air is carried outside in regulated air-outlet conduits. When the fair is closed only one compressor is operating. This compressor alone produces a pressure that safeguards the hall against high wind loadings. If the internal pressure falls – due to a pane of glass cracking, for example – then a second compressor switches on automatically and off again after replacement of the damaged part. Even hurricane gusts cannot endanger the hall, because the ventilation system reacts by means of wind speed gauges. In the event of a power failure an internal combustion engine starts automatically and takes care of the run-on of one compressor. The heating installations provide fresh air heating and are designed for a temperature differential of about 20 °C. The heating output is 80,000 kcal/h. Heating is by town gas that has a lower calorific value of 4,000 kcal/Nm³. The equipment is connected to the cold air compressors.
(Bibl. 25.)

379

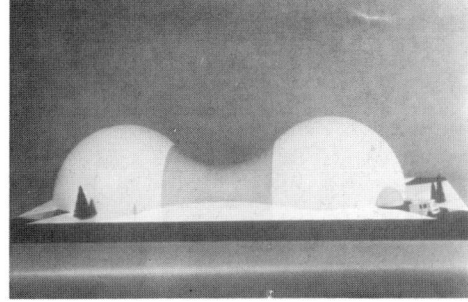

379. Combination of rotation surfaces and hemisphere
Design: Friedrich Krupp GmbH, Zentralinstitut für Forschung und Entwicklung, 1970

The model illustrates that it is also possible to form anticlastic surface areas in cases of large air supported halls with internal positive pressure. The project is ready for execution. Progressive firms are apparently making efforts to supply a greater variety of air supported halls.

380–384. Instant City
Design: José Miguel de Prada Poole, 1971

The initiative to build this pneumatic village, which accommodated young participants in the ICSID Congress of 1971, came from some architectural students in Barcelona. José Miguel de Prada Poole, who has already collected a good deal of experience in the field of pneumatic structures, developed a general plan and a "grammar of form" as a guide for manufacture and assembly. The lengths of transparent or coloured PVC were 0.3 mm thick and were fixed to each other by clamps. The ground anchorages were ballast pockets which were made by placing the membrane skirt in a 30×30 cm trench and loading it with the excavated earth.
The "grammar of form" was used very flexibly by the builders. When someone new arrived he built on as he liked – yet everyone kept to the prescribed structural details.
(Bibl. 112.)

380

381

382

383

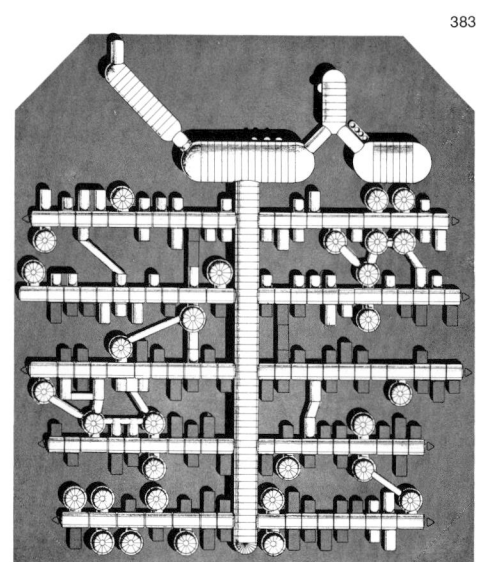

EJECUCION DE LAS ZANJAS DE ANCLAJE

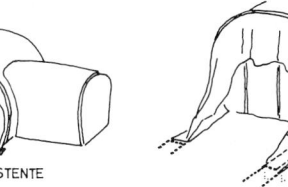

MODULO EXISTENTE

BORDE DEL LA ZANJA DEL MODULO EXISTENTE

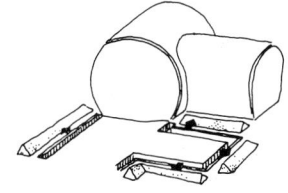

EJECUCION DE LA ZANJA DE 30X30 cms.

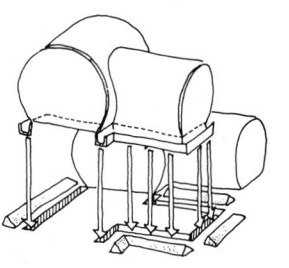

COLOCACION DEL BORDE DEL NUEVO MODULO EN LA ZANJA

RELLENADO Y APISONADO DE LA ZANJA

AXONOMETRIA DEL CONJUNTO

REPERTORIO GENERAL DE FORMAS NEUMATICAS SIMPLES

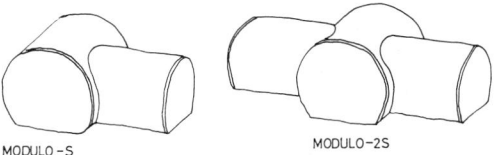

MODULO-S

MODULO-2S

constan del elemento propia-
mente dicho , mas la parte
proporcional de las zonas
comunes.

MODULO-C

MODULO-E

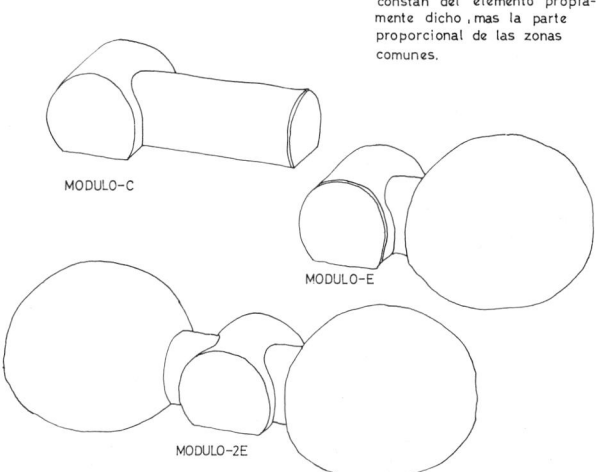

MODULO-2E

REPERTORIO GENERAL DE LA GRAMATICA DE USO

Comprende los siguientes elementos:

SIMPLES	DOBLES	MIXTOS
M-S	M-2S	M-SC
M-C	M-2C	M-SE
M-E	M-2E	M-CE

COMPOSICION DE LAS PIEZAS

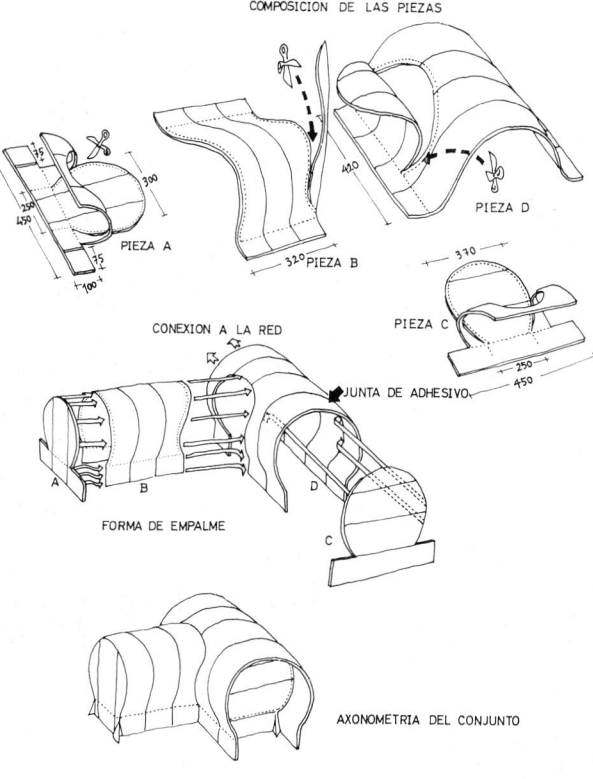

PIEZA A

PIEZA B

PIEZA D

PIEZA C

CONEXION A LA RED

JUNTA DE ADHESIVO

FORMA DE EMPALME

AXONOMETRIA DEL CONJUNTO

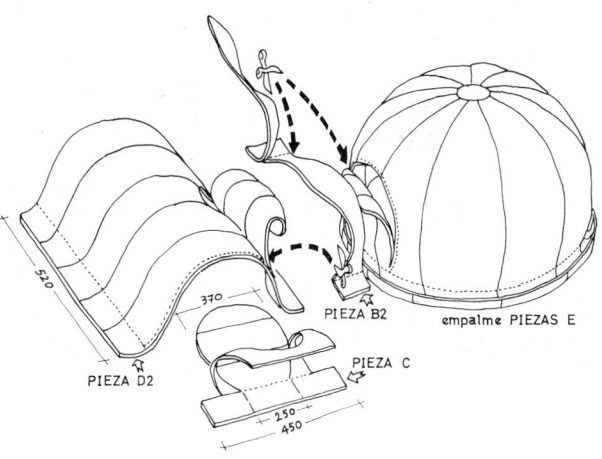

PIEZA D2

PIEZA B2

empalme PIEZAS E

PIEZA C

COMPOSICION DE LAS PIEZAS QUE
FORMAN EL MODULO E (6 PERSONAS)

AXONOMETRIA DEL CONJUNTO

JUNTA DE ENLACE
DE LA PIEZA C

BORDE QUE SE INTRODUCE
EN LA ZANJA DE CIMENTACION

BORDE QUE SE INTRODUCE EN
LA ZANJA DE CIMENTACION

EMPALME A LA RED

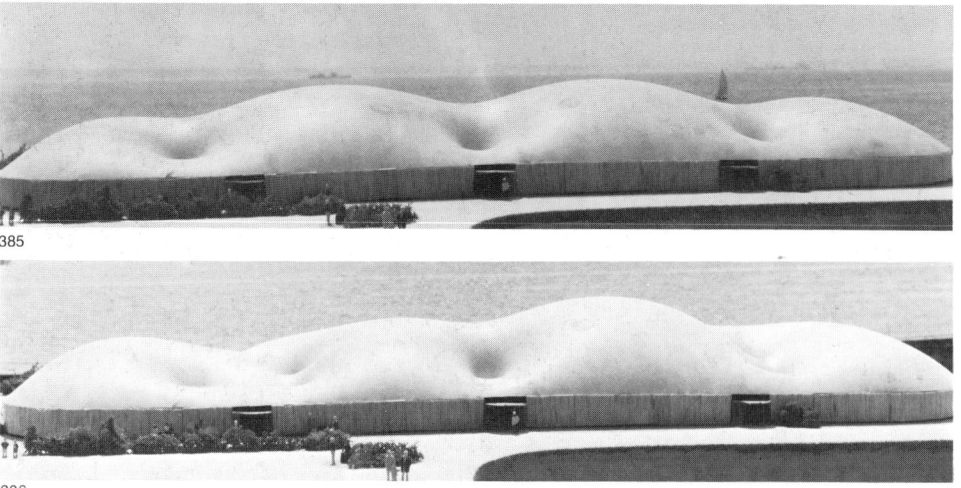

385

386

385, 386. Multi-purpose hall on the English South Coast

Design: Gernot Minke with Croucher and Salt, 1969

As far as the author knows, this is the first time anyone has thought of using additional point support on a structure in which a main direction of extension is dominant. The rubber membrane of the model was guyed first at five and then at eight points. The hall was to have a ground surface of 2,000 m²; the membrane would consist of PVC coated polyester fabric and be fastened to a 3 m high base.

387. Hemispherical combination

Design: Friedrich Krupp GmbH, Zentralinstitut für Forschung und Entwicklung, 1970

The central spherical element is cut so that two quarter circles appear on the outside contour when viewed from above. Such a definition achieves an equal division in the encircling arcade. (Bibl. 91, p. 28.)

388, 389. Roof over an open air swimming pool in Wolfsburg

Design and manufacture: Krupp Universalbau, 1971

The difficulty in this roof lay in the transition area between the rectangular hall and the spherical section covering the diving pool. It was designed with the help of model studies so that the cutting line in the loaded state remains approximately as an arc. The tensile stresses concentrated here are taken up by a steel cable and diverted to the foundations. A "soft" transition would also have been possible, but for various reasons was not chosen.
Air vent valves are put in various places to ensure tolerable temperatures in the event that the structure is not dismantled in summer.
(Bibl. 28.)

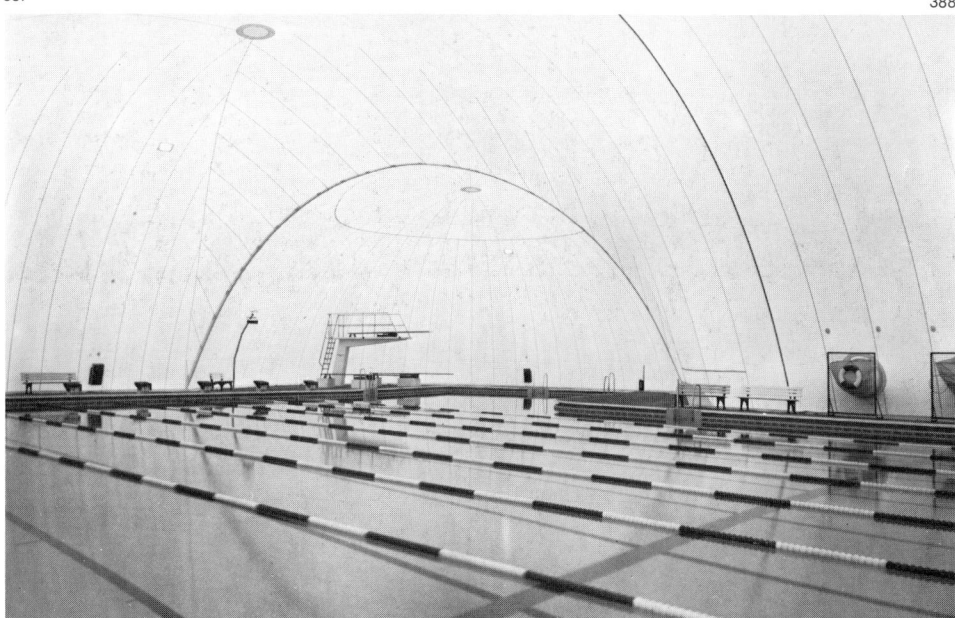

387

388

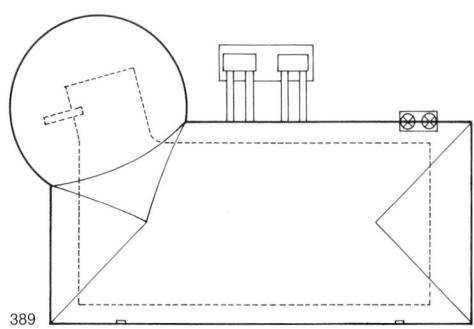

389

390–394. Modern Art Museum, Munich

Design and manufacture: J. B. Sanders & Söhne, 1971

The building is 35 m long and 15 m wide in the centre. From a central hemisphere emerge two half cylinders, at the ends of which are placed quarter spheres in which in turn are cut the entrances. At the transition areas to the half cylinders and to the vehicle lock the membrane forces of the main dome are transferred to the foundations by means of sewn in cables. The formation of wrinkles at the transition areas between the various structural parts shows that the choice of cutting pattern was not completely successful.

390

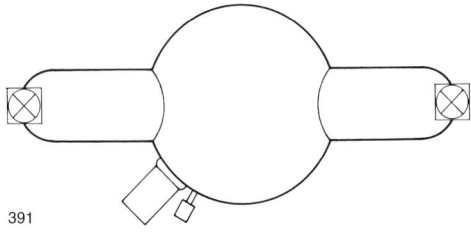

391

393

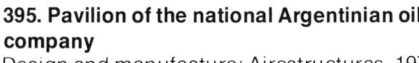

392

394

395

395. Pavilion of the national Argentinian oil company

Design and manufacture: Airestructuras, 1972

The pavilion acts as a demonstration room for a travelling exhibition. In contrast to the originally intended constriction of the main room by means of a cable, in production the area was designed as a band that is bounded by a cable on both sides. The fan is neatly placed at its foot. As well as the design of the constriction the cutting pattern of the membrane is remarkable. The membrane sections do not converge at the top but at the connection points or access doors. Thus not only is assembling simplified, but the spatial impression is favourably affected in as far as in the spherical sections the zenith has no special emphasis.

396

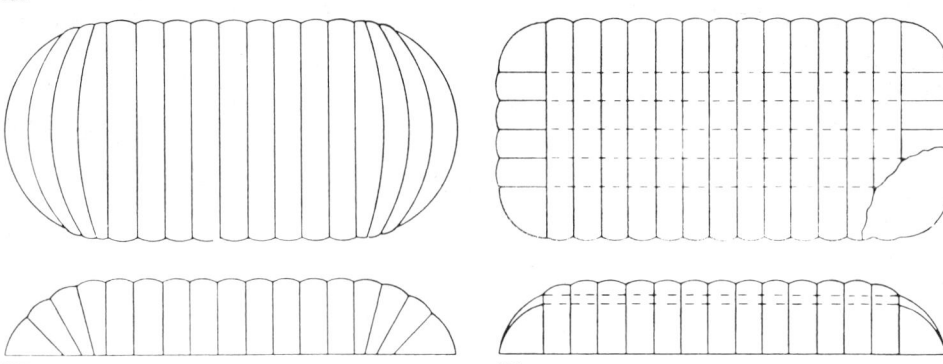

397   398

399

396—398. Cabled fieldhouse
Design and manufacture: Birdair Structures, Inc., 1972

Cables which reduce the curvature of the skin and considerably improve the internal acoustics run at 6 m intervals on the inside of the membrane.
The manufacturer offers two different forms of end section (Fig. 397, 398).

399. Cable reinforced ship cover
Design and manufacture: Birdair Structures, Inc., 1971

The 140 m long by 25 m wide prototype helps to investigate a new process, that of "mothballing" a ship. It is a great advantage to be able, by means of suitable adjustment of the internal climate, to keep the corrosion of the superstructure under considerable control.

400—406. Pneumatically stabilised plastic foil greenhouses
Design and manufacture: H. Brügge, Abt. Zugbeanspruchte Konstruktionen

Normal plastic foil greenhouses are already very cheap. When they are also pneumatically stabilised they can scarcely be beaten for price. A net made of polyamide fibres or coated steel cables was laid over a film of PVC or – as here – polyethelene. More accurate cutting patterns are of no interest in this type of use and for this reason no effort is made to avoid the formation of wrinkles.
The green houses shown in Fig. 400 are each 80 m long and 12 m wide. Figs. 401 to 406 demonstrate the erection.

400

401

403

405

402

404

406

407–410. Conference pavilion, Arnhem, Holland

Design: Eventstructure Research Group
Manufacture: Hoogerwerff B. V., 1971

The Arnhem conference pavilion had two levels that were connected by means of a spiral staircase with a wooden platform. The lower level served as a foyer, the upper as a conference room. The positive pressure amounted to 60 mm of water pressure below, but only 15 mm above, so that the ceiling between the two levels was pressed upwards. It consisted, as did the upper dome, of yellow PVC coated polyester fabric and was held in place at 1 m centres by cables to which sand bags were fastened below. Thus it was point supported so that seats were formed above. Up to thirty people could be seated on this surface under the above mentioned pressure differential. The upper dome, which was fastened all round by a zip fastener, could be detached in fine weather. The outer skin of the foyer consisted of transparent 0.8 mm thick PVC which, because of its great elasticity, was linearly reinforced by polyester bands. As a comparison of the building with the section shows, deformation of the membrane was different from what the designers had presumed.

The building was 9 m high and had a diameter of 18 m. The wooden stage lay 2 m above ground level and had a diameter of 5 m.

407

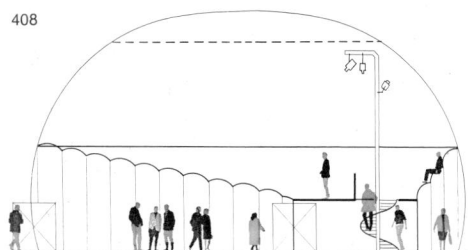

408

409

410

411

412

411–414. Interchangeable air cushion roof over a swimming pool in Rülzheim, near Germersheim
Design: Kleine und Richrath with Krupp Universalbau
Manufacture: Krupp Universalbau, 1972

This combination of air supported structure and air cushion offers interesting possibilities.
In the simplest form the membrane of a normal air supported hall is designed in two layers and the intermediate space is also under pressure. By using suitable cutting patterns different forms of buildings can be produced, as for instance structures which, with horizontal upper and vertical side cushions, are similar in outline to conventional air supported halls. If in pneumatic buildings of this kind the roof cushion is suspended on masts, the side membranes can be rolled up giving free access to the outside. In the example illustrated here two swimming pools lying one behind the other are covered by such a structure.
(Bibl. 29.)

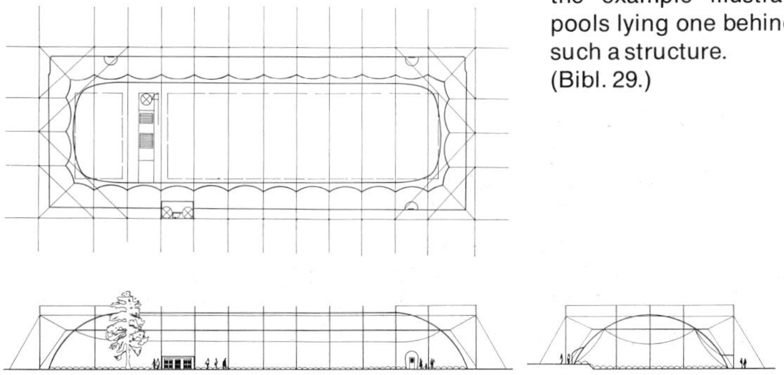

414

413

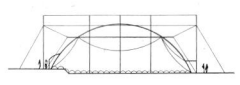

415. Mobile multi-purpose system
Design: Simon Conolly and Mark Fisher, 1970

Pneumatic forms of all kinds are extended from a transport vehicle to form an alterable environment.
(Bibl. 111.)

416–419. Pavilion "Atoms for Peace"
Design: Victor Lundy
Structural calculations: Severud-Elstad-Krueger Ass. and Birdair Structures, Inc.
Manufacture: Birdair Structures, Inc., 1960

The pavilion served as a travelling exhibition for the United States Atomic Energy Commission which was sent throughout Central and South America. It combined a refined technical concept, which has not been bettered in any way since, with an architectural efficiency of the highest rank.

The envelope was formed by two membranes made of vinyl coated polyamide fabric with a 1.20 m air space between which was divided into several compartments. The positive pressure in this air space was 40 mm of water pressure; inside the pavilion, which housed a cinema with 300 seats, workrooms and a test reactor covered by another air supported dome, the pressure was 50 mm of water pressure.

The entrances were rigid frameworks with revolving doors. While the exhibits were moved in and out of the building the entrance canopies,

415

416

which were linearly supported by internal lamellae, together with these entrance frames and temporary outer membranes, which were later dismantled again, functioned as airlocks.

The building was 100 m long, 40 m wide, 18 m high and covered a surface of some 2,000 m². The weight of the outer envelope was less than 6,000 kg. Erection lasted 3 to 4 days and was carried out by 12 men.

(Bibl. 44, p. 41 ff.)

418

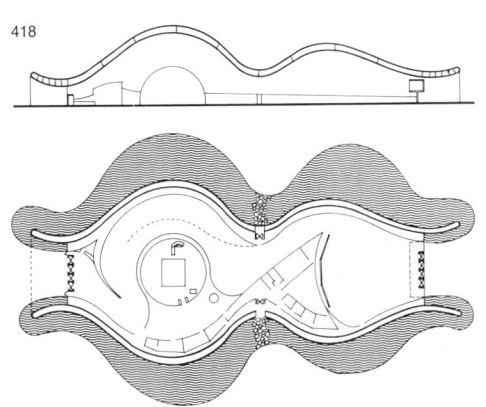

Sc, Mo + Mc / 3 di + 2 di + 1 di

420

421

422

423

424

427

425

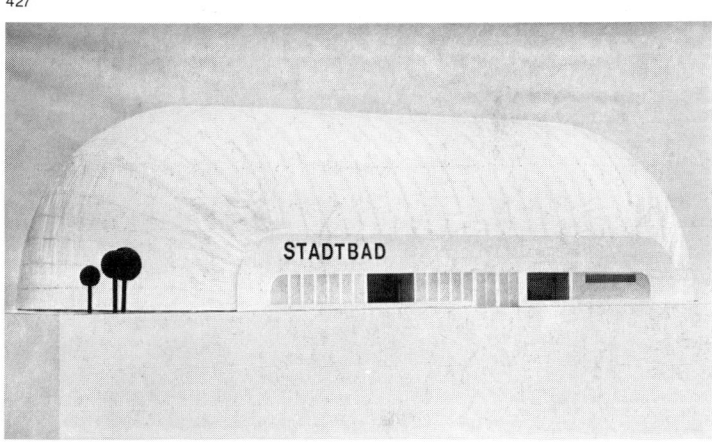

428

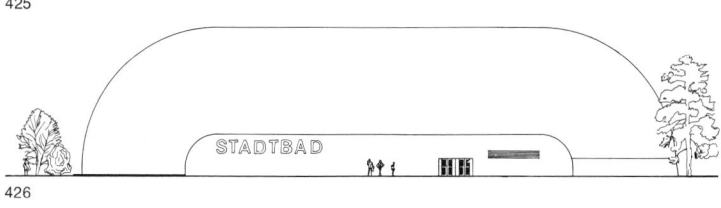

426

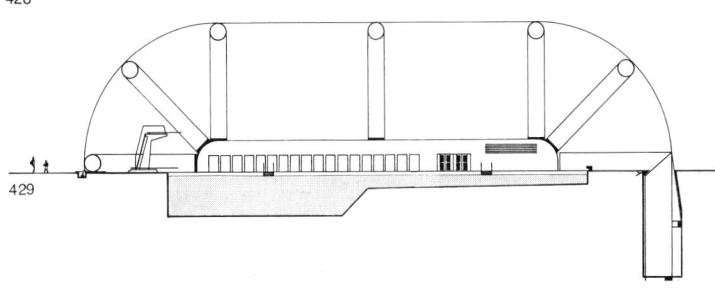

429

420, 421. The Bubble

Design and manufacture: Research Development Establishment

The device is used for carrying goods or land vehicles over water, swamps, etc. A tubular closed membrane with a diameter of 60 cm functions as a floating body. Its internal positive pressure is 1,400 mm of water pressure. Attached to it is an open membrane which forms a canopy over the goods to be transported. On the inside of this canopy cables are fixed by means of loops and the goods are fastened to these cables. If the inner space is put under positive pressure the goods are raised up. An air cushion is created on which the vehicle can be moved. Sealing to the ground or water surface is provided by an encircling membrane apron. An outboard motor is used for travelling over water. The cargo is loaded or unloaded while air is let

out on both sides in the central portion of the tube, thus creating a joint which allows the structure to hinge open and shut.

Because of its low transport volume the craft, together with inflation equipment, can be conveyed on a normal vehicle trailer.

Fig. 421 shows the transport of a lorry with a dead weight of 4,250 kg and a cargo of 3,750 kg.

422, 423. The Ark

Design: Research Development Establishment

This is a smaller design of lifting vehicle. It can be stowed away in a trunk and weighs only about 100 kg, while the weight of the transported car amounts to 1,500 kg.

424–429. Alterable air supported hall with air tubes

Design: Krupp Universalbau, 1971

Until now air tubes have seldom been installed as loadbearing structures for large spans. They must have large diameters and high internal pressure to withstand wind forces. However, they are highly suitable as erection aids for light membranes. The illustrations show a 50 m swimming pool with diving tower that can be used with or without the encircling envelope according to the weather conditions. The air tubes support the membrane only when it is being pulled across and tilted; they do not need to be designed for the usual wind loadings as the hall is only opened or closed in calm weather. When closed the hall is operated with internal positive pressure, i.e. as a pure air supported hall. (Bibl. 28.)

5. Pneumatic structures – technical details

5.1. Material properties of the membrane

5.1.1. Survey of materials

Tests have been carried out on all kinds of materials. However, requirements for tensile strength, flexibility and durability heavily restrict the range of suitable materials. A summary is given by the following classification which is divided up into isotropic and anisotropic materials. (Isotropic materials show the same strength and stretch in all directions; anisotropic materials have direction orientated properties.)

5.1.1.1. Isotropic materials

Plastic films
Plastic films are primarily produced from PVC, polyethelene, polyester, polyamide, polypropylene, polyvinylfluoride, polyterephthalate or synthetic rubbers (polisobuthylene, chloroprene).

Fabrics
This covers fabrics made of glass fibres or synthetic fibres which are coated in a PVC, polyester or polyurethane film.

Rubber membranes
Rubber membranes are very flexible. They are particularly useful as test specimens, as a wide range of forms can be constructed from them without a complicated cutting pattern. They are not very suitable for permanent pneumatic membranes of larger dimensions because of their low modulus of elasticity and their low weatherability. They are used particularly in special pneumatic concrete and synthetic forming processes, which make use of their great ability to deform. (Bibl. 119, p. 168.)

Metal foils
Metal foils possess a very high gas diffusion resistance and their tensile strengths range up to 90 kg/mm². However, they have very low breaking loads and can only be used for pneumatic structures when they are fairly ductile. Aluminium foils are mainly used in outer space because of their high reflective properties.
Rocket research in recent years has provided information on the application of chrome-nickel steels for membrane structures. At the moment investigations are being made on how far such materials can be economically applied in architectural situations. The author does not know of any completed projects.
One of the major problems in the use of metal foils is the need to produce very exact cutting patterns.

5.1.1.2. Anisotropic materials

Woven fabrics
Woven fabrics have two main directions of weave initially at right angles, but some angular displacement between threads is possible. They can be made of:

organic fibres, e.g. wool, cotton, hemp or silk,
mineral fibres, e.g. fibre glass,
metal fibres, e.g. thin steel wires, or
synthetic fibres, e.g. polyamides (Perlon, Nylon, Dederon), polyesters (Dacron, Diolen, Grisuten, Terylene, Trevira), polyacrylnitrilene (Dralon, Dolan, Orlon, Redan) or polyvinyl (Rhovyl).
Organic fibres are seldom used today for pneumatic structures. Deterioration processes can be largely prevented by additive materials; however they have a considerably lower durability and a less favourable elasticity after outside exposure than mineral or synthetic fibres.
The lowest elasticity under loading is shown at the moment by the fibre glass fabrics (breaking strengths of 3 to 4%). However, because of the low elasticity their spatial deformability is relatively low despite the angle displacement of the fabric threads and thus their ability to adapt to synclastic or anticlastic surfaces (which frequently occur in pneumatic structures) is also low, so that a very accurate cutting pattern is necessary. It is for this reason too that metal fibres have hardly ever been used in practice despite their great strength.
A great variety of materials is used for coating these fabrics, in particular rubber, bitumen, paraffin, polyester, acrylester, polyacrylic acid (Plexigum), plasticised PVC, polychloropor, (Neoprene, Perbunan), chlorysulfonated polyethelene (Hypalon), alkathene, athylene/propylene-Terpolymerisat (Holstapren), butyl and polyurethane. (Bibl. 15; Bibl. 57, p 3 ff; Bibl. 104; Bibl. 119, p 168; Bibl. 123, p 38; Bibl. 144, p 102.) They must be specially applied and provided with additives for ultra violet protection and flame resistance.

Gridded fabrics
Gridded fabrics are coarse weaves made of organic, mineral or synthetic fibres or metallic networks with mesh sizes of 3 to 20 mm. They are often embedded in several layers of synthetic films by means of casting or rolling processes. Gridded fabrics are particularly recommended when maximum light transmission plus high strength is required.

5.1.2. Synthetic films

Synthetic films are especially suited for use in pneumatic structures because of their high gas diffusion resistance and their great flexibility. They also offer advantages in their easy workability (welding and cementing) and their high light transmission.
Like all synthetic materials, however, under constant load they show an increased tendency to stretch so that they only maintain their form over a period of time if relatively low loads of less than 10% of the short term tear resistance are applied.
Furthermore, limits are placed on the use of films on the basis of their relatively low tensile strength of 3 to 20 kg/mm², their low tear-propagation resistance and their low weather resistance. The strength of thin films is affected primarily by ultra violet rays in conjunction with

Collaborators to this chapter: Rainer Hascher, Claudia Häfele and Verena Herzog-Loibl.

water vapour and can be greatly reduced within the first few months after an object was erected. The durability can be considerably increased by the use of absorbent carbon coatings or vaporisation or lamination of metals (above all aluminium) for protection against ultra violet rays – although at the expense of transparency, which is in any case greatly reduced by dirt in the course of time. If high light penetration is required it is often most economical to use the cheapest possible material (to which category most films belong) and not to clean this when it gets dirty but simply to change it.

5.1.3. Coated woven fabrics

5.1.3.1. Interaction of fabric and coating

As already stated, woven fabrics are anisotropic surface forms with two right angled preferential directions whose angles can be displaced. In the weaving process the threads are stretched more strongly in the warp direction (Fig. 1); thus this has a lower elasticity in comparison with the weft. In an extreme case the warp lies flat in the weave. In a fabric with the same form in both directions there is a somewhat lower stretch or higher modulus of elasticity due to the coating in the warp direction.

Other properties of the uncoated fabric, such as tear-propagation resistance and flexibility are also directly affected by the coating. For example, if the coating has a very high adhesive strength, then the long term tearing resistance is reduced.

However, one can state basically that the uncoated fabric is primarily responsible for strength and elasticity. Further important requirements, such as high gas diffusion resistance, flame resistance, resistance to ultra violet rays, insensitivity to mechanical influences and chemicals, must be provided by the coating.

Coatings vary greatly with regard to colour and light penetration.

5.1.3.2. Strength properties

Comparison with other construction materials
A comparison of the strength properties of these materials stressed only by tensile forces with other structural materials is offered by the tension length. It is measured on any constant cross-section and signifies the length at which a vertically hanging thread or rod breaks off by its own weight at the point of suspension. The size and shape of the cross-section are unimportant. The tension length is specific to each material and is expressed in terms of the breaking strength and the specific weight. It signifies the capacity of a material in terms of the possible spans within a structure.

The concept has long been used in the textile industry, because textile threads, which have no clearly measurable cross-section, can be best defined by their performance in material tests by means of weighing and tearing.

A thread or rod breaks when its weight cor-

responds to the breaking load. This is equal to the maximum weight (GMax). As:

$$R = \frac{V \, max}{F} \quad and$$

$$F = \frac{G \, max}{\sigma} \quad and$$

$$\gamma = \frac{G \, max}{V \, max} \quad equals$$

$$R = \frac{\sigma}{\gamma},$$

where:
R = the breaking length in km,
V = the volume of the rod in cm^3,
γ = the specific weight in g/cm^3,
F = the cross-sectional surface in mm^2,
σ = ultimate stress of rupture of the material in kg/mm^2.

Table 1 shows tension lengths of freely suspended profiles with prismatic cross-sections (Bibl. 137, p. 5). From this it can be seen that, disregarding the duration of load application, temperature and other limiting factors, the best values are achieved by those materials which have low volumetric weight and are able to take up high tensions. Under short term loading natural and synthetic fibres have a greater tension length than high strength steel wires (St 200/220), as they have a considerably lower specific weight. Today tension lengths of over 80 km can be achieved with high strength synthetic fibres as well as with glass fibres; even higher values can be attained by monocristal fibres (e.g. silizium-carbide), so-called whiskers, which have tensile strengths of 2000 kp/mm^2 and more.

In theory considerably larger spans can be achieved with Perlon wires than with metal wires. Because of the relatively low modulus of elasticity, the flow behaviour under permanent load, the comparatively low fatigue strength and the considerably higher safety factors which are therefore required, Perlon is still inferior to steel in terms of the maximum spans which can actually be achieved.

Using a safety factor of 5 on a typical fabric as used today, the maximum span of a flat dome without additional stabilisation is about 40 m.

Tensile strength and stretch
The actual strength of the cross-section of fabric membrane is not clearly definable; construction and non-homogeneity of the materials used make an exact definition of the tension in kp/cm^2 impossible.

The tensile strength of woven fabrics is usually given in kp/5 cm. To determine the tension length samples of 50 mm widths (in accordance with DIN 53354) are torn with a free clamping length of 300 mm and a speed of 300 mm/min. Up until now the strength behaviour of coated fabrics has generally been tested successively in the weft and warp directions.

The actual tensile strength of a woven fabric depends on the number of threads per cm, on the thread denier and on the type of weave.

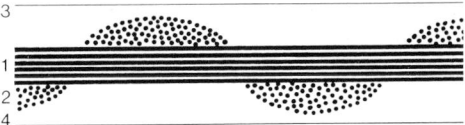

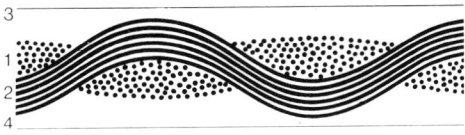

1. Diagrammatic cross-section through a coated fabric in the warp and weft direction. Key: 1 warp, 2 weft, 3 top side, 4 bottom side.

Material	σ (kg/mm^2)	γ (g/cm^3)	R (km)
lead	1.7	11.4	0.15
aluminium wire	17	2.7	6.5
structural steel St 52	52	7.8	6.7
duraluminium	50	2.8	18
pine wood	10	0.5	20
steel wire	220	7.8	28
silk	–	–	45
cotton	–	–	26–40
perlon wire	57	1.14	50
aluminium oxide whisker	2000	3.3	606
graphite whisker	2,100	1.4	1,500

Table 1. Tension lengths of different materials.

(Denier expressed by "den" is the weight of a 9000 m long thread in grammes. Some time ago the unit of measurement "tex" was also introduced. 1000 den corresponds to approx. 110 tex, or 1100 dtex.)

High pressure structures require woven fabrics with strength values of over 1000 kg/5cm and thicknesses of several millimetres; in low pressure structures, which make considerably lower demands on strength and stretch behaviour, strength factors of 200 to 600 kp/5 cm and thicknesses of 0.7 to 1.2 are usually sufficient.

Table 2 shows the results of uniaxial tensile tests on an uncoated and a PVC coated polyester fabric. The higher elongation in the weft is characteristic of coated polyester fabrics. It can be either smaller or larger depending on the method of production. It depends primarily on the shrinkage allowance during the coating process.

In contrast to these uniaxial tensile tests the biaxial ones offer a better insight into tension/elongation behaviour in practice, where the membranes are nearly always equally tensioned in both directions.

The bursting test, in accordance with DIN 53869, is made to establish the strength and elongation of laminar textiles during bulging due to air or fluid pressure up to bursting point of the ring clamped sample. The bursting pressure is generally given in kp/cm². For example, for a polyester fabric that has a tensile strength of 400 kp/5 cm in both the weft and warp, it amounts to 20 kp/cm². (Bibl. 103.)

Table 3 shows an evaluation of the cross-test described by M. I. Petrowkin (Bibl. 144, p. 103).

For larger mechanical stress so far polyamide fabrics (e.g. Nylon, Perlon, Dederon) have been employed. These elastic, friction resistant fibres are used especially in barrage dam construction. For normal pneumatic structures polyesters are almost exclusively used in Europe today because of their high dimensional stability. Polyester fibres have somewhat lower elongation and strength than polyamides, although the difference, especially in strength, between polyester and polyamides is now small.

Polyester threads can be produced in normal strengths and high strengths. For the manufacture of coated and rubberised fabrics the high strength, low shrink types are used almost exclusively. (Bibl 103).

Tear-propagation resistance and adhesion

Tests to determine the tear-propagation resistance give information regarding the "tear-propagation load" (kp) at which a sample, which is already notched on one edge, tears on. This kind of measurement of tear-propagation resistance, in accordance with DIN 53356, is disputed and attempts are being made to find other methods. On the other hand the present method is easy to carry out.

The adhesion strength expressed in kp/cm specifies the resistance of adhesion of the coating to mechanical separation from the woven fabric generated by tensile forces (not by shear forces). For testing purposes two freshly coated strips of fabric 5 cm wide were laminated under light pressure; the force which is required to separate the two layers of fabric is given as the adhesive strength in kp/5 cm (DIN 53357). The test procedure is called a "Peeling Test".

In contrast to woven fabrics made of rough fibres, thick woven fabrics made of smooth fibres offer the coating a mechanical anchorage that is inadequate for the usual fields of use.

Without an adhesive agent the adhesive strength of polyester fabric 1000, den 60 T/m, 9.2/9.7 threads/cm, for example amounts to only 4 to 5 kp/5 cm. It is different in the case of open and lattice weaves, where an adhesive agent is not usually necessary, as here a welding of the coatings can take place by means of the formation of small plugs.

In the case of PVC coatings on polyester fabrics, isocyanates in conjunction with polyesters have proved successful as adhesive agents. Increased additions of isocyanate do not further improve adhesion and usually lead to a stiffening of the fabric.

In the case of polyurethane coating on polyester the addition of adhesive agents is unnecessary as the isocyanate itself is present in a very concentrated form.

Today adhesive strengths of up to 25 kp/5 cm can be achieved. However, as a decrease in the tear-propagation resistance occurs simultaneously with an increase in the adhesive strength, 12 kp/5 cm cannot usually be exceeded.

The tear-propagation resistance is also influenced by the fabric form, the type of weave, the formula of the coating paste and the twist of the thread.

In woven fabrics with a denier of 1000 the tear-propagation resistance is usually between 20 and 30 kp. The lowest tear-propagation resistance – here approximately 7 to 8 kp – occurs when each thread tears separately and the threads remain separated from one another.

For low pressure structures woven fabrics are most frequently used at the moment in basket or mat weave. Regardless of the value of the tensile strength, in general one can estimate the tear-propagation resistance at 10 to 15% of the tensile strength. (Bibl. 27; Bibl. 103.)

Relationship between strength behaviour and load duration, temperature and humidity

The long term tensile strength of coated fabrics is considerably lower than the tensile strength which can be determined from a short term test in accordance with DIN 53354.

Tests with PVC coated polyester fabrics showed that under permanent loads of 80% of the tearing strength some samples tore after only hours, many after several days.

The samples resisted continuous loads up to 10,000 hours at 50% of the tearing strength. In a subsequent short term test no reduction in strength worth mentioning could be distinguished. (Bibl. 104.)

Table 4 shows the stretch of the fabric in both thread directions under different loads and the increase in stretch under load continuing for some time. The increase in length stopped when a specific final value for the load in question had been reached.

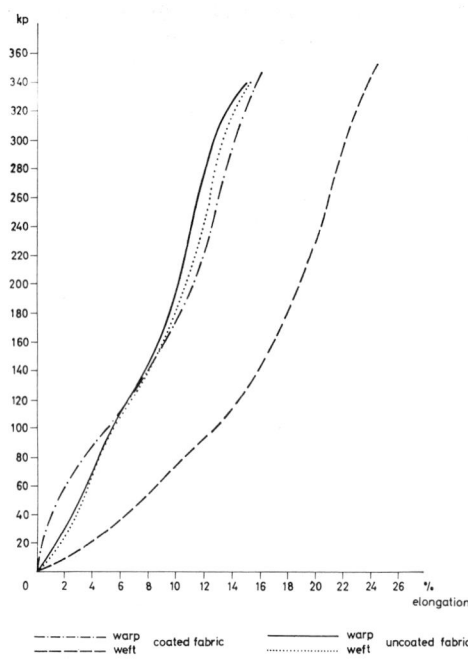

Table 2. Diagram of a uniaxial tensile test on an uncoated and PVC coated polyester fabric 1000 den, 9/10 threads/cm.

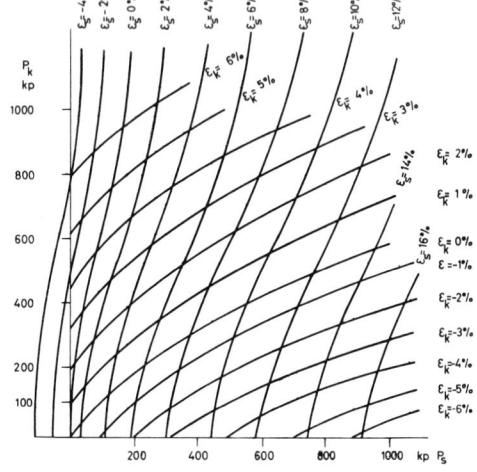

Table 3. Diagram of a biaxial tensile test.

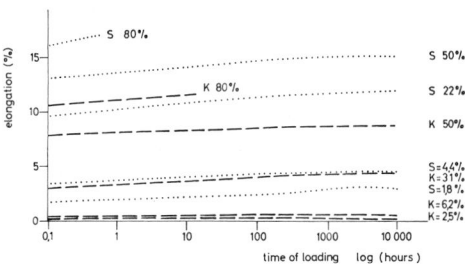

Table 4. Progress of elongation of a PVC coated polyester fabric in warp (k) and weft (s) direction under permanent load (loadings in % of breaking load).

Breaking elongations of 14 to 20% in the weft or warp direction were rarely attained in low pressure structures because of the usual factor of safety of 5 to 6.

Strength and stretch are dependent on temperature. With an increase in temperature, stretch increases and strength decreases. The material reacts conversely when it is cold. In any case one can state that temperatures of –25 °C to +70 °C are normal for coated and rubberised synthetic fabrics and within this range the influence on strength and stretch behaviour is low. (Bibl. 27.)

Even in temperatures of under –30 °C such fabrics are still stable; however flexibility is considerably reduced so that tears could be initiated by a sharp kink.

In Tables 5 to 8 the stretch behaviour in the warp and weft direction is shown for PVC coated polyester fabric under different temperatures (Bibl. 104).

If polyester fabrics are vulcanised, then even after several hours storage in the drying cupboard at +180 °C no loss of strength in the material can be established.

The influence of humidity is also important for the dimensional stability of a woven fabric. The better dimensional stability of polyester fabrics under the influence of temperature and humidity as against polyamide fabrics (in damp conditions polyamides show a tendency to lengthen and have a higher humidity absorption than polyester) is itself a decisive reason why polyester fabrics are used in preference today. (Table 9; Bibl. 27.)

5.1.3.3. The different coatings

PVC coating

Good weather stability, favourable price, easy workability and above all variation in respect of colour and transparency account for the fact that in over 90% of cases in Europe synthetic fabrics are PVC coated.

Usually it is stipulated that a PVC coating should have a thickness of at least 0.2 mm over the crossing points of the threads. For reasons of price and weight it is not in any case economical to greatly exceed this figure.

The support fabric hardly ages when the coating is strongly pigmented (e.g. with carbon).

If transparency or translucence is desired, it is usual to try to make good the lack of colouring by adding ultra violet absorbents (e.g. Tinuvin or Uvinul). (Table 10.)

Single membrane structures are often completely made out of translucent material, whereby in certain conditions additional transparent plastic sheets or acrylic glass windows can be inserted. Although translucent material only has a light transmittance of 5 to 10%, with an external brightness of 1200 lx light values of 120 lx are achieved inside such buildings. This corresponds to the brightness in a conventional office under the same external conditions. (Bibl. 104.)

A light transmittance of over 50% with a strength of approximately 300 kp/5 cm is provided by a

special polyester lattice fabric with transparent coating.

Rotting due to mould or bacteria cannot occur in synthetic coatings or in synthetic woven fabrics. A reddish colouring which was observed in the membrane was found after investigation to be due to metabolic products of a fungus type whose nutritive substratum was probably fine dust. Even this could be prevented by special additives in the PVC. (Bibl. 104.)

Even atmospheric elements (water, oxygen, ozone, industrial gases) have hardly any effect on the PVC coating. The coating can if necessary be additionally provided with very thin rolled metal foil or even be damped with aluminium. A final coat of varnish, which for ultra violet protection can be strengthened with the finest aluminium lamellae, mica or quartz granules, prevents plasticiser dispersion and thereby sticking or smearing on the surface and, in the case of coloured coatings, diffusion of coloured pigments. As these varnishes are very smooth, they also reduce the effects of pollution and allow water to drain away more quickly.

Rubberising

In contrast to PVC coatings the durability of synthetic rubbers is considerably affected by industrial gases.

Translucence can also be achieved in rubberised coatings, as in the use of chlorosulfonic polyethelene (e.g. Hypalon) and ethylenepropylene rubber; polychloroprene (e.g. Neoprene, Perbunan) is only suitable for dark coatings because of its tendency to yellowing. (Bibl. 103.)

Rubber coatings are practical especially when a greater mechanical tensioning has to be taken into account, e.g. in pneumatic barrage dams.

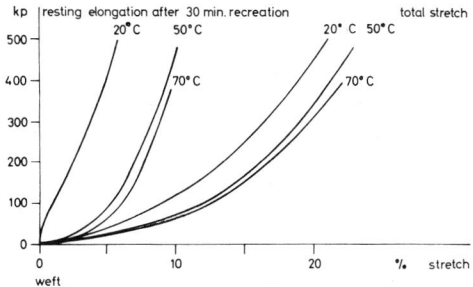
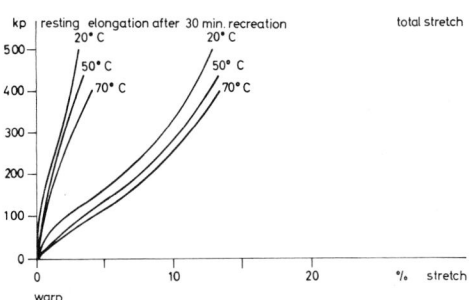

Table 5, 6. Stretch behaviour of PVC polyester fabric (high strength Trevira) at different temperatures.

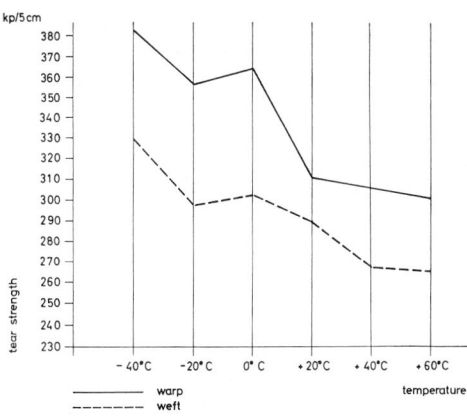
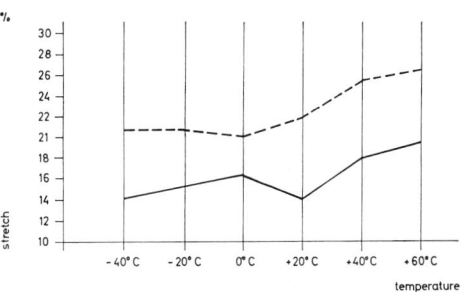

Tables 7, 8. Strength and stretch behaviour of a rubberised polyester fabric (high strength Trevira) at different temperatures.

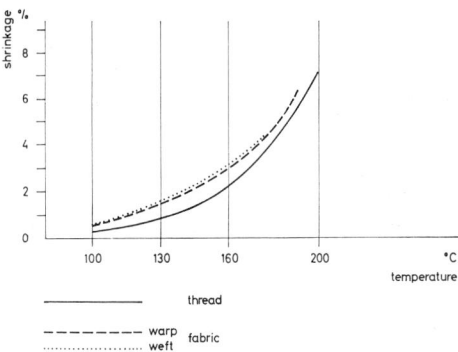

Table 9. Shrinkage behaviour of a low shrink polyester fabric (high strength Trevira 1000 den).

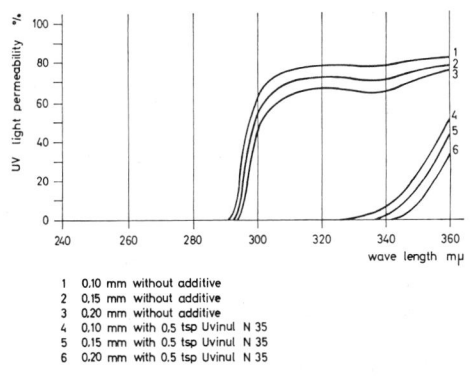

1 0.10 mm without additive
2 0.15 mm without additive
3 0.20 mm without additive
4 0.10 mm with 0.5 tsp Uvinul N 35
5 0.15 mm with 0.5 tsp Uvinul N 35
6 0.20 mm with 0.5 tsp Uvinul N 35

with Titanoxide 2 tsp TiO or 0.5 tsp Tinuvin P no transparency

Table 10. Ultra violet permeability of PVC foils

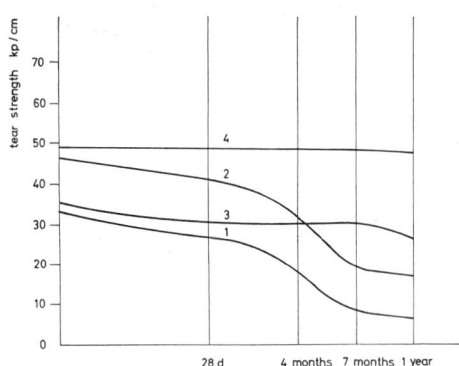

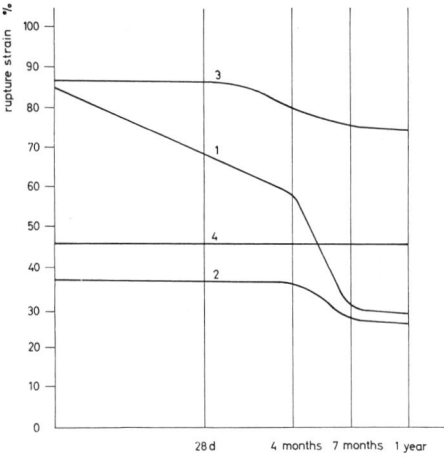

1 transparent material in weft direction
2 transparent material in weft direction
3 dark material in weft direction
4 dark material in weft direction

Table 11, 12. Aging of polyamide rayon fibre fabric DC15 coated with PVC on both sides.

Other coatings
Despite various advantages such as high flexibility, abrasion resistance and resistance to cold, polyurethane coatings have not much been used due to their high price and the fact that in the long run they are being attacked by hydrolytic decomposition (Bibl. 57).
The most weather resistant coating at the present time can be achieved with polytetrafluorethelene (Hostavan, Mylar). However, due to its high price and difficult workability it is only used in special cases.

5.1.4. Safety factors and data on durability

In coated woven fabrics the fatigue strength is considerably less than the tensile strength, the seam strength is usually lower than the material strength and a gradual decomposition through ultra violet rays cannot be completely eliminated. Therefore in low pressure structures today a safety factor of 5 to 6 is generally applied for the tensile strength of high performance woven fabric structures.
In tests to determine the suitability of specific materials more significance is given to the results of external weathering, in so far as this is carried out over a sufficiently long period of time, than to those of artificial weathering. The durability of synthetic materials is particularly affected by ultra violet rays. In the case of uncoated fabrics a higher degree of failure is usually expected in the weft direction, as in the cross-woven areas the weft threads are above the warp threads. (Bibl. 104.)
Tables 11 and 12 show the effect of aging on the strength and stretch behaviour of polyamide fabric DC 15 which is coated with PVC on both sides (Bibl. 144, p. 103); Table 13 gives information on the expected durability of coated high strength polyester fabrics in relation to the coating (coating strength as generally used in practice). (Bibl. 86; Bibl. 104.)
From the economic viewpoint the durability of coated woven fabrics is completely satisfactory, if it is borne in mind that any slight damage can easily be repaired in situ by a piece of replacement material and special adhesive agent and that apart from occasional cleaning to improve the light transmission, no maintenance of the envelope is necessary. (In this connection the caution of the German manufacturers, who at the moment will only furnish guarantee periods of 2 years maximum even for standard air supported halls, is difficult to comprehend.)

5.1.5. Behaviour in fire

Previous evidence of the behaviour of pneumatically stressed membranes has been gained from tests on single membrane structures with internal positive pressure ("air supported halls"). The information gained should not therefore be directly generalised.
Subsequent tests have been carried out by the Institut für Baustoffe and the Institut für Grubensicherheit in Weimar (Bibl. 153). The tests were made on a hemispherical membrane of 25 m diameter made of polyamide fabric with an inflammable PVC coating. During the tests the following weather conditions prevailed:
The air temperature was −6 °C and humidity was 52%; the wind speed on the ground was 1.8 m/sec. and at 14 m height 3.85 m/sec.
1. At a distance of 1 m from the membrane a row of wax torches was installed inside at varying heights. After 12 minutes the first hole appeared in the membrane. It increased in size only slightly due to the fact that the flame was drawn out by the air pressure differential and extinguished. It was deduced from this that the dimensions of burn holes depend only on the size and heat of the furnace.
2. In the centre of the hall 50 litres of benzine was ignited on a surface of 3 sq m. After 2.5 minutes the interior was completely filled with smoke; after less than a further half minute as a result of the increase in pressure caused by the increase in volume, a hole appeared at the top of the membrane through which the smoke was drawn; after 18 minutes the building began to sink in. The greatest heat inside occurred after 5 minutes.
3. At a distance of 1 m from the inside of the membrane a pile of wood with a ground surface of 60 × 60 cm and a height of 1 m was lit over a container with 10 litres of benzine. 3 to 4 m high flames licked at the membrane but did not ignite

Coating	Durability
PVC, transparent, 0.5% UV absorber	3 – 8 years
PVC, translucent, 2 to 6% TiO$_2$, 0.5% UV absorber	8 – 12 years
PVC, opaque, pigmented	10 – 18 years
chloroprene	8 – 12 years
chlorosulfonated polyethylene	12 – 20 years
polyurethane	4 – 6 years

Table 13. Durability of coated high strength polyester fabric in relation to the coating.

Country	Material	Weight g/m²	Thickness mm	Tearing resistance kp/5 cm	
				warp	weft
FRG	Hostaphan	53	0.04	45	36
USA	Mylar	16	0.015	8.2	5.5
USA	polyterphthalic acid ester, aluminised	110	0.127	~ 15	~ 11
USA	polyvinyl chloride	285	0.25	36	32
USSR	polyethylene	53	0.065	4.5	4.5

Table 14. Technical data for different foils.

it. After 3.25 minutes heat radiation caused a hole in the membrane; after 4.1 minutes the pile of burning wood was extinguished; after 7 minutes the fan switched off – the building began to sink in. The greatest internal heat arose between the 4th and 5th minute.

4. At a distance of 1 m from the inside of the membrane four piles of wood, each with 5 litres of benzine, and another in the centre of the hall were ignited. After 0.6 minutes the increase in pressure caused a hole at the top of the membrane; after 1.8 minutes heat radiation caused holes in the region of two furnaces; after 2.8 minutes the third hole appeared; after 3.3 minutes the fourth hole appeared; after 10 minutes the fan switched off; although two of the openings made were several square metres in size the hall did not collapse for 25 minutes. The greatest internal heat came after 6 minutes.

Similar combustion tests were undertaken by the Swedish Research Laboratory (Bibl. 131, p. 33 ff.).

The first test was undertaken on an air supported hall made of polyester woven fabric with PVC coating. The wind speed was 5 to 7 m/sec.

At the edge of the hall inside a furnace was installed. After 2.7 minutes a hole appeared in the membrane. The flames penetrated through to the outside without directly touching the membrane. No further damage occurred.

The second test was carried out on a hall made of polyethelene lattice foil with polyamide fibres. The wind speed was 2.2 to 2.7 m/sec.

Here also a small furnace was installed at the edge of the hall. After 5 minutes a large hole appeared and the hall collapsed so severely that the furnace lay in the open air.

In the Krupp work reports the following test was described (Bibl. 28): "In order to test a water spraying apparatus 8 plates with various sprays, toothpastes, tins, etc. were set on fire by means of remote ignition. The hall was a hemisphere of 11 m diameter. The sprinkler was set to switch on at 68 °C and reacted with most success after 57 seconds. The pressure rose from 27 mm water column to over 50 mm water column and the temperature from 10 °C to 22 °C; the oxygen content was almost constant. The rise in pressure caused the nonreturn valve in the draft channel to close automatically".

The results of these tests showed that the combustion danger for air supported halls made of rubberised and coated fabrics was very low, as long as they were not used for storing inflammable goods, and that the same fire protection conditions could be applied as for solid structures. Smoking is permitted. Welding work can also be carried out as long as the usual safety precautions are fulfilled. (Bibl. 28; Bibl. 104; Bibl. 153.)

Safety measures are specific (Bibl. 102; Bibl. 131, p. 34, 68):

– use of flame resistant material;
– adequate escape routes (maximum intervals 30 m);

Table 15. Technical data for different coated fabrics.

Country	Material	Weight g/m²	Thickness mm	Tearing resistance kp/5 cm		Tear propagation resistance kp		Breaking elongation %	
				warp	weft	warp	weft	warp	weft
FRG	PVC coated high-strength Trevira fabric, dtex 1100 f 200, 9.5/9.5 threads/cm, linen weave 1/1	850	0.75	330	310	35	35	14	20
FRG	PVC coated high-strength Trevira fabric, dtex 1100 f 200, 11/12 threads/cm, mat weave 2/2	1,000	0.9	400	400	70	70	15	21
FRG	PVC coated high-strength Trevira fabric, dtex 1100 f 200, 14/15 threads/cm, mat weave 2/2	1,100	1.0	500	500	70	70	15	23
FRG	high-strength Trevira lattice fabric, foil coated on both sides, translucent, dtex 1100, 5/5 threads/cm, linen weave 1/1	420		200	190	56	56	16	16
FRG	PVC coated super-strength Diolen fabric, 12/11.6 threads/cm, mat weave	873		451	362	43	56.3	17.3	21.5
FRG	PVC coated super-strength Diolen lattice fabric, dtex 1100, 17/18 threads/cm	1,250	1.02	550	625	70	62	20	28
CSSR	PVC coated Atmotol 800/TE 516, 7/7 threads/cm	691		226	234			20	28
CSSR	polyurethane coated Chemlon TE 514-TH 370, translucent, 14/13 threads/cm	532		562	479.6			43.2	31
CSSR	butyl rubber coated Chemlon TE 522, 11/10 threads/cm	770		406	326			25.4	39.2
GDR	PVC coated polyamide fabric	760	0.76	375	370				
UK	PVC coated polyester fabric	540	0.51	180	115			52	
France	PVC coated polyester fabric	650		230	200				
France	PVC coated polyester fabric	730		300	270				
France	PVC coated polyester fabric	850		400	360				
Japan	Hypalon coated PVA fabric	980	0.92	770	720	48			
Japan	PVC coated "Vinylon KV 71,000" PVA fabric	1,000	0.95	600	575	45	60	24	24
Japan	PVC coated "Vinylon KV 70,600" PVA fabric	600	0.6	325	300	12	14	25	20
Japan	PVC coated "Tetoron TT 51,000" polyester fabric	600	0.45	375	850	50	50	25	30
Japan	"Vinylon KV 40,629" PVA fabric, Hypalon coated on the outside and PVC coated on the inside, double ply	3,200	3.5	c. 2,000	c. 2,000			37	33
Japan	PVC coated "Cordoglass X-340/864" fibre glass fabric	1,000	1.2	550	500	30	26	4	15
Japan	PVC coated "Tetoron TT 4,000" polyester fabric, double ply	3,700	3.2	c. 630	c. 920	300	350	25	30
Japan	PVC coated "Tetoron TT 55,000" polyester fabric	1,100	0.7	c. 750	c. 750	60	60	25	30
Sweden	type 199 PVC coated nylon fabric	950	0.8	490	410	60	60	16	28
Sweden	type 196 PVC coated nylon fabric	708	0.6	300	260	35	30		
Sweden	PVC coated nylon fabric, dtex 840, 5/5 threads/cm	750	0.8	201	170	56	74	25	30
USA	Hypalon coated dacron fabric, double ply below 45°	2,380	1.78	910	910	180			
USA	silicone rubber coated dacron fabric	540	0.51	270	270	41			
USA	PVC coated nylon fabric	610	0.66	360	305	50			
USSR	rubber coated polyamide fabric No. 24	1,200		270	216			32.9	39.6
USSR	rubber coated polyamide fabric No. 806	1,200		180	130			28	28

– emergency exits in the form of single doors opening to the outside; however, these have the disadvantage that the high loss of pressure which occurs when they are opened hastens the collapse of the building;

– provision of spare fans, to maintain support pressure for as long as possible;

– rigid support structures preventing the membrane from collapsing completely;

– fire extinguishers;

– smoke vents at the top of the building;

– sprinkler equipment.

In Federal Germany it is usually stipulated that the envelope should comply with the combustion chamber test in accordance with DIN 4102. In the FRG guidelines for the construction and operation of air supported halls (Bibl. 136) brought in in 1971 proof of this was also required.

In order to comply with the strict demands of this Standard the coating must be made sufficiently flame resistant. This is achieved by means of flame resistant plasticisers (e.g. alkylarylphosphate) and other additives.

The air supported hall materials produced in Federal Germany are flame resistant materials Class B. 1. It is frequently said that this test is too stringent for coated and rubberised fabrics. Thus for the development of flame resistant materials initial tests are carried out in accordance with DIN 53 906 and DIN 53 907 (determination of the flame resistance of textiles with vertically and horizontally clamped specimens).

The decisive criteria for the evaluation of combustion tests are undecomposed surface lengths of the sample, combustion time, afterburn time, after-glow time and flue gas temperature. The least stringent test is that complying with DIN 53 907, where surface decomposition and after-burning are lowest. The combustion chamber process (DIN 4102) compared with the combustion tunnel process developed by the construction materials testing office in Hamburg, has a somewhat lower proportion of undecomposed surface length and is therefore the more stringent test. DIN 53 906 provides the hardest test. Because of longer after-burn times the sample material is often completely destroyed. However a longer after-burn period can also give proof of low combustibility of the material. (Bibl. 57.)

H. L. Malhotra at the International Symposium on Pneumatic Structures in Delft in 1972 reported on a series of fire tests carried out mainly in England and Sweden. The observations made there deviate slightly from the test result described so far (Bibl. 102). It became clear how problematic generalisations of individual test results are and that as regards fire behaviour one is often obliged to deal with prognoses.

It is comforting to know that the number of cases of combustion of single membrane structures is very low (approximately 0.5% in the USA). For all other kinds of pneumatically stabilised membrane structures no evaluated test results were known at the time of writing.

For low pressure double membrane structures it can, however, be stated that destruction will normally take place more quickly because the internal air volume is less than in single mem-

brane structures and because in most cases the loadbearing structure is stressed by bending, torsion or buckling. If the loadbearing structure consists of individual compartments whose air spaces are not connected to each other, then in certain circumstances only parts are destroyed. In this case it is, however, important that the membrane material does not continue burning of its own accord.

In the case of high pressure loadbearing structures damage to the membrane, including fire damage, could cause an immediate collapse.

5.1.6. Summary tables

Tables 14 and 15 show the most important technical values of the most commonly used membrane materials as at Spring 1972 (Bibl. 44, p 94, 98; Bibl. 82; Bibl. 103; Bibl. 104; Bibl. 131, p 168, 169; Bibl. 144, p 104). The indication of a country in which a material is manufactured or used does not preclude the fact that this or a similar material may be manufactured or used in other countries, but means simply that the technical details originate there.

5.2. Development of the membrane envelope

5.2.1. Structural design

In almost all cases single layer membranes are used. In places where concentration of forces occurs (anchorages, connections to other structural parts or triangulations) reinforcement is provided by superimposed flanges, so that the forces are conducted away from the membrane surface (Fig. 2; Bibl. 131, p. 235).

Doubled membranes are structurally necessary over the whole surface when – perhaps because of an above average internal pressure (e.g. in special high pressure structures) or of a lower than average membrane curvature – the forces in the envelope become so large that they can no longer be taken up by a single layer of the membrane material available.

It was not to increase strength, but in order to avoid corrosion through ultra violet rays, that the physicist Dr. Laing developed a three layer membrane with an outer coating of glass fibre fabric, a middle one of aluminium foil and an inner one made of plastic foil to which were attached revolving aluminium lamellae (Bibl. 96, p. 163; see also p. 162).

A multi-layered membrane is also produced by spraying insulating synthetic foam or by sticking soft lamellae on the membrane. Such measures bring about an additional stiffening. If the coating is so stiff that the pneumatically stressed membrane is structurally superfluous, then this can subsequently be removed. Then the building can naturally no longer be referred to as a pneumatic design.

There are two main possible designs for pneumatic envelopes made of two membrane

layers separated from one another by a support medium:

1. The space to be utilised receives the same atmospheric pressure as the exterior (Figs. 3, 5).

2. The space to be utilised receives a higher or lower atmospheric pressure than the exterior space (Figs. 4, 6). In the cases illustrated here the support pressure is reduced from layer to layer towards the outside so that the inner membrane is stressed by the support pressure of the utilised space.

In Example 1, the most simple design, the membranes touch only at the ends. Lenslike cross-sections are formed (Fig. 5). If the membrane surface is damaged, the support air escapes and the membranes become slack, although the structural part will not collapse. Only in the event of severe applied forces need subsequent damage be expected. Lower heights with smaller radii of curvature in the membrane are achieved with positive pressure by means of guys (Figs. 7–11) or with negative pressure by means of supports (Fig. 12, 13). A large number of proposals which have not yet been executed can be found in Bibls. 83, 84 and 119. In the case of larger spans it is recommended that the space between the membranes be divided into individual compartments so that if the skin is damaged only single parts of it fail (Figs. 14–16).

The main advantage of a structure such as Example 2 is that in the event of external damage there is still a reserve membrane available (Fig. 17, 18).

A combination of both principles is represented in "pneumatic sandwich plates". Fig. 19 shows an English design in which both layers are joined together by narrow, parallel threads. The internal positive pressure is between 0.2 and 0.8 excess atmospheric pressure. The lamellae are produced in Great Britain as well as the USA and are available in different thicknesses. They are characterised by a high thermal insulation capacity as well as a high resistance to wind loading. Because of their relatively high price they have been used almost exclusively in the military field.

5.2.2. Cutting pattern

The commercial semi-finished material for membranes in the case of foils and coated fabrics is lengths which in Europe are usually supplied in widths of 150 cm, less often 120, 140, 160 or 200 cm. At the sides overlaps of 2 to 4 cm are required, so that an untreated width of 150 cm for simply curved surfaces gives finished widths of 146 to 148 cm according to the type of bonding used.

Overlaps are formed by localised linear reinforcements of the skin, which, depending on the cutting pattern and construction of the seam, can lead to a reduction in the radius of curvature of the skin – similar to that in the use of cables as additional structural elements.

With a doubled skin (See Section 5.2.1) the joints of the membrane lengths in the individual mem-

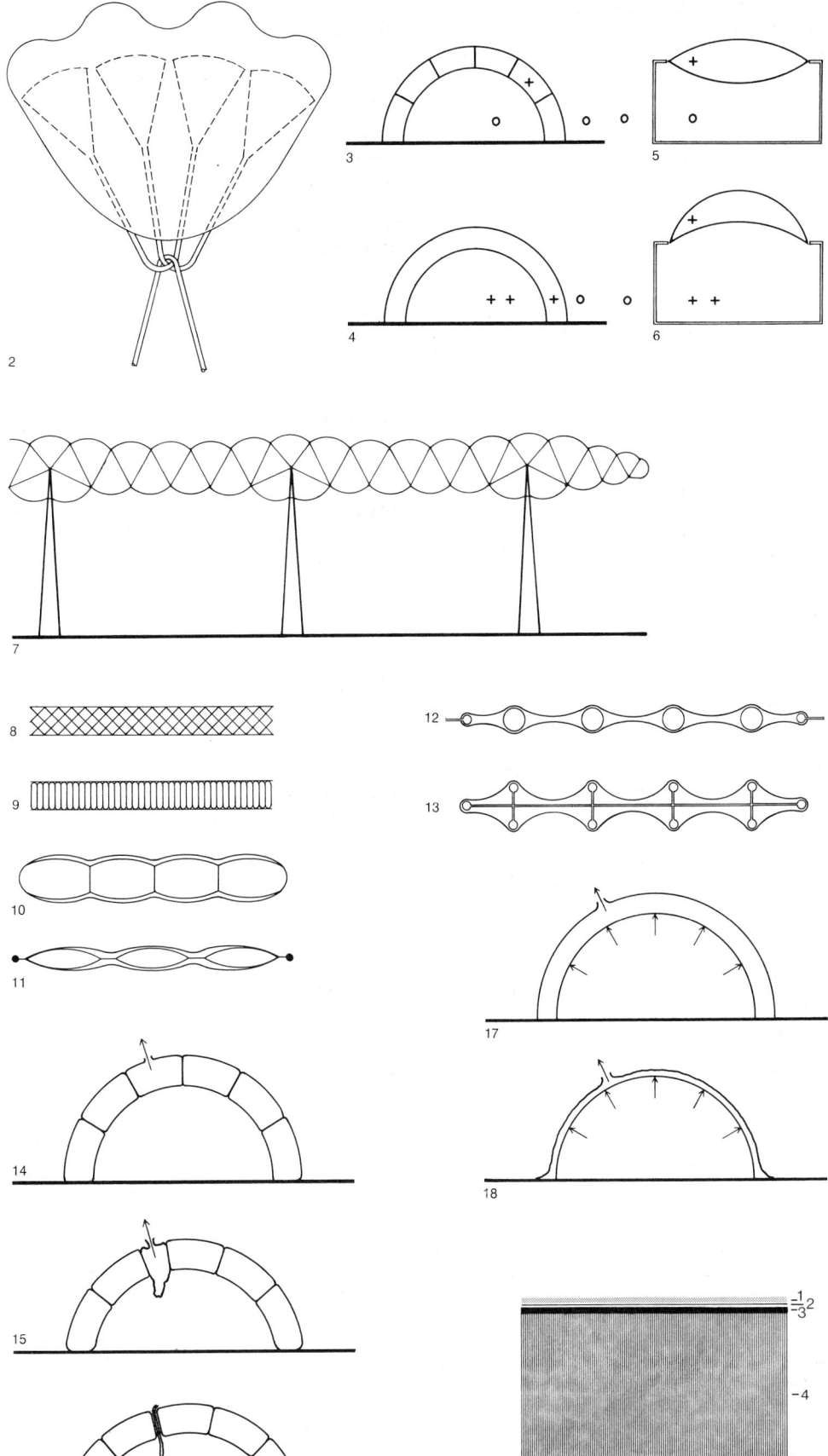

2. Surface diversion of forces for triangulations.

3, 5. Double membrane structures with the same pressure in the space to be utilised and outside.

4, 6. Double membrane structures with different pressures in the space to be utilised and outside.

7–11. Sandwich type positive pressure structures.

12, 13. Sandwich type negative pressure structures.

14–16. In the case of double membrane structures with the same pressure in the space to be utilised and outside, it is recommended that the space between the membranes be divided into individual compartments so that only parts of it fail if the skin is damaged.

17, 18. If the utilised space of double membrane structures is also under positive pressure then a reserve skin is always available in the event of external damage.

19. Pneumatic sandwich plates, M. L. Aviation Co. Ltd. Key: 1 Hypalon, 2 ply, 3 air sealed layer, 4 nylon threads.

brane layers can be staggered so that the stability is practically the same in all directions.

With synclastic and anticlastic surfaces the lengths must be cut in rounded contours according to their form. The less flexible the material and the less the possibility of generous deformation, the smaller the surface elements must be which comprise the total form.

In the case of Fig. 20 the seams increase – in relation to the same surface unit – towards the top of the dome. The reinforcement of the membrane towards the top leads to unequal elongation on the skin and thus to a deviation from the ideal form of the spherical section.

Fig. 21 shows another cutting pattern in which the seams are fairly equally distributed over the whole surface. However, manufacturing costs are higher in this case; furthermore there is a greater danger of failure under aerodynamic stress.

Figs. 22 to 26 show typical cutting patterns for standard halls by different manufacturers.

When geometrically different partial forms are put together, then various membrane tensions occur which can easily lead to wrinkles forming at the transition areas. In order to predetermine such tensions and to adapt the cutting pattern accordingly, exact dimensioning models are necessary.

Cutting pattern drawings of standard halls are today being prepared by programme controlled plotters. Further simplification will be brought by automatic pattern cutters, which are directly linked with computers to avoid the intermediate step "drawing".

5.2.3. Jointing techniques

5.2.3.1. Inseparable joints

The production methods used in envelope manufacture depend primarily on the basic material and sometimes on its coating. The requirements made of the joints are:
– a strength equal to that of the main material;
– as great a flexibility as possible, so that no kinks or break points occur in inflating and deflating the envelope;
– a density which will prevent the escape of air and penetration of surface water;
– a low relief so that there is little susceptibility to meteorological factors.

For the production of inseparable joints the possibilities are as follows:
– sewing,
– cementing,
– vulcanising,
– welding,
– riveting,
– clamping.

Sewing

Suitable forms of seam are the double and multiple sewn simple and double overlaps. There are also various special forms (Fig. 27; Bibl. 144, p. 113).

In the case of natural fibre fabrics the seams seal themselves through swelling of the damp fibres; in the case of synthetic materials methods of seam sealing must depend on the actual material used.

In the case of coated fabrics wrinkles occasionally occur in the seams; the use of saddle making machines with combined lower, upper and needle carriage as well as thread with contra-rotation and chromium plated round pointed needles is therefore recommended. (Bibl. 57, pp. 6, 8.)

The strength of sewing threads is decreased by ultra violet rays. It is therefore best if the point with the highest loading, the overlap between the upper and lower thread, can lie deep in the material. Effective protection of the seam can be achieved by coating the seam area with a pigmented protective varnish or a pigmented foil. (Bibl. 104.)

The strength of a seam depends largely on the strength of the seam thread and the number of stitches. With too many stitches tears occur in the perforated skin; with too few stitches and weak sewing thread tears occur in the seam.

Cementing

Cementings usually have very high strengths. In peeling tests the cemented area is often stronger than the bond between fabric and coating. In the course of time the adhesive strength of soft PVC cementings falls considerably when affected by higher temperatures – probably due to plasticising diffusion in the cement.

Cementing is fairly complicated and comparatively expensive. Therefore it is only worthwhile for very high grade materials such as butyl rubber, neoprene or Hypalon, for repairs in rather inaccessible places and in the production of complicated forms.

Vulcanising

Vulcanisation can be used for joining together rubber skins or rubberised fabrics.

Welding

From the point of view of the time involved, welding, which can be carried out on all thermoplastic synthetics, is one of the best bonding techniques. In the case of PVC coated fabrics with adequate adhesive strength the bondings are stronger than the basic material. The main requirement is that not more than 60% of the coating should be on one side of the basic fabric. (Bibl. 20, Table 3; Bibl. 44, p. 37.)

There are three different processes: the hot key, the hot air and the high frequency welding process.

In the hot key process the materials to be bonded are fused to the seams by means of a heated key on the surface and bonded under pressure by means of two pressure rollers. Portable hot key welding machines exist which under favourable conditions can reach a working speed of 5 m/min. Welded seams 30 mm wide are now possible; usually however a seam width of 20 mm is adequate. (Bibl. 57, p. 9 ff.)

In the hot air process the membrane material is fused by means of jets of hot air. Here also bonding occurs through pressure of rollers.

In the high frequency welding process a high frequency field is set up between electrodes which are usually ridge-shaped. This high frequency field heats the parts to be welded to the necessary temperature at the area of bonding. The welded joint is achieved by means of simultaneous pressure of the electrodes on the seam. The process works discontinuously. The seam length depends on the actual length of the electrodes. After each individual welding stage the whole machine moves one electrode length further. The working speed can, according to machine size and width of seam, go up 3 m/min including insertion time. The advantages of high frequency welding are that more than two layers as well as very thick materials can be bonded in one stage.

Riveting

Sealing is achieved using "pop" rivets placed at short intervals, while the inner membrane is pressed against the outer at the point of overlap. The process is little used.

Clamping

In this process metal clamps which look like large wire staples and which are equally deformed when applied, are shot in with air pistols at short intervals. The process is new and was first used on some temporary structures.

5.2.3.2. Separable joints

Separable joints can be necessary:
– in order to be able to insert movable parts within a section of the envelope;
– in order to be able to exchange parts of the envelope for other or new parts;
– in order to be able to separate large envelopes into parts suitable for transport and erection;
– in order to be able to manufacture large individual sections as standard elements and to combine them into buildings according to individual requirements;
– in order to be able to achieve a compound structural effect with individual pneumatic parts;
– in order to be able, in the case of expansion or reduction of the building, to add or remove sections.

With single membrane structures the possibility of changing size can only be planned in a few places for inseparable seams are considerably cheaper than any separation mechanism.

Separable seams, when they lie between zones of different pressure, must be as airtight as possible. If several individual pneumatic designs are joined together and the same pressure applies inside the building as outside, then the bondings only have to satisfy mechanical requirements.

From the point of view of technical practicality the following are available:
– zip fasteners,
– press fasteners,
– lacings,
– peg joints,
– connecting strips,
– different combinations of clamps, springs,

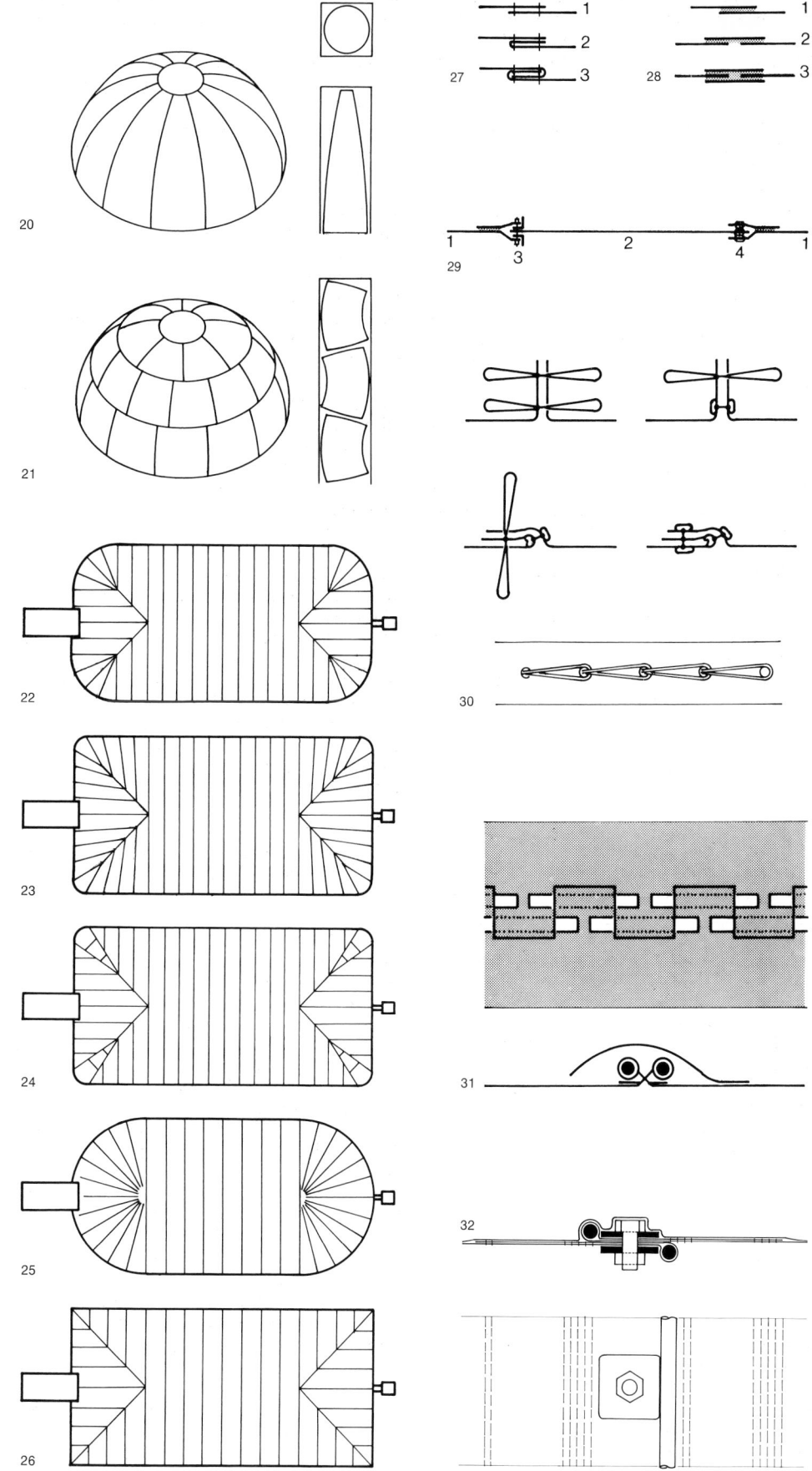

20, 21. Different cutting patterns for a hemisphere.

22–26. Cutting patterns of standard halls.

27. Sewn seams. Key: 1 simple lap joint, 2 simple overlapping seam, 3 double overlapping seam.

28. Welded and glued joints. Key: 1 simple lap joint, 2 connection with single cross-section, 3 connection with double cross-section.

29. Interchangeable window. Key: 1 envelope, 2 window, 3 L profile, screwed, 4 press stud.

30. Simple tent lacing ("Dutch lacing").

31. Peg joint.

32. Separable clamp joint with metal plates. Fuji Pavilion, Expo '70, Osaka.

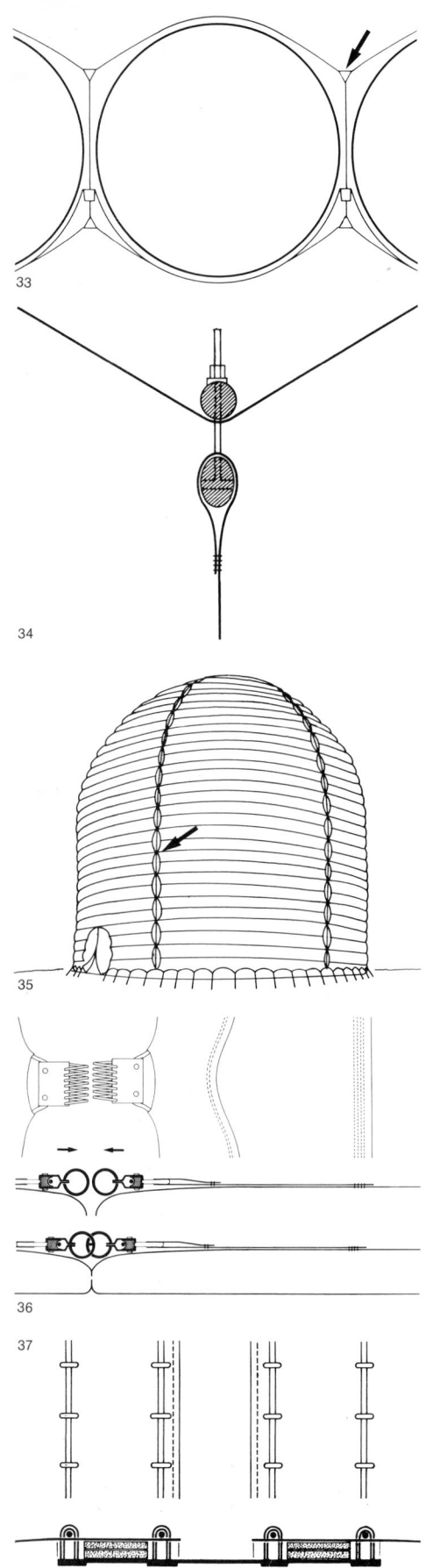

33

34

35

36

37

rings, material loops or membrane belts with inserted cables, link chains, etc.

There are zip fasteners for pneumatic structures in gas and watertight as well as normal specifications. Products with rubber or PVC coated support bonds (according to the envelope material) are also used.

Press fasteners are only used in "shear" tensioned structures (and even then the tension must not be very high). They serve to secure membrane sections to openings (Fig. 29) or to mount membrane flaps on the skin surface.

Simple tent lacings are usually carried out by "Dutch lacing". Hooks run along one side and cable loops along the other. Fig. 30 shows the phases of lacing. This is a type of joint which can be performed in one direction only at each seam.

Peg joints consist of membrane loops linked together and stabilised in position by round pegs which are pushed in sideways. They form a simple and safe joint which can be protected against meteorological factors by an overlapping apron which is welded onto one side during manufacture (Fig. 31).

A clamp joint with metal plates is shown in Fig. 32. In order to distribute the tensile forces of the membrane within the screwed joints and to prevent the skin from slipping through, continuous round profiles (steel or plastic rods, cables or pipes made of steel or plastic) are inserted at the edge of the membrane. The detail shown here is from the Fuji pavilion in Osaka (see pp. 76–78). The seams were covered on the outside by a protective paint.

Figs. 33 and 34 show the jointing of the high pressure tubes of the Fuji pavilion. 50 cm wide girdles were used which encircled the whole building horizontally both inside and out and were guyed against each other. The girdles were jointed outside the guying by the principle of the peg joints.

Figs. 35 and 36 show a further process of jointing closed pneumatic forms. Metal springs are fixed on to flaps and fastened to sewn-in cables. When the steel springs are pushed into one another, the pneumatic forms are pressed together. A cable is pushed through and holds the joint in position. (In accordance with DBP 1 559 112.)

The separable joint shown on the single membrane structure in Fig. 37 is also formed by a cable being pushed through. Brackets placed on a strip of synthetic material are inserted through the hooks of the membrane sections and four cables are drawn through the upper part of the bracket. Sealing is by means of strips of expanded rubber. (In accordance with DBP 1 172 025.)

33, 34. Separable joint for high pressure tubes. Fuji Pavilion, Expo '70, Osaka.

35, 36. Separable joint for flat pneumatic forms. DBP 1 559 112.

37. Separable membrane joint. DBP 1 172 025.

5.3. The anchorage of pneumatic envelopes to the ground

5.3.1. General

The anchorage has the task of conducting to the foundations the vertical and horizontal forces carried by the membranes. These forces result from the internal positive or negative pressure and the external loading. Figs. 38 to 40 show the dependence of the vertical and horizontal force components, which stress the anchorage, on the tangential angle at the base of the membrane. These are tensile forces.

To take up the vertical forces the deadweight of the anchor can be used or an applied load.

To take up the horizontal forces either the "passive earth pressure" of the directly adjacent ground is used or the friction generated along the anchorage surface.

Most anchorages represent mixtures of different types. Fig. 41 gives a schematic survey (Bibl. 94). The most important types of anchorages are briefly described below. In some diagrams the connection to the membrane is also illustrated for further clarification. Detailed illustrations of tensile anchorages can also be found in the Bibliography 119, p. 279 ff. and in the technical literature on foundation engineering and soil mechanics.

5.3.2. Anchorage structures

5.3.2.1. Ballast anchors

In the case of ballast anchors lying on the ground, the size of the permissible tensile force Z is dependent on the deadweight of the anchor as long as it is only vertically tensioned. In the case of transverse tensioning the vertical force components are taken up by a friction force which acts horizontally. Their size is dependent on the roughness of the friction surface and the condition of the subsoil.

Cast-in-situ concrete strip foundations (Fig. 42) are usually only used in large permanent pneumatic structures.

Special precast concrete parts (Figs. 43, 44) can be formed in such a manner that structures with curved (polygonal) as well as those with rectilinear outlines can be anchored. Standard components such as concrete pipes (Fig. 45), road building slabs (Fig. 46), canal shells (Fig. 47) and barrels filled with gravel or sand (Fig. 48) are sunk as appropriate; after the dismantling of the building they can be redirected to their original uses.

Ballast pockets or ballast containers are filled with bulk material or water. Figs. 49 and 50 show the use of sandbags (Bibl. 20; Bibl. 144, p. 110). In Fig. 51 a very primitive form of pocket for small buildings is illustrated, formed by wrapping the lower end of the membrane skin around stones (Bibl. 3). The method shown in Fig. 52, in which a box made of steel plate with brackets welded on is used, is more solid (Bibl. 144, p. 110). In Fig. 53 the use of water tanks is shown (Bibl. 20). Ballast pockets have proved especial-

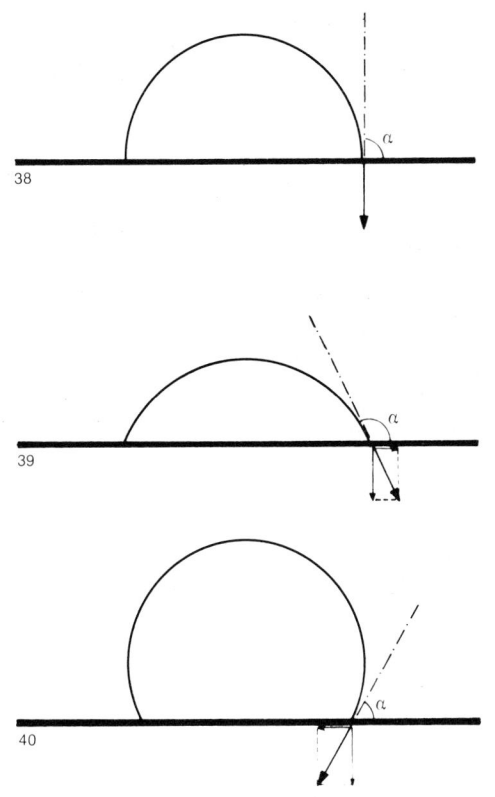

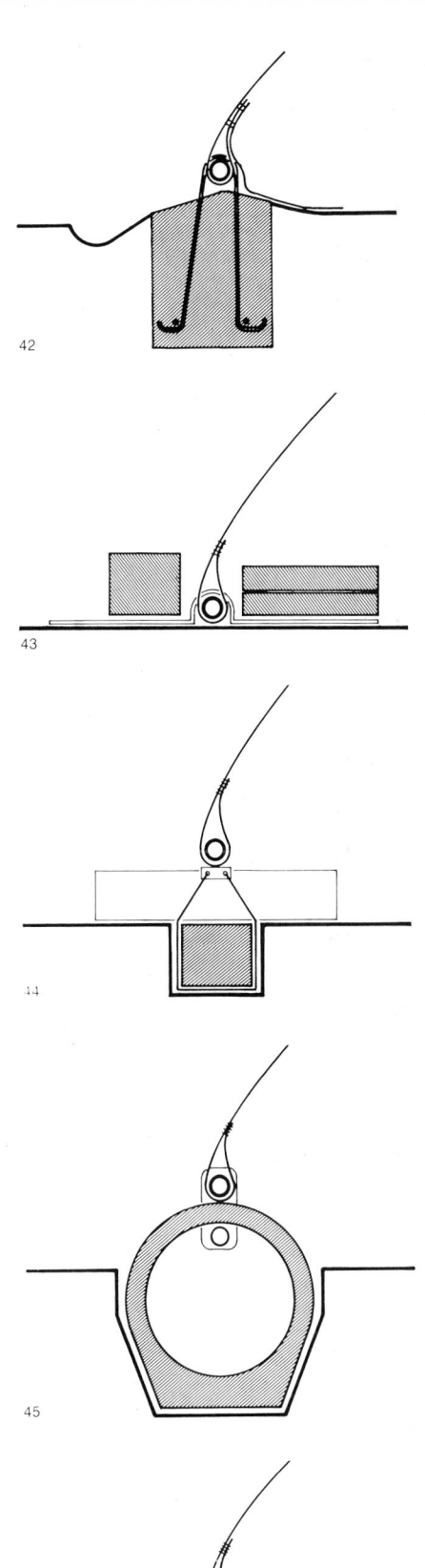

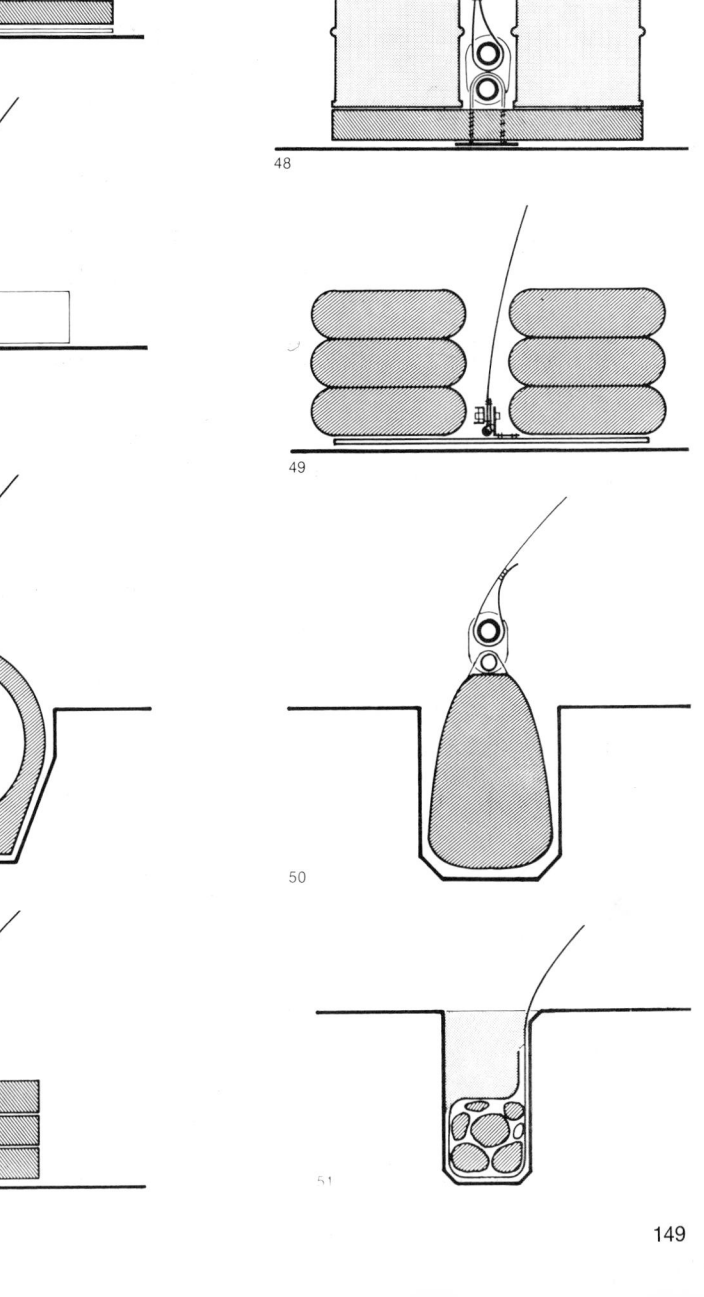

38–40. Force components at the membrane edge in relation to the tangential angle at the base of the membrane.

41. Schematic survey of the various types of anchorages.

42. Ballast anchorage – cast-in-situ concrete strip foundation.

43, 44. Ballast anchorage – precast concrete parts.

45. Ballast anchorage – concrete pipes.

46. Ballast anchorage – road building slabs.

47. Ballast anchorage – canal shells.

48. Ballast anchorage – barrels filled with gravel or sand.

49, 50. Ballast anchorage – sandbags.

51. Ballast anchorage – stones around which the lower end of the membrane is wrapped.

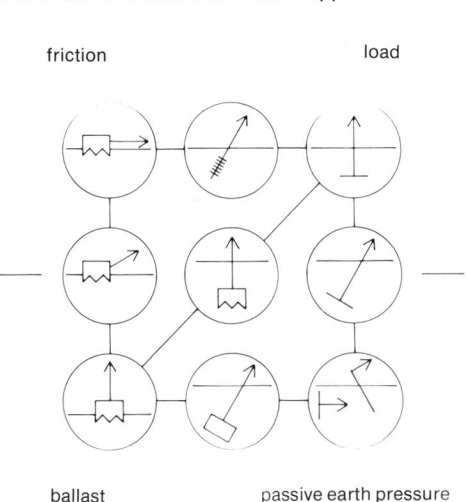

friction   load

ballast   passive earth pressure

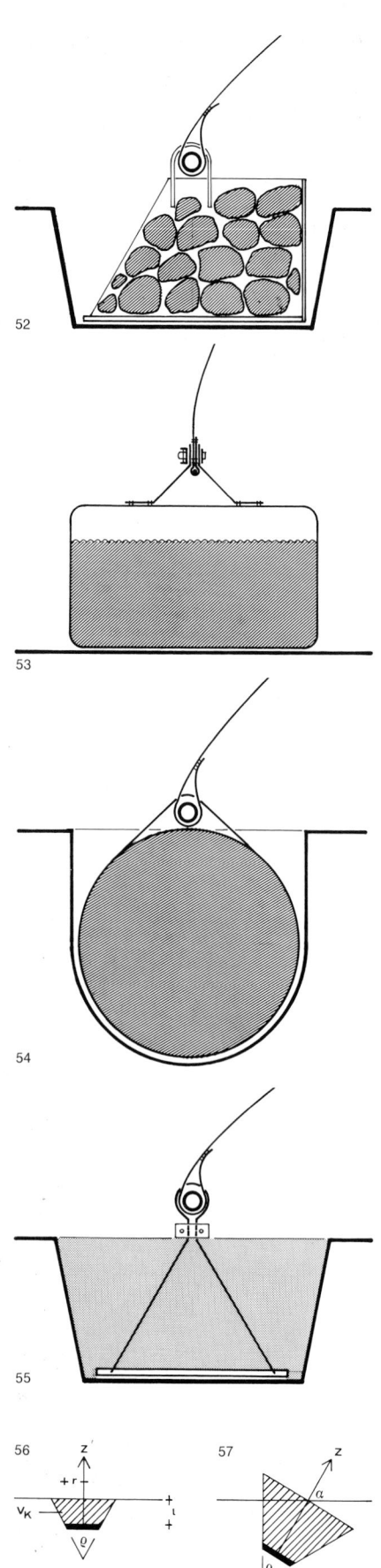

52

53

54

55

56

57

ly good for use in average sized structures with a span of 15 to 20 m.

Circular flexible tube foundations are particularly suitable for small structures. They are filled with water or sand and, like the skin of a pneumatic structure, are made of membranes and to a certain extent are part of the envelope. Fig. 54 shows an embedded water tube (Bibl. 15); yet flexible tubes, which can adapt up to a certain degree of unevennes in the site, are just as suitable for setting on top of the ground.

5.3.2.2. Ground-load anchors

Anchor plates (Fig. 55) are usually made of steel or reinforced concrete. In the case of vertical application of stress the possible tensile force results from the deadweight of the anchor and the weight of the earth frustum above it. If the application of stress is not vertical (as in Fig. 56) but oblique (as in Fig. 57), then as the angle between the direction of stress and the horizontal under constant stress is reduced, the horizontal component becomes larger and the vertical component smaller. For an approximate calculation of the earth resistance one can also take as a basis the weight of the earth frustum and ignore consideration of the skin friction of the frustum or the passive earth pressure, as long as α is not less than $90° - {}^2/_3 \varrho$. Anchor plates are a simple, often very cheap type of anchorage (Bibl. 144, p. 110). A variation is provided by anchor beams which are dug deep in order to guarantee the necessary load application.

Screw anchors can be tapped or driven in according to profile. The anchor must have sufficient inherent rigidity against the buckling and torsion stress which arises. There are screw anchors where the shaft and the boring blade are securely connected and those in which an anchor line is connected to the boring blade. The shaft serves to install the anchor which cannot then be withdrawn. Anchor lengths of 60–300 cm are usual with plate diameters from 10 to 35 cm. Fig. 58 shows various possibilities for the design of the screwing blade. The short boring blade is particularly suitable for permanent anchorages, while those with long threaded screws are more appropriate for short term anchorages.

Screw anchors are capable of alteration in use. The force which is to be conducted into the ground can, in most cases, be transferred into the anchor axis by an adequate thread direction. In very dense soils, which the anchor cannot penetrate, a hole must first be augered; this is later backfilled after the anchor has been installed, e.g. with tamped concrete. In order to compress the earth, stakes can be driven in around the anchorage.

Large anchor screw depths do not necessarily mean greater grip. This is only the case when firmer earth strata lie below.

Because of the possible loosening of the subsoil when screwing down, a safety factor of at least 3 should be incorporated as the basis of the calculation of loadbearing capacity of screw anchors. In an area of ground water in cohesive soils a negative effect of frictional resistance has to be taken into account, and with uniform sandy soils a positive effect through the washing effect of the ground water.

Table 16. Holding power of screw anchors in relation to type of soil, disc size and reach of the screw. According to Lange/Glienke.

Disc diameter (mm)	Reach of screw (m)	Rich clay, dried-up, also containing stones	Well-graduated gravel and sand mixtures; equal-grained gravel with few fine components	Sand and gravel, coarse-grained, close-packed	Loam, firm (easily to hardly kneadable)	Filling-up, unconsolidated, loosely bound; sand, fine-grained
1	2	3	4	5	6	7
100	0.70	615	520	420	320	250
	1.00	950	800	650	500	400
130	0.70	910	780	600	450	380
	1.00	1,400	1,200	950	750	600
	1.50	2,550	2,300	2,100	1,500	1,000
150	0.70	1,100	950	700	600	450
	1.00	1,700	1,450	1,100	880	720
	1.50	3,000	2,800	2,400	1,800	1,200
	2.00	5,500	4,000	3,200	2,500	2,200
200	0.70	1,600	1,450	1,100	900	700
	1.00	2,500	2,200	1,700	1,300	1,100
	1.50	4,500	4,000	3,600	2,700	1,800
	2.00	7,350	6,000	4,800	3,800	3,500
250	0.70	1,500	1,400	1,100	800	700
	1.00	2,300	2,000	1,600	1,200	1,000
	1.50	6,600	5,500	3,300	2,500	1,700
	2.00	7,500	5,800	4,300	3,300	2,500
300	1.00	3,300	2,800	2,200	1,800	1,400
	1.50	5,800	5,200	4,600	3,500	2,300
	2.00	9,000	7,500	5,700	4,300	3,300
350	1.00	4,200	3,600	2,800	2,200	1,800
	1.50	7,500	6,800	6,000	4,500	3,000
	2.00	12,000	10,000	8,000	6,300	5,700

The loadbearing capacity of screw anchors is calculated as follows (Bibl. 94):

$$P_z = \frac{1}{V_s} \cdot q_r \cdot t \cdot d_s \cdot \pi.$$

where

P_z = permissible bearing capacity (kg),
V_s = safety factor (3),
q_r = specific frictional resistance (kg/cm²),
t = screw length (cm),
d_s = screw diameter (cm).

Screw anchors can be used individually or in combination (Fig. 59). They are cheap and can be moved with simple equipment or machinery. Their possible applications depend very much on the type of soil.

Table 16 shows the holding power calculated by one manufacturer (Lange/Glienke) in relation to type of soil, disc size and reach of the screw.

Driven-in anchors (Fig. 60) have shovel-like blades which are joined together by means of a hinge. They are driven a fair way into the ground and contract further under tension. (Bibl. 144.)

Spreading anchors (Fig. 61), which are driven into the ground like stakes, under tension spread barbs. They can usually be used only once and thus are relatively expensive. (Bibl. 119, p. 302.)

Large spreading anchors, for which a hole has first to be augered in the site, consist of a round anchor plate of a diameter similar to that of the hole, a screwed-in tensile bar and two, three or four "resisting" surfaces, which folded together are set up over the round plate and connected by hinges. Pressure is applied by means of a tube or special tool, which grips the central tensile bar, to the hinge of the anchor plate which is then expanded and pressed into the earth. Finally the hole is backfilled.

Hinged anchors (Fig. 62) consist of a hinged tip, which is rammed into the earth by means of a tube which is subsequently withdrawn, and cables which transfer the tensile forces. A further cable regulates the position of the hinged tip and recovers it.

Injection anchors act as frictional resistance anchors. So-called "needle anchors" are frequently used. These are tubes, often provided with barbs or pins, which are open at the lower end. Using these under a pressure of 30 to 40 excess atmospheric pressure, cement or other hardening agents or agents effecting petrifaction of the earth are pressed in. Thus around the needle anchors a solidified, hardened mass arises, which offers sufficient frictional resistance in its tension loading.

The process is particularly suitable for difficult unconsolidated soils. Sand or gravel strata are especially suitable for the use of injection anchors. The cavities of a gravel structure can easily be injected with cement so that in the area of force entry or compression a solid anchorage is formed. Even when cement grout cannot penetrate into a sandy soil then a spread footing forms as a result of the high pressure and the subsequent local compression of the sand. The size of this footing is to a large extent dependent on the soil conditions, especially on the thick-

52. Ballast anchorage – metal boxes filled with stones.
53. Ballast anchorage – tanks filled with water.
54. Ballast anchorage – flexible tube filled with water.
55. Ground-load anchorage – anchor plates weighted with sand.
56, 57. Operation of a ground-load anchorage consisting of entrenched anchor plates.
58, 59. Screw anchors.
60. Driven-in anchor.
61. Spreading anchor.
62. Hinged anchor.
63. Screw plug with screw socket.
64. Screw plug with glued socket.

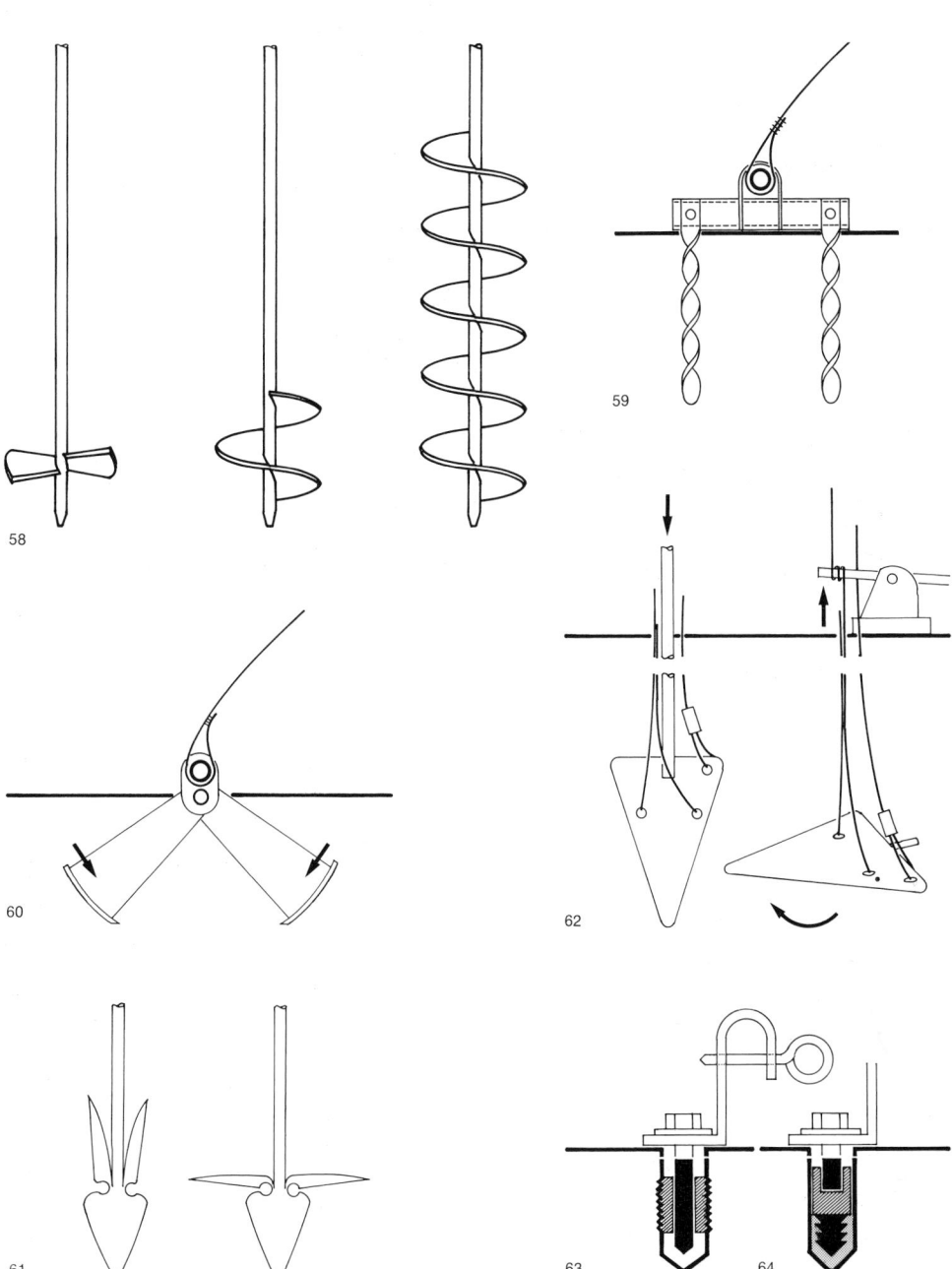

ness of the sand stratum. In the case of cohesive soils certain differences have to be considered with regard to the type of soil (distribution of granules) and the water content (consistency). Although cohesive soils cannot take up any compression mass, with the aid of an additional work process, injection anchors can still be produced in clay, loam and glacial marl of a firm to soft consistency and in silt of a firm to stiff consistency. The maximum working load is 25 to 30 Mp per anchor. (Bibl 94.)

Socket anchors are required for fixing on very hard ground such as rock or concrete slabs. Screw sockets with screw plugs (Fig. 63) are used as much as glued sockets (Fig. 64). The connection to the membrane is frequently made by means of shackles (Fig. 63). (Bibl. 20.)

The anchorage chosen in each individual case depends on the condition of the site. When using frictional resistance anchors it is advisable to carry out extraction trials on the intended site. Naturally the applied load is also governed by the distance between the anchors which in the case of individual anchors is usually 60 to 90 cm.

If different tensions occur in the membrane, perhaps because of the geometry of its surface, then the anchorages are correspondingly tensioned. They must be dimensioned according to the strongest tension or be adapted to the different tensions.

5.3.3. Fixing of the membranes to anchorage structures

In order to conduct the membrane forces into the anchorage a hem is made into which a cable, rod or tube is pushed and this transfers the tensile forces of the membrane to the individual anchorages. In order to prevent air escaping an air apron in the form of a membrane flap is attached to the inside. In the case of continuous anchorages the hem of the envelope is clamped by clamps, splints or cover plates. An airtight lower seal is usually automatically guaranteed in this kind of anchorage.

5.3.3.1. Single membrane structures

Securing of point anchorages
In a "one pipe system" the pipe running through the envelope hem is connected to the foundation by means of bolts (Fig. 65; Bibl. 15), hooks (Fig. 66; Bibl 92) or brackets (Fig. 67; Bibl. 54). There are notches in the hem where the anchorages grip.

The so-called "two pipe system" is somewhat more independent of the accuracy of erection of the anchor. As Fig. 68 shows, one pipe runs through the hem of the envelope and a second between the hem and the foundations. (Bibl. 92.) When cables are inserted a "garland line" arises between the individual anchors, which has to be taken into account in the cutting pattern for the envelope. As to the cutting in the case of small foil structures the cable can be assumed to be straight as the foil stretches to correspond with the tension. A simple project is shown in

Fig. 69 (Bibl. 20) and a more expensive one in Fig. 70 (see p. 104, Figs. 292, 293). In the case of Fig. 71 the cable takes on the membrane forces at the periphery of the building and conducts them into the individual anchorages while the cables running across the building to the opposite side serve to reduce the membrane forces on the surface (Bibl. 131, p. 118).

Cables can be sewn in during manufacture. Because of the possibility of simple and quick erection and dismantling, such structures are particularly suitable for "flying buildings".

Figures 72 to 74 show the boundary point of the USA Pavilion at the world exhibition in Osaka (see pp. 116, 117). A cable was sewn into the hem at the edge of the membrane, which was anchored to a huge concrete ring foundation by means of tensile cables.

Securing of continuous anchorages
The envelope hem is usually compressed with clamps against a continuous anchorage profile and a profile sewn into the hem prevents the membrane from slipping out. The disadvantage of greater erection costs contrasts with the advantages of this type of fixing, namely uniform conduction of forces, airtight sealing and good control of the drainage of the skin surface.

In Figs. 75 to 77, connections to profiles consisting of wood, concrete and steel are shown.

In the fixing detail of a radome skin to a steel cylinder shown in Fig. 78 clamps are used to produce the bearing pressure (see p. 93).

The detail shown in Fig. 79 illustrates a combination of continuous and point anchoring: a tube which runs in the hem and bridges the distance between the clamps is connected with individual clamps to the steel profile.

5.3.3.2. Double membrane structures

From the details given so far analogous solutions can be deduced for multi-membrane structures. The provision of an inner membrane apron for sealing is, however, no longer required.

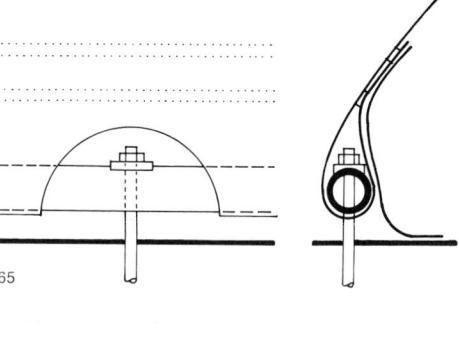
65

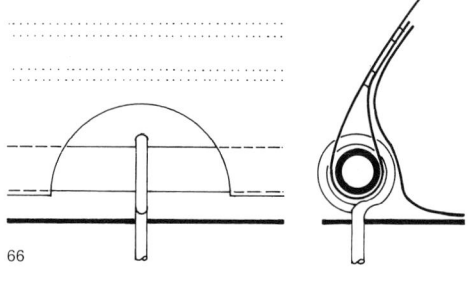
66

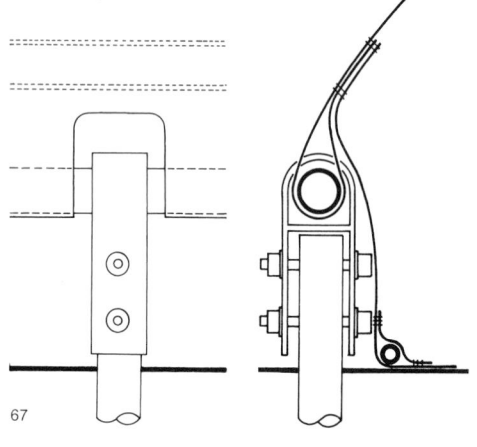
67

65–67. Fixing of membrane to point anchorage by means of single pipe ("one pipe system").

68. Fixing of the membrane to point anchorage by means of two pipes ("two pipe system").

69–71. Fixing of the membrane to point anchorages by means of boundary cables.

72–74. Fixing of the boundary cable to a concrete ring foundation. USA Pavilion, Expo '70, Osaka.

75. Fixing of the membrane to a continuous wooden anchorage profile.

76. Fixing of the membrane to a continuous concrete anchorage profile by means of a steel clamp.

77. Fixing of the membrane to a continuous steel anchorage profile by means of a steel clamping rail.

78. Fixing of the membrane to a steel cylinder by means of a steel clamping rail pressed on to it by screwed clamps. DBP 1 225 908.

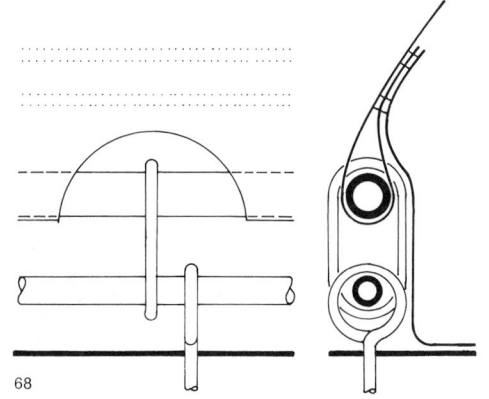
68

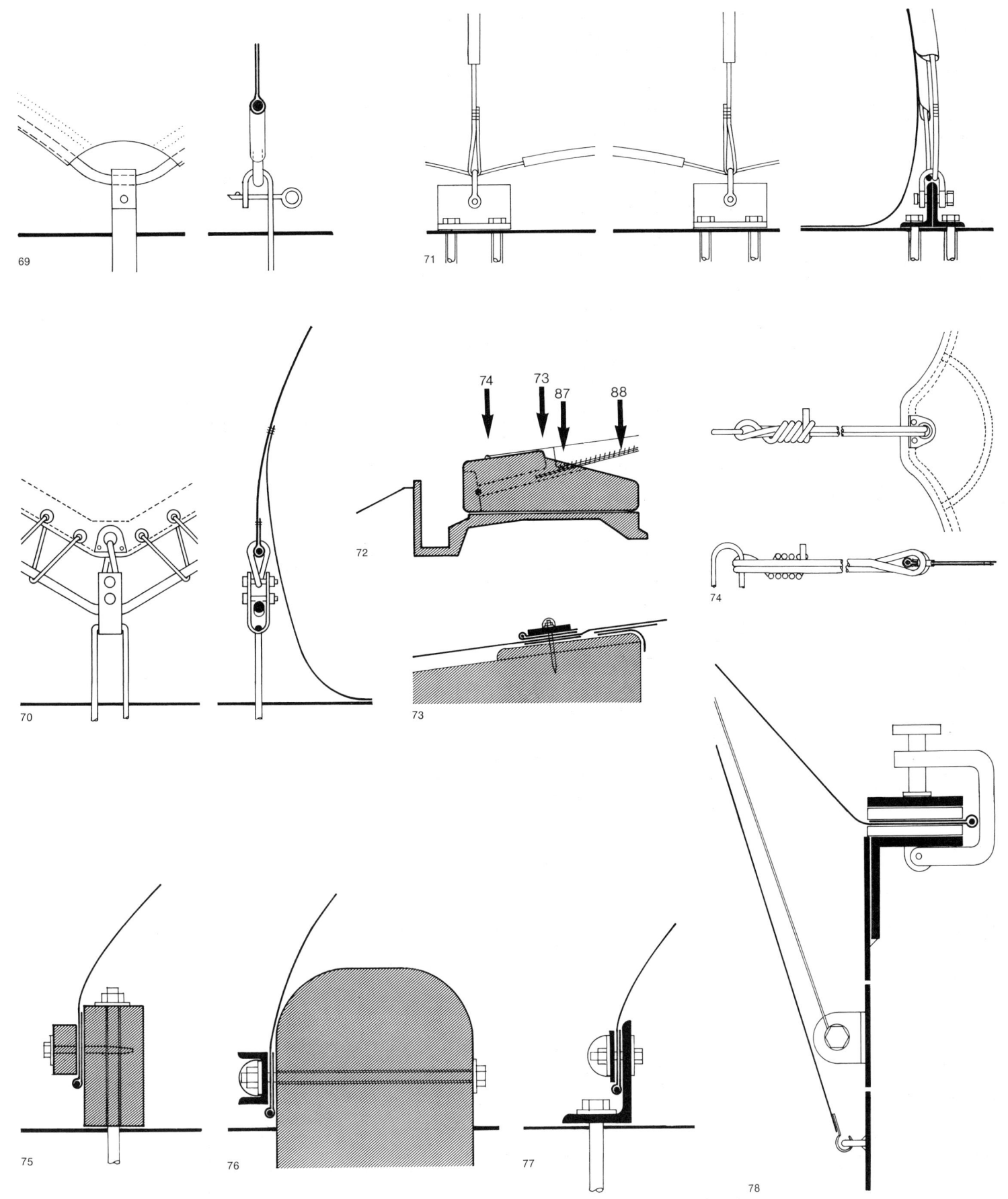

69

71

70

72

74

73

75

76

77

78

79. Fixing of the membrane to a steel profile by clamps.

80. Boundary formation of a mobile membrane apron. "Floating Theatre", Expo '70, Osaka.

81. Fixing of the membrane of a high pressure structure to the anchorage by means of closely spaced flat steel clamps. Fuji Pavilion, Expo '70, Osaka.

82, 83. Fixing of the membrane to the anchorage structure by means of flaps held by press studs. DBP 1 684 972.

84. Simple fixing of the membrane to a wooden framework.

85. 86. Fixing of the membrane to a steel ring. Information Pavilion, Expo '70, Osaka.

87, 88. Fixing of the membrane to a cable net construction by means of skirts of membrane material. USA Pavilion, Expo '70, Osaka.

89. Fixing of the membrane to a cable net construction by means of clamps.

90. Joining of individual membrane sections by means of sewn-in boundary cables. DBP 1 434 630.

91, 92. Fixing of the membrane to a cable net construction by means of loops; the cables are held together at the points of intersection by means of positioning clamps. Cabledome.

93. Internal drainage: fixing of the membrane to a rainpipe anchored in the ground.

94. Internal drainage: fixing of the membrane to a circular pipe guyed by cables to a concrete foundation.

95. Internal drainage: fixing of the membrane to a rainwater collection vessel.

96. Internal drainage: fixing of the membrane to a special precast component.

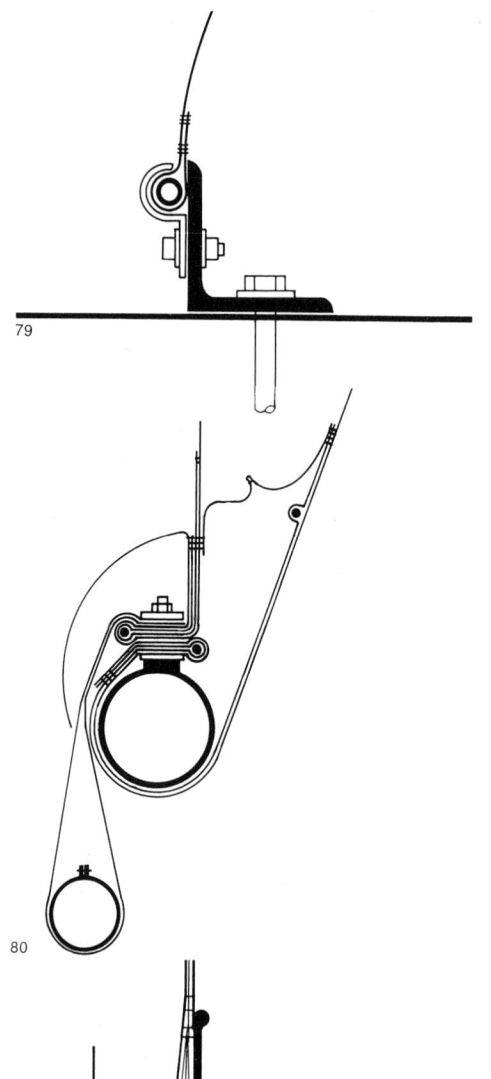

79

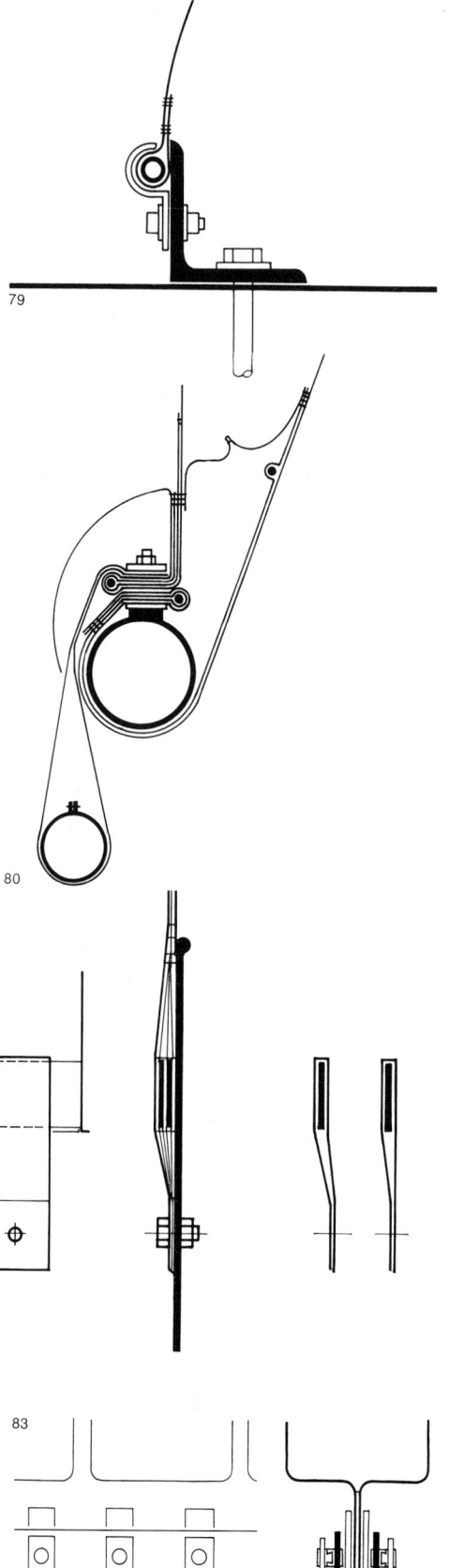

80

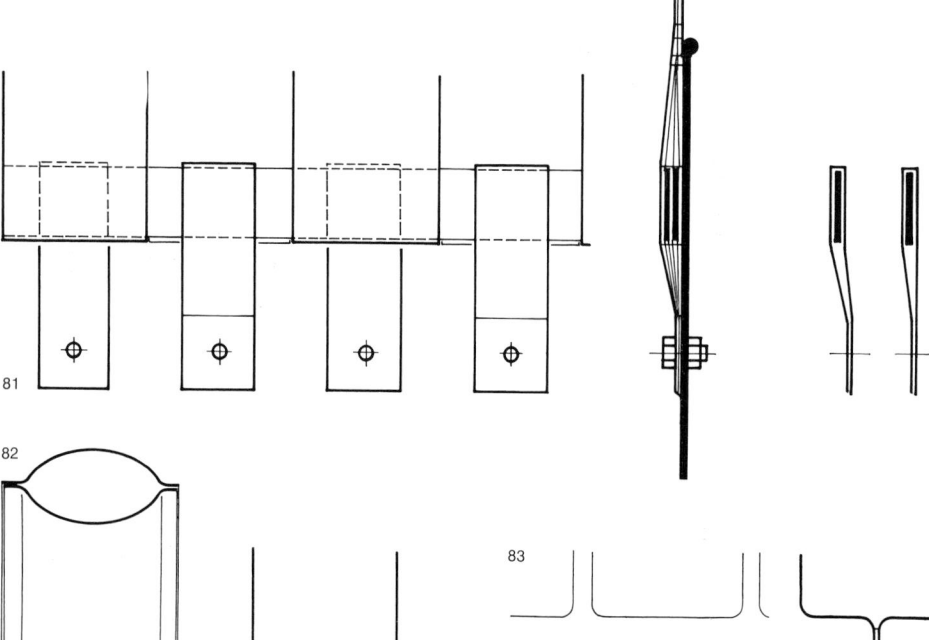

81

82

83

In addition some special forms are illustrated here:

Fig. 80 illustrates the "mobile" anchorage of the "Floating Theatre" at the world exhibition in Osaka (see pp. 88, 89). The two tubes were flexibly fixed at each end. The upper tube followed exactly the curvature of the envelope; the lower was the connection point on the ground. As there was negative pressure between the membranes, the tube arches were raised up so that they released a large opening which acted as an entrance and exit for visitors to the theatre. In order to close the opening, the lower tube and thus the whole envelope was drawn down.

The securing of the membranes of the high pressure tubes on the Fuji Pavilion (see pp. 76–78; see also pp. 147, 148) had to withstand an extremely high loading. Closely positioned flat steel clamps held two flat steel rings which ran through the uniformly divided hem of the envelope (Fig. 81). In this way the membrane was not weakened at any place on the hem.

In the solutions suggested in Figs. 82 and 83 the membranes are held by flaps which are fastened with press studs (DBP 1 684 972). The membrane tensions produced by the internal pressure are not transferred to the fastening. It is affected only by external forces.

5.4. Fixing of pneumatic envelopes to other structures

5.4.1. General

In the same way as they are fastened to ground anchorages, membrane ends are also fastened to other structural components. The main task is always that of conducting the membrane forces into the other structural elements.

5.4.2. Single membrane structures

The fixings to a simple wooden framework (Fig. 84) or to a steel ring (Figs. 85, 86; see also p. 39, Figs. 7, 8) show the relationship to the ground anchorages detailed above: the membrane is compressed on to a guide profile by clamps and also secured under heavy loading by cables or synthetic profiles sewn into the hem (Bibl. 46; Bibl. 119).

The skin of the USA Pavilion in Osaka (see pp. 116, 117; see also p. 153) was subtended by a diamond grid cable network. The membrane forces were conducted into the net by means of narrow vertical fabric skirts attached to the cables (Fig. 87). As the bands of cables lay on two planes because of the intersections (Fig. 88), these fabric skirts had to have different heights. (Bibl. 46.)

A similar construction was carried out in 1972 on two warehouses in France. Here also the membrane was additionally stabilised by subtended cables. Fig. 89 shows how the membrane forces are transferred into the cables by means of two clamps and fabric loops. According to the manufacturer this detail is considerably cheaper than that on the USA Pavilion.

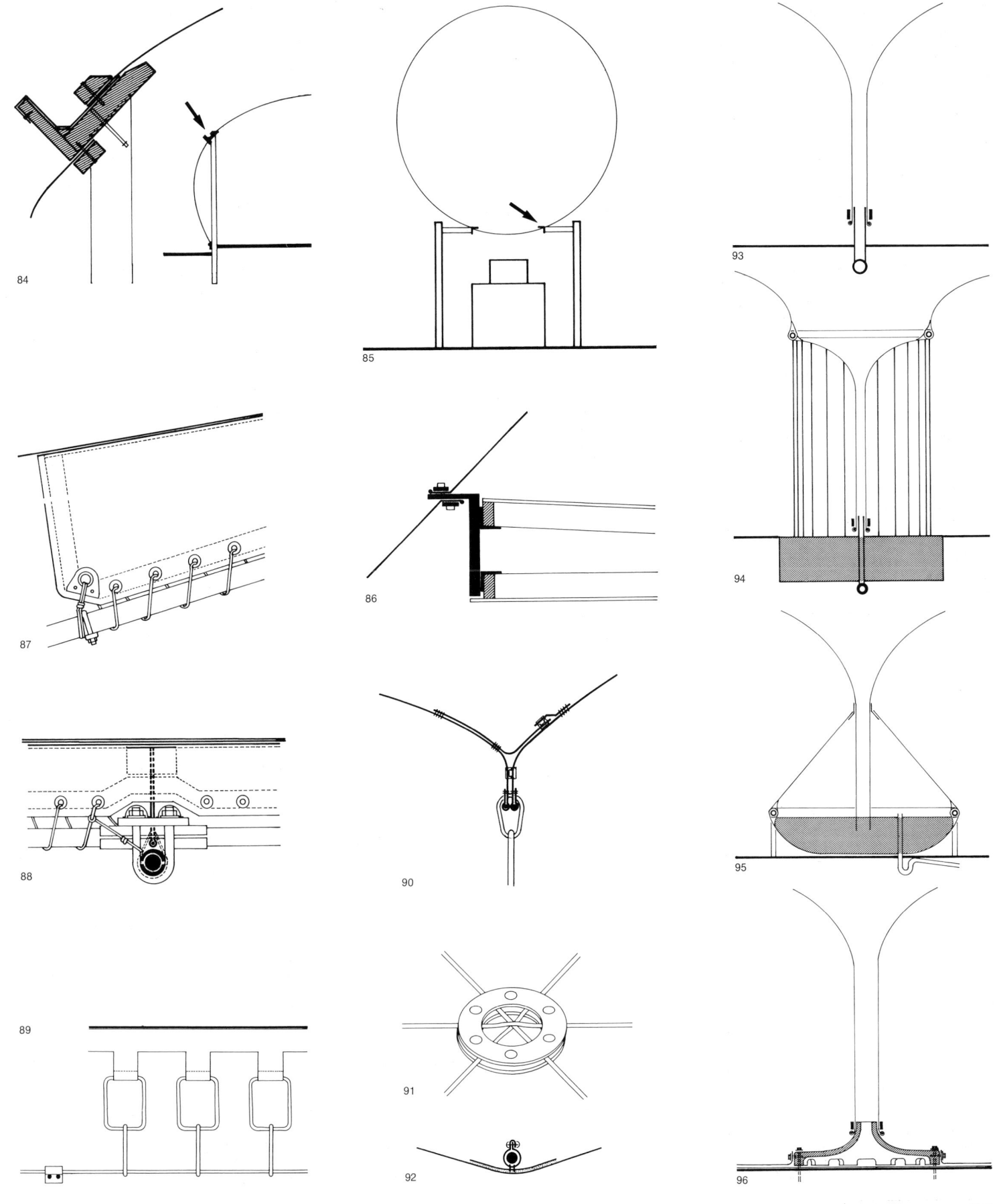

84

85

86

87

88

89

90

91

92

93

94

95

96

97, 98. Simple connections of cushion structures to steel profiles.

99, 100. Connections, sealed with synthetic profiles, of cushion structures to steel profiles. DBP 1 937 998.

101, 102. Fixing of a cushion structure to a flat steel profile. Forum Steglitz, Berlin.

103, 104. Fixing of individual cushions to a tubular frame. Festival Plaza, Expo '70, Osaka.

105. Cable guyed cushion framework.

106. Cushion structure guyed against an arched framework.

107–109. Connection of a cushion structure to a framework of horizontal I profiles. Boston Arts Centre Theatre.

110. Connection of two cushions to a pipe.

111. Rectilinear cable net as support for a cushion structure.

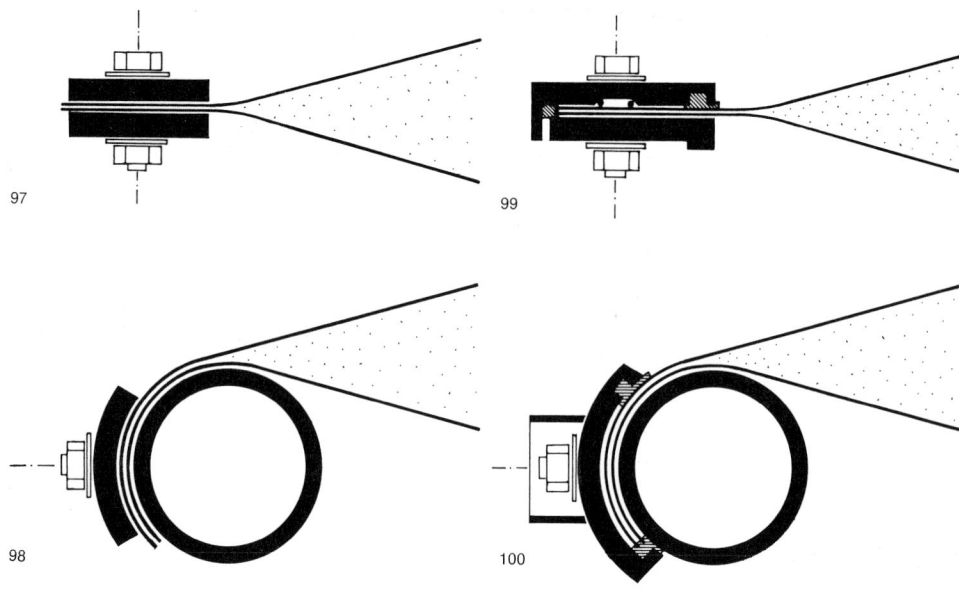

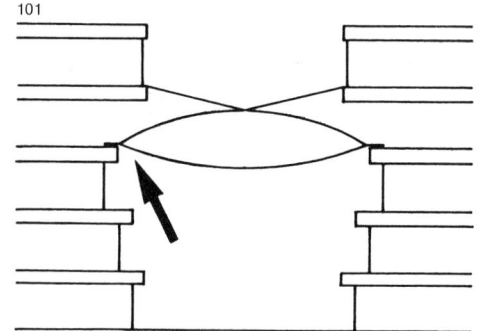

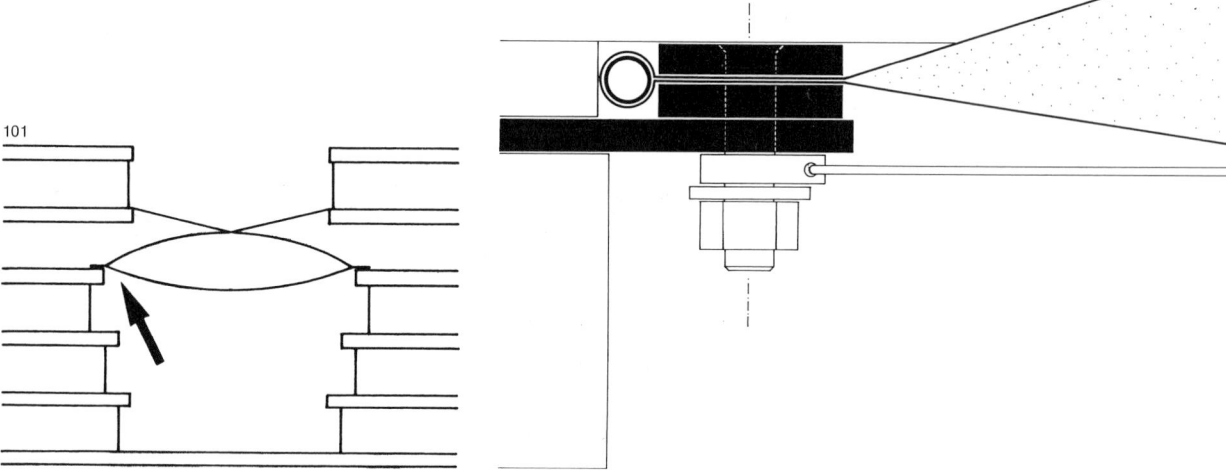

Another solution is shown in Fig. 90. Here individual membrane sections are joined together by means of sewn-in boundary cables. The resulting groove is protected against rainwater by a flap which is sewn on to one side and attached to the other by press studs. (DBP 1 434 630.)

For the "Cabledome", a development by Birdair in 1971 (see p. 121), a very simple detail was found for joining the cables to each other. The cables, which are arranged in three sheaves on the outside of the membrane, pass through from base to base and are held in the intersections by positioning clamps (Fig. 91). Loops serve to fasten the membrane to the cables (Fig. 92).

Compared with a reinforcement of the membrane by a cable net lying on the outside a subtended net has clear advantages:

– the connection points between membrane and

cable are not exposed to weather and pollution by the atmosphere;

– the connections do not need to be rainproofed;

– the cables do not chafe on the skin when this moves under the effect of wind;

– erection is simplified as the membrane lies on the cable net instead of being suspended from it.

If the rainwater is drained off inside the building, then there are additional places at which the membrane must be connected to other structural components.

Fig. 93 shows the direct connection to a rainpipe anchored in the ground. In Fig. 94 the membrane forces are taken up by a circular pipe which is guyed by cables to a concrete foundation. Further possibilities are the connection to a water collection vessel with overflow (Fig. 95) or

to a precast component (Fig. 96) which channels the water into a double membrane which forms the floor. (Bibl. 119.)

5.4.3. Double membrane structures

Patented connections to steel profiles in simple solutions such as those illustrated in Figs. 97 and 98 (DBP 1 937 998) have already been used several times.

The central hall of Forum Steglitz in Berlin (see p. 58) is covered by two large air cushions. Fig. 101 shows the position, Fig. 102 the details of the boundary connection. The cable net, which is used for safety reasons, is attached to the bolts that hold the flat steel profiles in place. (Bibl. 138.)

The large air cushions of the Festival Plaza in

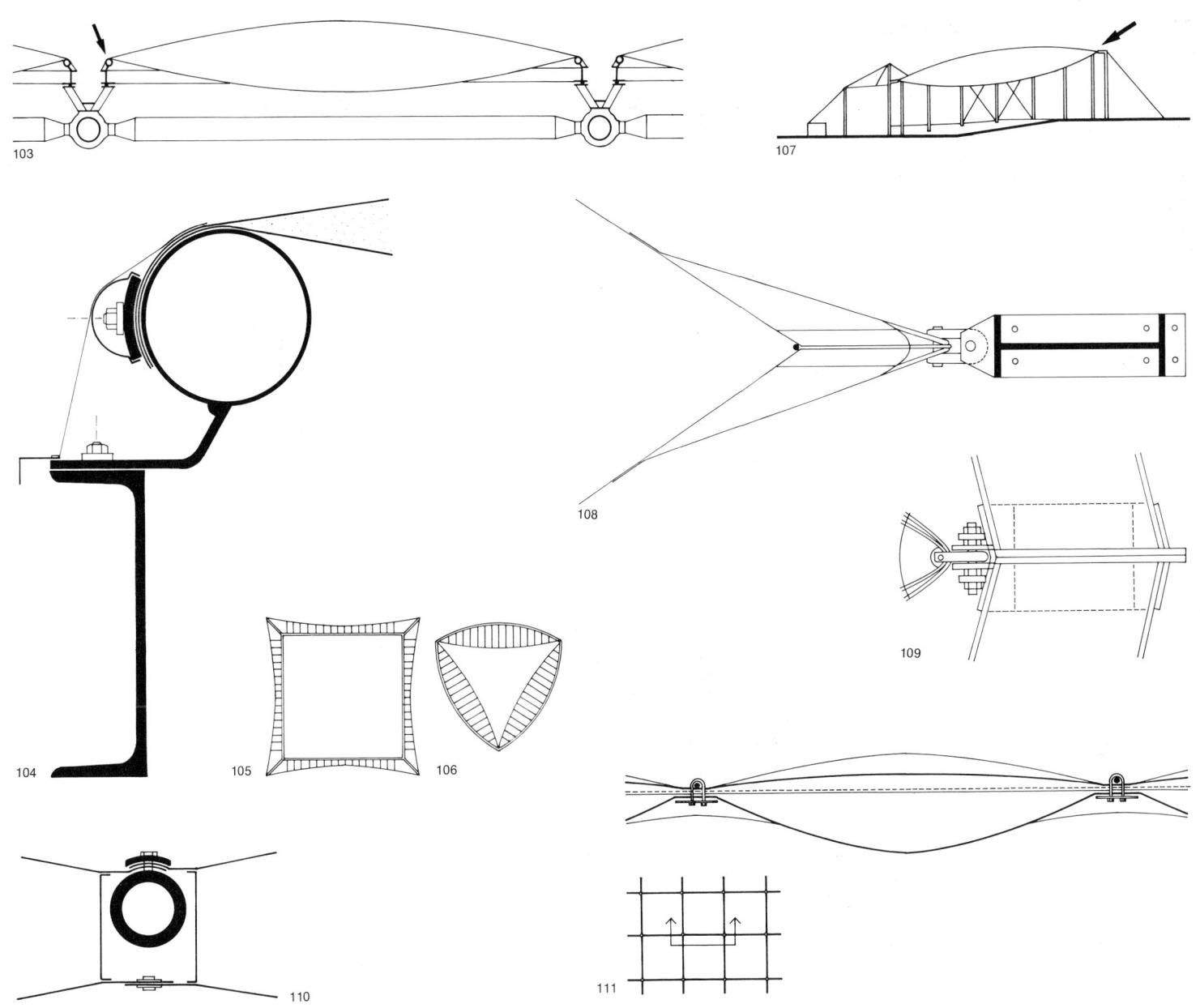

103

107

108

109

104 105 106

110 111

Osaka (see pp. 48, 49) were erected on a tubular framework which in turn sat on a framework made of [profiles (Figs. 103, 104). Aprons were drawn over the screwed joints to drain off the rainwater.

Where greater sealing is demanded the connections can be further improved by the insertion of flexible profiles (DBP 1 937 998) (Figs. 99, 100).

Frameworks for cushion structures can be guyed by cables on the outside, as in Fig. 105, in order to reduce the strong bending moments in the frame profiles. In Fig. 106 the membrane forces are conducted by means of cables into arched frame elements mainly under compression stress. (Bibl. 119.)

In a theatre in Boston (see p. 47) in 1959 the combination of an encircling framework of horizontal I profiles with cables which resisted the membrane forces had already been realised.

(Figs. 107–109.) The framework is polygonal in plan form. Three cables are sewn in to the edges of the cushion and these run together into the corners of the polygon. Thus bending tension in the framework elements as a result of membrane forces is avoided.

Figs. 110 and 111 show combinations of pipes and cable nets with double membranes (Bibl. 119). Neither idea has been executed so far.

5.5. Access constructions

5.5.1. General

Buildings whose utilisation space has a higher or lower pressure than the exterior need special access structures which are as airtight as possible when closed and keep the leakage of air as

low as possible during the passage of persons or materials.

5.5.2. Passage of persons

Alongside the rigid conventional doors which are generally used in building, there are some special constructions for installation in membranes.

Trapdoors, which are situated in cable reinforced round sections of the envelope (Fig. 112), are kept in balance under a central axis of rotation and low internal pressure of the building (Bibl. 119). Under an eccentric axis of rotation the regulation of pressure must be achieved by means of springs or weights. The loss of air is controlled by the period of opening. According to Bibl. 136 trapdoors are only permitted as sup-

157

112. Trapdoor.
113. Simple membrane apron.
114. Slipping-through door.
115. Lip door.
116. Cushion door.
117. Revolving door for personnel.
118, 119. Air locks.
120, 121. Air locks with revolving doors. DBP 1559208.

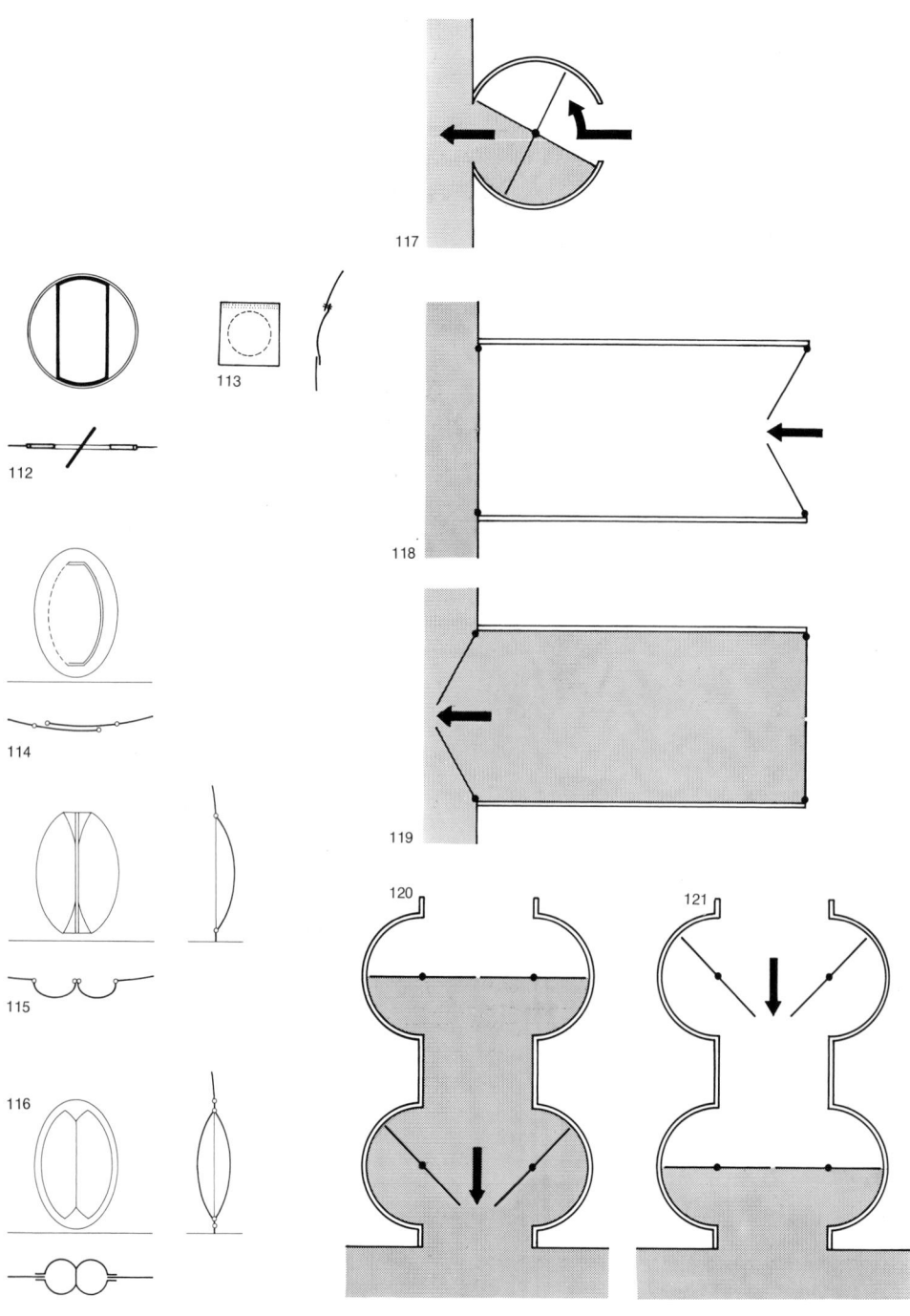

112

113

114

115

116

117

118

119

120

121

plementary doors; simple trapdoors, such as the one illustrated in Fig. 113, are at the most allowed in primitive "bubbles" which do not require authorisation. The same applies to all other access openings not intended as emergency exits.

Slipping-through doors consist of two membranes of which the inner one is pressed against the outer one (Fig. 114). They are difficult to open from both sides and the apex and base are in danger of splitting.

With lip doors two lip-shaped pouches are pressed against one another by the internal pressure (Fig. 115). Lip doors are easy to open from the outside, but less easy from the inside. Here also the apex and base are at risk.

With cushion doors two elongated rolls are pressed against each other by their internal pressure (Fig. 116). Usually no satisfactory sealing is achieved at the base and apex. The positive pressure in the rolls must be separately maintained independent of the pressure inside the building; it must be higher than the pressure inside the building.

Revolving doors (Fig. 117) are always under stable balance. They are the most frequently used type of access and permit constant through traffic in both directions without great losses in pressure.

5.5.3. Passage of material

Because of the high loss of pressure from large openings, single doors can only be used in the case of an automatically controlled short period of opening, or in very large structures.

Air locks (Figs. 118, 119) must have doors at both ends of the air lock which open and close alternately. They are particularly suitable for the transport of bulk goods and the use of vehicles within the building. (Bibl. 144, p. 114.)

A particularly expensive air lock is illustrated in Figs. 120 and 121 in which the doors are positioned in the centre (DBP 1559208). In order to produce the necessary pressure inside the air lock, small supplementary fans can be installed in the wall of the air lock. The side walls of the air lock can consist of stiff material or of a framework with a skin covering.

The connections between the flexible membrane and the stiff access construction are particularly difficult. Abrupt tension differentials always occur in the membrane when no transition elements are provided. Therefore the tension from the membrane must be properly intercepted, i.e. with a sewn-in boundary cable, then a membrane collar (for example) must form the joint from the cable to the access construction.

According to the FRG guidelines (Bibl. 136) every pneumatic structure must have at least two exits situated as far apart from each other as possible, which must be easily and safely accessible and may be no further than 35 m from any point in the building. In addition, near to air locks and revolving doors, outward opening doors leading directly outside can be required.

These access constructions show very clearly

how little aesthetic interests have concentrated up until now on pneumatic structures and how much architects have ignored this field.

In the last decade thousands of pneumatic structures have been produced and yet, from the aesthetic point of view, it will hardly be possible to find among them any satisfactory solutions for those structural elements which are not part of the membrane.

5.6. Stabilisation measures

5.6.1. General

The stabilising pressure of the membrane is an important structural element in a pneumatic structure. Its production and maintenance as well as its constant control require the provision of special technical equipment.

In the case of buildings and building elements the stabilising pressure is usually produced by inflation devices which are arranged on the outside of the structure.

As an additional measure for positive pressure structures, especially those with large spans, it has been repeatedly suggested that the wind pressure be intercepted by large funnels and conducted into the interior of the pneumatic structure. Thus under external wind loading one would also at the same time be able to produce a higher internal pressure. However this is not practical for very flat forms where the surface is stressed only by wind suction. (However the author knows of no instance of this idea being put into practice.)

Further possibilities are seen in the use of gases (helium, coal gas, hydrogen) for multi-membrane structures. Such solutions will, however, always remain special cases, as gases are either too expensive or too explosive.

More realistic is the exploitation of the uplift forces of warm air in buildings with large spans. These are effective both for single membrane structures which are situated in cold climatic zones and for buildings which are heated in winter.

For smaller structures, say up to the size of small camping tents, the use of high pressure gas cylinders for inflation has proved effective. The process is, however, only practicable for hermetically sealed elements. The same applies for bellows and hand pumps.

5.6.2. Mechanical equipment

A distinction is made between axially, radially and tangentially working inflation devices according to the control of the air current.

In the case of *axially* working devices (Fig. 122) the air current flows in the direction of the axis of the device, whereby several devices can be arranged in series. The noise at high speeds is greater than with other devices. The capacity can be increased by the arrangement of propellers rotating in opposite directions. The direction of the air current can be easily reversed by reversing the direction of the propeller.

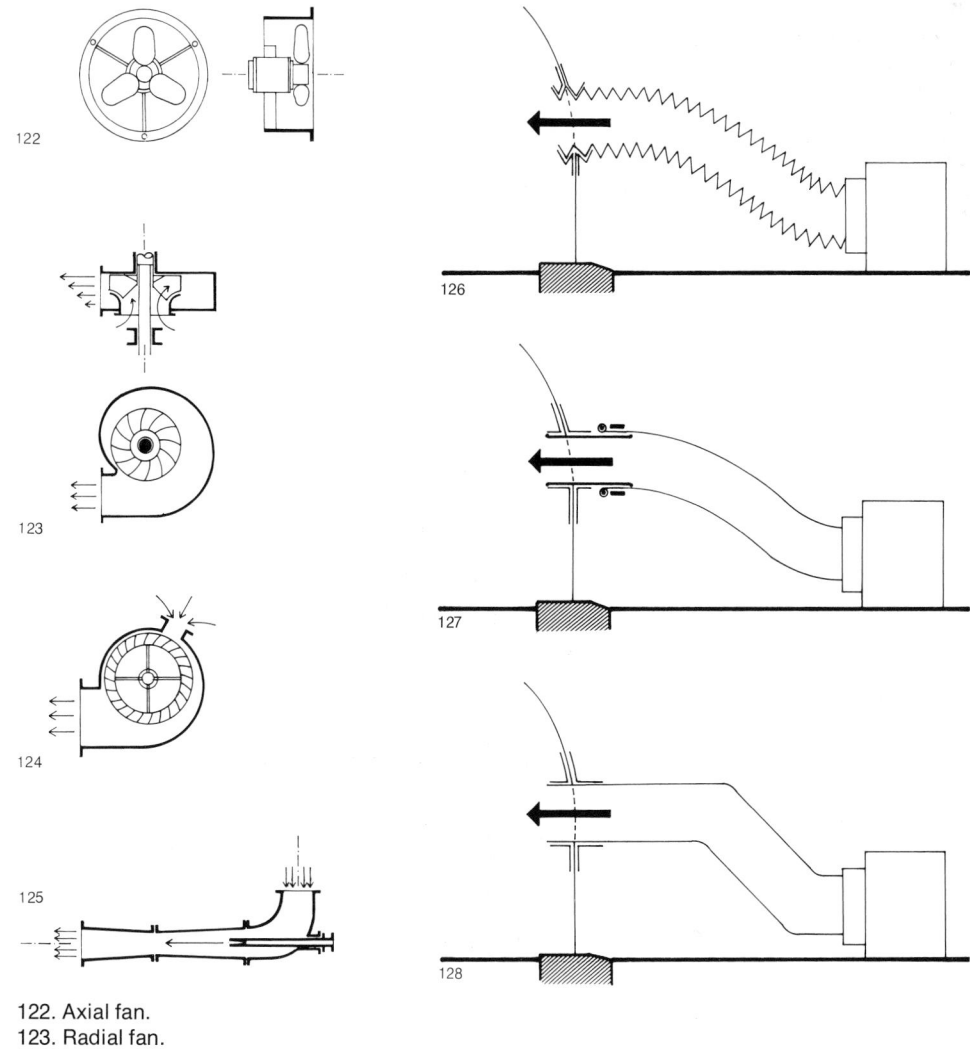

122. Axial fan.
123. Radial fan.
124. Tangential fan.
125. Jet compressor.
126–128. Diagrams of various possible joints between fan and envelope.

Table 17. Relationship between fan capacity and building volume or ground surface of building in the case of standard air supported halls.

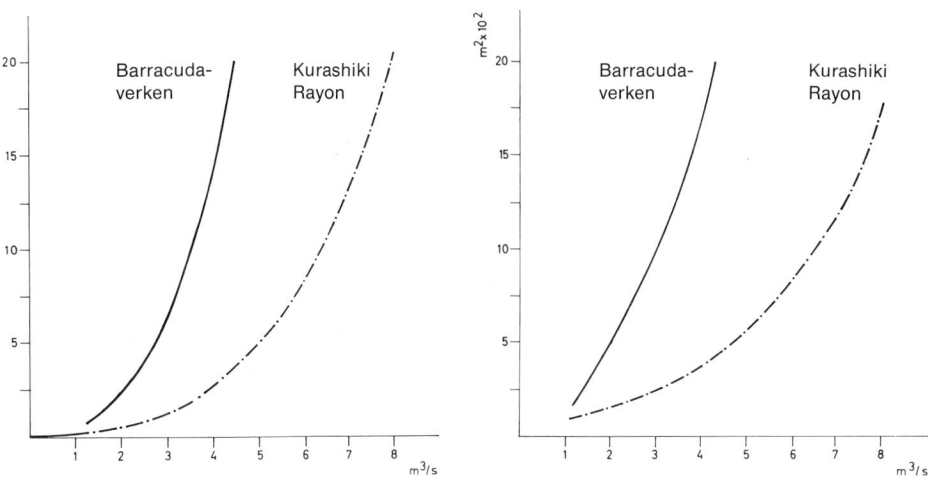

In the case of *radially* working devices (Fig. 123) the air is sucked in sideways, forced outwards centrifugally by the rotation of a cylinder and blown out at right angles to the air intake. Because of air deflection the devices cannot easily be added to each other. High running speeds are necessary to produce the air pressure. The direction of flow cannot be reversed.

In the case of *tangentially* working devices (Fig. 124) a shaft with fins which act as impellers revolves inside a cylinder. The air is sucked in tangentially in the area of the cylinder casing and also blown out tangentially. This device is usually only produced in small sizes. It is used in preference where increased demands for operational quietness are made. They are especially suitable for producing negative pressure. However, the direction of flow is not reversible.

According to the air capacity a distinction is drawn between fans, blowers and compressors.

A *fan* is an apparatus with a low pressure capacity but large cross-sections for the air current. An (axial) propeller fan needs the most power when producing a low air current under high pressure (e.g. at the end of the inflation of a building), a (radial) centrifugal fan needs the most power producing a strong air current under low pressure (e.g. at the beginning of the inflation of a building).

A slightly higher pressure can be produced with *blowers* than with fans. Radial blowers can only be installed under certain conditions because they have only a moderate volumetric capacity. The best energy yield is offered by circulating displacement blowers, in which no internal compression develops. They work with minimal pressure and gradually develop only as much pressure as is required by the counter pressure of the pneumatic structure. Their pressure capacity adapts and is only a little above the actual internal pressure of the structure at any time.

Compressors, as the name suggests, are appliances with high compression and therefore high pressure production. Aerodynamic compressors are expensive high speed machines of which the main disadvantage is the high temperature of the discharged air (over 200 °C); the output is over 3 excess atmospheric pressure.

With positive displacement compressors, which have at their disposal a fixed internal compression ratio and a constant delivery flow, pressures of up to 7 bar can be produced which makes them suitable for use in high pressure structures. In the case of low pressure structures such a high pressure output is not desired because the air expands again after entering the structure; furthermore, the enormous force of the air current can lead to damage to the membrane. If one still wishes to work with compressors on low pressure structures, one can add to the compressors jets, in whose air vent area air from the surroundings is carried forward by negative pressure (Fig. 125). In this way the positive pressure of the main air current is reduced and a larger quantity of air is transported with the same amount of energy. (Bibl. 163.)

In contrast to the distinction made here, in line with general usage the term "fan" will be used henceforth even when blowers or compressors are being discussed.

5.6.3. Fan specifications and their control mechanisms

Determination of the required fan capacity results from the anticipated quantity of air leakage as well as the characteristics of the fan and the supply channels.

In the case of single membrane structures the need for support air does not increase linearly with increased building volume conditioned by the relatively low proportion of openings. Table 17 shows the relationship between fan capacity and volume of building or building ground surface. (Bibl. 44, p. 100.)

The average current consumption can be estimated according to the following formula (Bibl. 25):

$$\text{KW required} = \frac{\text{covered surface (m}^2)}{200 \text{ to } 250}.$$

Occasional fluctuations in support pressure can become very dangerous in the presence of snow or wind loading. Therefore in the first place the support air in winter should always be warmed (approx. +12 °C is sufficient inside, measured at the apex), so that no snow remains. In the second place it should be possible to adjust the fan capacity to the actual wind pressure. In the case of buildings with almost equal main directions of extension, the internal pressure must be raised when the wind loading increases in order to further stabilise the membrane in position. With flat structures the wind loading acts in the form of suction forces. Here the internal pressure must be reduced by decompression or automatically controlled positive pressure valves, so that the membrane tension is not increased too much.

If the interior of a pneumatic structure is used by living beings, then the fan equipment has the function of air renewal as well as the task of producing support pressure. In small buildings, because of leaks the air supply must be maintained high enough to ensure adequate ventilation. With large halls one calculates 1 to 2 changes of air per hour.

To ensure uniform distribution of the inflowing air current and to reduce the air speed, the provision of "air current distributors" is recommended, perhaps in the form of an air distributing ring system with a great number of small discharge openings (Bibl. 44, p. 104).

In Figs. 126 to 128 different kinds of air supply lines are illustrated. Fig. 126 shows a concertina type tube whose length is variable; Fig. 127 shows a membrane tube, whose connection to the building envelope is reinforced by a piece of pipe. A stiff supply system is shown in Fig. 128; it has the advantage of an exactly controllable air supply but a considerable disadvantage in that it cannot yield to movements in the building without transition members which can lead to tension peaks in the membrane.

For conventional air supported halls the following rules can be compiled for the establishment of stabilisation equipment (Bibl. 44, p. 101; Bibl. 144, p. 114).

1. The fan installations must be suitable for continuous operation and be designed in such a way that they meet even maximum demands. The maximum stabilisation pressure must be quickly and reliably attainable.

2. Two or more fans, where under normal conditions only one would be operated, guarantee that in the event of a mechanical defect another fan can be switched on. Valves prevent loss of air due to non-functioning fans.

3. Care must be taken that no snow can collect in the area of the fan intakes so that the supply of airs is not reduced.

4. The injection pressure should be just above the internal pressure. This will make the best use of the energy supplied.

Further detailed directions can be found in Bibl. 136.

5.7. Transport and erection

How small the transport volume in relation to the final volume and the weight of the pneumatic structure is and of how few individual parts it consists becomes especially evident in the transport and erection procedure; this aspect also makes pneumatic structures particularly suitable for use in remote, e.g. polar, areas and in outer space.

In all structures the erection and dismantling procedure has to be carefully thought out for each individual case, as in this phase the membrane is not stabilised in position and is especially susceptible to damage.

In particular the following properties have a favourable effect on transport and erection (Bibl. 44, p. 121 ff.; Bibl. 144, p. 105 ff.).

1. With the exception of anchoring on site all parts of the structure are normally prefabricated. The anchoring too can be formed from prefabricated parts.

2. The weight of the structure is extremely low. It is essentially defined by the weight of the membrane, which is about 3 kg/m² including localised reinforcements.

3. The proportion of erection costs to total expenditure is only 8 to 25%.

4. The structure can be transferred to other sites and also be used after lengthy storage.

5. The erection and dismantling times are very short. Depending on the size of the building, ground conditions, type of anchorages, size of erection team – usually 5 to 8 men – erection for a standard hall lasts 1 to 4 days, dismantling less than 1 day including packing for transport.

5.8. Structural/physical factors

5.8.1. Humidity control

Almost completely airtight and watertight membranes are logically only used for high pressure structures, while in buildings whose positive or negative pressure zones are occupied by men a certain air and moisture penetration is even desirable (Bibl. 119, p. 168).

If highly impervious synthetic foils are used in single membrane structures, ventilation flaps are indispensable as an aid to greater circulation of air (Bibl. 131, p. 33). By using coated fabrics such flaps can usually be dispensed with if there are at least 4 to 5 changes of air per hour.

Even in pneumatic swimming pool roofs condensation formation can be brought under control to a large extent and even completely eliminated by means of additional measures. Thorough investigations into this were made by a German air supported hall manufacturer and the results are reproduced below (Bibl. 28): If the water temperature is 23° to 24 °C then the air temperature should be about 26 °C. This does not only mean that the room temperature is pleasant, but also limits the amount of water evaporating from the surface.

If the fresh air in the inblown air volume is so proportioned that almost total absence of condensation is assured, a temperature difference of at most 30 °C can be bridged with the usual fans so that at an internal temperature of +26 °C the hall membrane remains dry down to an external temperature of −4 °C. If the proportion of fresh air is reduced and the proportion of recirculated air is correspondingly increased, then a temperature difference of 40 °C is attainable, so that an internal temperature of +26 °C can be maintained down to an external temperature of −14 °C. With the reduction of the proportion of fresh air the drying effect, e.g. the absorption capacity for humidity, is reduced; however, the film of condensation on the interior of the membrane is usually insignificant. Moreover, the temperature stratification within the room, i.e. the situation that the warmer air rises to the upper zones, has a favourable effect.

As the heat storage capacity of the water is considerably greater than that of the air, it is more practical to make allowance for a slight cooling of the water during the night than to continually draw off warm air to the outside. Even with a night temperature outside of 0 °C to +5 °C and corresponding air temperatures in the hall of +15 °C to +18 °C, considerable amounts of condensation occur on the inside of the membrane when the pool water is constantly kept at bathing temperature. If the pool is covered at night with a sheet of foil, then the hall membrane will remain completely dry down to an external temperature of some −30 °C and a corresponding internal air temperature of up to +10 °C.

5.8.2. Thermal control

It is clear that such thin and light materials as foils and woven fabrics can only offer extremely low thermal insulation and heat storage. PVC coated polyester weaves have at 4.0 to 4.5 more or less the same calorific value as a normal window pane.

Considerably lower coefficients of heat transmission can be achieved by laminating the membrane with insulating material. Flexible foams made of rubber, polystyrol, PVC and polyurethane are most frequently used and these are either stuck on in sheets or sprayed on direct as a foam. The calorific value attainable here is about 3.0 (Bibl. 123, p. 53). In Southern France air supported halls were used for refrigeration as well as for quicker ripening of fruit and vegetables. The required internal temperatures of +3 °C to 27 °C could be achieved without difficulty by pasting on 80 mm thick sheets of polyurethane foam.
(Bibl. 60, p. 8.)

The pasting on of stiff and relatively thick foam sheets results in a shell-like stiffening. With such stiffened halls a very high internal pressure is necessary in order to prevent the changes in form due to aerodynamic loading. If the air supported hall frequently has to be erected and dismantled, this method cannot be used for the sheets can only be pasted on when the structure is already erected and the insulating layer must be removed again before the structure is dismantled.

Another solution is offered by the membrane sandwich construction described on p. 145 which has a calorific value of 2.6.

A further possibility of achieving a higher thermal insulation is the use of multi-layered membranes with enclosed air cushions. An example of this is the American exhibition pavilion "Atoms for peace" by Victor Lundy (pp. 134, 135). However, it should be remembered that with a membrane distance of over 5 cm an increased heat transfer can take place by means of convection in the vertical direction within the individual chambers.

In contrast to conventional buildings whose great thermal inertia demands constant heating, pneumatic structures only need to be heated while in use, so that short term heating systems, such as hot air or radiation systems, are the most suitable.

Propane gas burners are not recommended, as they are dangerous to health and increase the formation of condensation on the inside of the membrane (Bibl. 44, p. 124). In preference to these, heat exchange systems should always be chosen. A very uniform temperature distribution is obtained when the ventilated warmth caused by the loss of air is transmitted through the ventilation ducts or in their immediate proximity. The necessary transmission warmth is expediently supplied by heating elements arranged regularly in the interior of the hall. (Bibl. 144, p. 115.)

The warm air fans most frequently installed are provided with directly lit heat converters run off heating oil. Hot water or steam heating is only occasionally used. Heating by electricity is uneconomical for the most part.

In the cushion structure of Forum Steglitz in Berlin warm air heated up to 45 °C is blown in so that an accumulation of snow, which is not only undesirable on structural grounds but also reduces translucency, is prevented.

The heating of the air inside air supported halls in strong sunlight is only a problem when the hall is low in height. Large halls with a height of 10 to 20 m even in strong sunlight have a temperature in the main hall only a little above that outside, although the membrane temperature at the top of the hall could be +40 °C and more; this is because thermal convection causes the temperature to reduce from top to bottom at the rate of about 1 °C per metre (Bibl. 25).

The higher temperature in the upper area is anyway convenient in so far as it helps in melting snow and ice in winter. Thus in order to achieve the melting temperature of +12 °C laid down in the FRG guidelines, it is sufficient in the case of an external temperature of −10 °C and a 10 m high hall for the air volume at 1 m (height) to be heated up slightly above 0 °C. It is recommended that the proportion of fresh air be kept as low as possible and that the fans run with more recirculated air in order to save thermal energy. (Bibl. 28.)

In the case of small halls, cooling the supplied air in summer is indispensable; tests show that in sunlight even near the floor temperatures of more than 10° above the external temperature can occur.

In the case of halls being less than 5.5 m high, cooling the supplied air is not feasible for economic reasons. However, here also favourable temperature conditions can be achieved by means of a higher rate of change of air, the possibility of ventilation in the crown of the membrane, local lifting of the internal skirt and water spraying apparatus from perforated tubes or pipes.
(Bibl. 28; Bibl. 104.)

The colouring and surface treatment of the hall membrane have a decisive influence on the heating of the hall interior by means of sun radiation. White and silver coatings show the highest reflection values and therefore result in the least heating. For this reason PVC coatings are frequently provided with an additional aluminium varnish. In the case of dark membrane surfaces the internal temperatures are 3 °C to 5 °C higher than with light membrane surfaces.

The significance of colour for reflection values can be seen in Table 18 (Bibl. 103); Table 19 shows the transmission and reflection curves of two white PVC coated polyester fabrics (Bibl. 104).

In 1967 N. Laing published a very interesting research work on the problem of the hothouse effect – great heating up of the air strata in the hall interior in sunshine and quick cooling when the sunshine goes and the temperature outside drops.

In his institute, variable transmission wall constructions of thin very light membrane material were developed, which in connection with radiation control systems use the sun for heating up and extra terrestrial space for cooling down. With minimal energy contributions for the servo system these wall elements can give frost temperatures in the Sahara and sub-tropical conditions in Newfoundland.

Some 20 different wall systems have been developed by which the internal climate of buildings can be widely controlled in respect of all classic climatic factors – radiation, air temperature, air humidity, precipitation and air current.

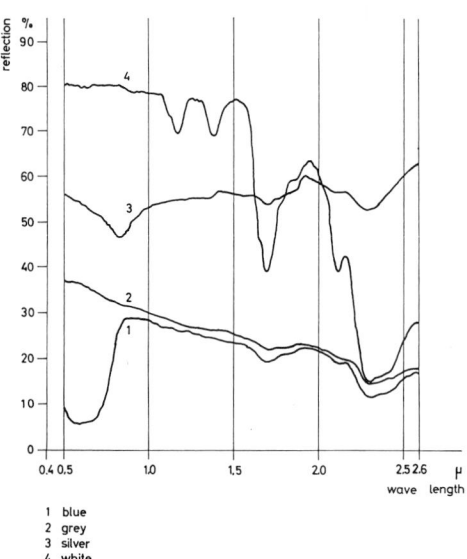

Table 18. Reflection curve of PVC coated polyester fabrics in various colours.

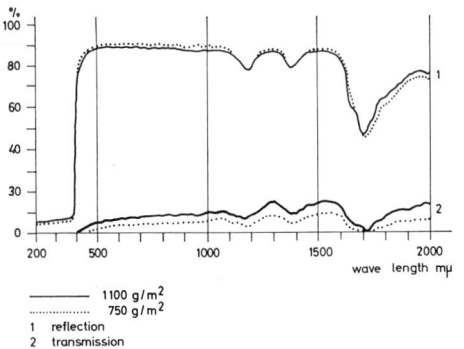

1100 g/m²
750 g/m²
1 reflection
2 transmission

Table 19. Reflection and transmission curves of two white PVC coated polyester fabrics.

Questions of manufacture and structural realisation as well as installation and maintenance expenses have already been extensively clarified.

In the simple wall construction shown in Fig. 129 an example of the operation of these variable transmission wall systems is explained (Bibl. 96).

The outer skin (A) consists of a fibre glass net; this is coated with a foil which has a certain optical quality. On the inside are channels made of very thin transparent film which are tightly filled with gas. Within these channels a partly metallised foil flap (K) is fixed which can be pneumatically moved to positions L and R.

The surfaces marked in black are made reflective on both sides by metal vaporisation. If the foil flap is in the position depicted on the left, then atmospheric radiation and ground reflection enter without hindrance through the wall. In the position depicted on the right the wall acts like a polished screen that prevents the entry of any radiation.

With such walls coefficients of heat transmission of less than 0.8 Kcalories/m²h°C have been achieved, which corresponds to the thermal insulation of a 30 cm thick cavity brick wall.

Dietz suggests another system for regulating the climate (Bibl. 47):

He believes that one should view the outer membrane as a kind of "solar trap" and provide blackened surfaces at intervals on the inside which can be heated up by sunlight. By means of appropriate fan equipment, air can be conducted past these surfaces which will transmit the heat to a central storage unit. From there it could be recalled for heating or cooling purposes by means of a heat converter.

The main problem lies in the choice of the storage medium, whose energy output must be controllable in small stages corresponding to changing demands. At the moment, appropriate model tests are being carried out at the Massachusetts Institute of Technology.

5.8.3. Acoustics

No completely satisfactory solutions for pneumatic structures have yet been found as regards acoustics. In the case of single shell structures a high weight per unit area and great rigidity are decisive for good sound attenuation, so that sound transmission is hardly affected in single membrane structures with their very light envelope systems in comparison with other materials.

In order to protect exhibition areas in pneumatic structures against noise, attempts were made to provide curtains made of leaded vinyl which, however, at the same time caused a greatly increased reflection of sound waves.

There are also suggestions for shielding pneumatic structures from outside noise by connecting a second sound absorbing skin to the envelope system or incorporating a sound absorbent material.

It is interesting that the sound insulation values of pneumatic structures did not prove to be as bad as had been expected. Detailed investigations into the influence of internal pressure in double and multi-membrane structures as well as investigations into the influence of tension conditions in the membrane have still to be carried out. Also inferences for acoustically improved envelope systems have still to be drawn. In this connection it is also worth mentioning the phenomenon that the intensity of sound waves from the noise of speaking is hardly reduced at all by a single pneumatically stressed membrane, but the sound waves are transposed so that on the other side no speech but only noises can be heard.

The acoustic behaviour of air supported halls also is generally better than was first expected by the experts. With correctly installed loudspeaker facilities, even lectures can be presented easily in very large halls. However, air supported halls are still unsuitable for musical performances with high acoustic demands.

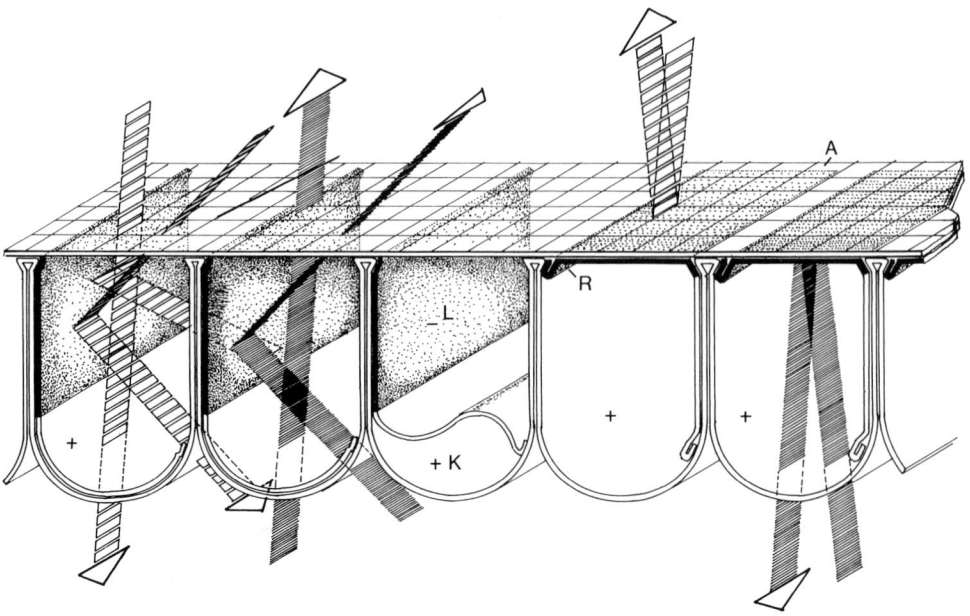

129. Variable transmission wall construction. N. Laing.

In a hall for 2,500 people with a covered surface of 1900 m² there was an average reverberation time of approx. 1.3 seconds over the total frequency range of 100 Hz to 1 Hz, while it should be 1.7 to 2 seconds in rooms of this size. When the internal pressure of the hall is reduced the reverberation time increases by 10% but is still not satisfactory. As a result of the acoustically transparent envelope music does not have an adequate sound volume. It sounds similar to an open air performance. In addition there are flutter echoes, whose fade-out time is sometimes longer than the reverberation time. The vaulted roof cross-section is the main cause of these flutter echoes. They are certainly moderated when the hall is full, but not suppressed. (Bibl. 28.)

Improvements in acoustics can be achieved by means of sound dispersing hanging surface elements, mobile sound absorbing curtains and by the pasting on of absorbent, soft porous materials. (Bibl. 123, p. 49).

6. On the statics and dimensioning of air supported structures

Hans Eggers

6.1. Introduction

For the last ten years air supported structures have been built in an increasing number as industrial, sports, store and exhibition halls.
The important stages in their production are:
– design and selection of geometry,
– proof of load safety,
– selection and calculation of cutting pattern,
– manufacture,
– erection.

Only the first four points are dealt with here. Special emphasis is laid on design, calculation and construction methods for relatively small, simple, air supported structures. More detailed investigations and model tests are required for very large or complicated structures.

Even in the choice of membrane form one has to take into account the fact that the envelope can only take up tensile forces. Following on from the soap membrane comparison, in Section 6.2 the design of an unwrinkled membrane over any ground plan is shown. Cylindrical and spherical forms have been especially successful on account of their simple cutting pattern and their trouble-free manufacture. Even for these simple forms the calculation process is relatively complicated when the large deformations of the membrane can no longer be ignored. The numerical effort involved in solving the non-linear equations is usually great. The loading – specifically the wind loading – is often known only in the order of magnitude. Therefore approximation methods, designed for practical use, are only given here which are sufficiently accurate to calculate the membrane forces and the displacements. The diagrams shown in Section 6.6 considerably simplify the calculations of spherical and cylindrical membranes. They are specifically laid down for the design load and proof of load safety given in Bibl. 136. Their practical use is shown in Section 6.7.

At this stage I would especially like to thank Professor Duddeck at whose suggestion this work was originated and who has greatly assisted it. Apart from that I thank all the companies who have helped me both with practical advice and by placing test results at my disposal, as well as Mrs. Lack for her very careful production of the drawings.

6.2. Membrane geometry

6.2.1. Soap film analogy

One of the simplest membrane forms is the soap film (Bibl. 119). It encloses a prescribed volume with the minimum surface area. Characteristic of the form law of the soap film is the identical disappearance of the shear forces:

$$n_{12} \equiv 0. \tag{1}$$

This condition is only fulfilled for any co-ordinate lines when the maximum value

$$\max n_{12} = \frac{1}{2} \sqrt{(n_{11} - n_{22})^2 + 4(n_{12})^2} = 0 \tag{2}$$

also disappears. From equations 1 and 2 it follows that only the hydrostatic membrane force state

$$n = n_{11} = n_{22} \tag{3}$$

can occur in the skin. If one ignores the minimal selfweight, then the membrane force n is constant in the whole skin (tangential equilibrium). The equilibrium, normal to the surface, is determined as follows:

$$b^\alpha_\alpha = \frac{p}{n} = \text{const.} \tag{4}$$

The mean curvature b^α_α ($= 1/R_1 + 1/R_2$ for rectangular co-ordinates) can be given for arbitrary co-ordinates with the help of the tensor calculation. For a given curvature b^α_α depends only on the relationship between internal pressure p and membrane force n. For the Cartesian co-ordinates according to Fig. 1 the equation 4 is converted to

$$\frac{z_{,xx}(1 + z_{,y}^2) + z_{,yy}(1 + z_{,x}^2) - 2z_{,xy} \cdot z_{,x} \cdot z_{,y}}{(1 + z_{,x}^2 + z_{,y}^2)^{3/2}} = \frac{p}{n} \tag{5}$$

where a comma denotes the partial derivatives

$$(\ )_{,x} = \frac{\partial(\ )}{\partial x}, \quad (\ )_{,y} = \frac{\partial(\ )}{\partial y}$$

Exact solutions of this partial non-linear differential equation are known only for special cases. Therefore it is solved in general iteratively. Because of the non-linear connection between the geometry and the curvature several equilibrium positions are possible. In the transition to a new state of equilibrium the membrane changes its form suddenly.

The soap film model is very clear and is usually replaced in the experiment by an inflated thin rubber membrane. Envelopes formed in accordance with the soap film analogy are uniformly tensioned by internal pressure and are free of wrinkles. For this loading – usually permanent load – the soap film is the ideal membrane form.

6.2.2. Choice of geometry

In general a pneumatic structure should be free from wrinkles. This condition can only be fulfilled for a "pneumatically formed geometry" in which only tensile forces occur under internal pressure. The membrane will not take compression forces. It changes the geometry through the formation of wrinkles, so that a tensile structure re-occurs. Even under short term wind and snow loadings the membrane should not bulge in order to avoid an uncontrolled increase in external loadings (see Section 6.3). Manufacturing discrepancies, distortions of the cutting pattern, as well as various stiffnesses of the seams and fabric in the weft and warp direction lead to unexpected membrane forces which are not usually taken into account in the calculation. They considerably influence the development of the membrane forces, however, as the measured bearing loads in Bibl. 114 and Bibl. 67 show. In order to include these influences approximately, a wrinkle-free membrane should satisfy the condition.

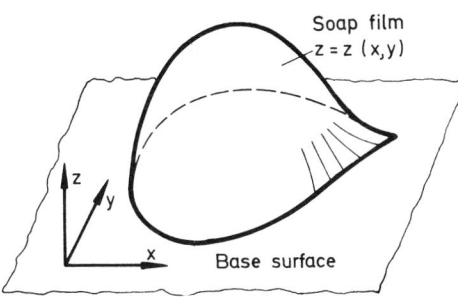

1. Soap film.

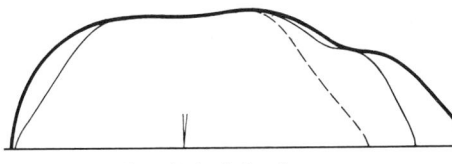

Side view with "arris lines"

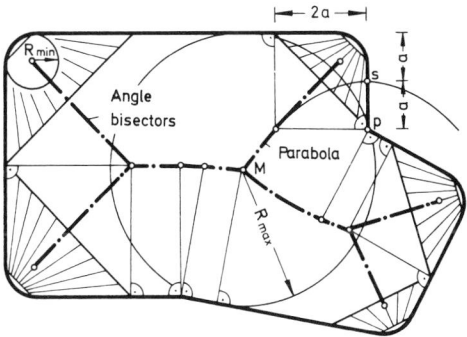

2. Envelope construction by inscribed spheres.

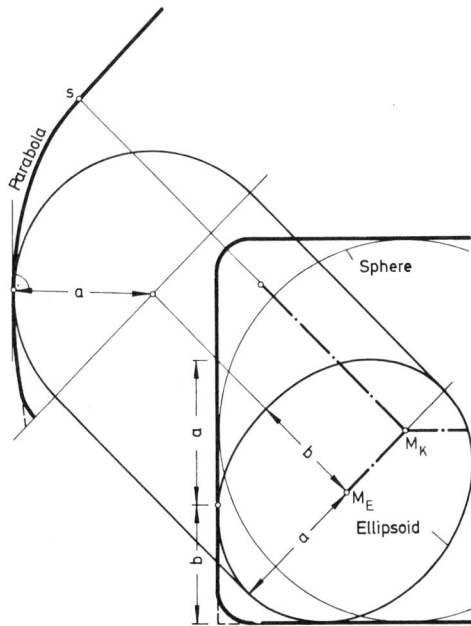

3. Angle construction by inscribed ellipsoids.

$$\min n(p) \geq 0.2 \cdot \max n(p). \qquad (6)$$

The curvature of a membrane must be steady. Even with complex geometries the curvatures must change over from one to another as uniformly as possible. Small jumps in the curvature are compensated for by the elasticity of the material and only rarely lead to the formation of wrinkles. Ridges, flutes and peaks can only be produced by additional forces. Theoretically conical apexes are formed by internal pressure, but are always rounded-off for reasons of stability. However, as $n_{11} = n_{22} = 0$ they do not comply with condition 6.

The geometry of a soap film is usually too complicated for practical application. Furthermore the flat corner areas are often undesirable. The soap film is, however, a very clear membrane form from which the geometry can be changed and simplified.

A very simple rule for the construction of wrinkle-free membranes is given by Frei Otto (Bibl. 119):

"A geometry can be formed pneumatically when spheres with a steadily changing radius can be included, the centre points of which lie on a curve and when at least one parallel of latitude of the generating sphere rests on the whole length of the membrane".

Condition 6 is approximately satisfied for spherical radii $R_{min} \geq 0.2 \cdot R_{max}$. In Fig. 2 the construction of a membrane according to Otto's rule is shown. In this example the centres of the sphere lie on the dotted line in the ground plane. The spheres touch the bearing line in at least two points. In the corner they transform into conical surfaces which are indicated by their generatrices. The base circles of the largest and smallest sphere are shown for comparison when $R_{min} = 0.2 \cdot R_{max}$. Furthermore the form of the membrane can also be affected by shifting the centre point of the sphere from the ground plane. In the vertical plane branchings of the central point lines are also possible.

The straight generatrices in the corner area do not look very attractive even when the tip of the cone is rounded-off by a sphere. A more pleasing form arises when the sphere is replaced by an ellipsoid that is tangential to the base lines and to a parabola in the direction of the angle bisectors (Fig. 3). Experience shows that even this geometry remains unwrinkled. The corner should always be rounded-off, as wrinkles can easily occur here otherwise.

The choice of geometry is difficult when the envelope is constricted by individual cables but there is still no cable net (Fig. 4). Then only the cable length is given, while the form of the cable lines are determined by the equilibrium conditions. In the first approximation the cable lines are estimated and then a membrane geometry is chosen following Otto's rule. Constricted membranes usually undergo much greater deformation than the simple smooth envelopes. Thus small errors are usually balanced out. With larger, more complicated structures the geometry and the membrane forces should always be defined by tests and more accurate calculations.

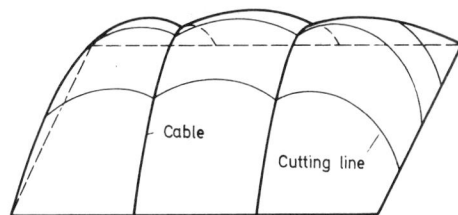

4. Membrane reinforced by cables.

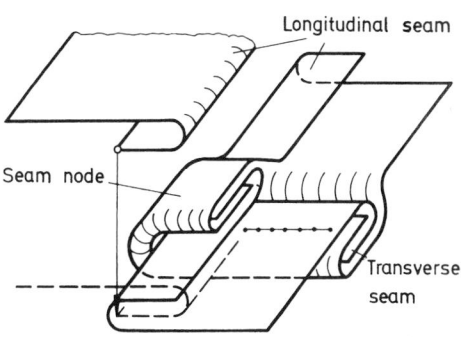

5. T-nodes on a folded seam.

6.2.3. Membrane cutting patterns

The envelope of a pneumatic structure is usually made of coated fabric or synthetic film. The material, which is supplied in a bale, is cut like a suit and either sewn, glued or welded. As simple and economic a cutting pattern as possible is sought for a given geometry. The seam pattern on the skin is chosen according to the following criteria:

– regular and large surfaced patterns are preferred to a 'patched' pattern because the seams in the inflated membrane can be clearly recognized;

– if seam nodes cannot be avoided, then as few seams as possible should converge at one point (in the T-node shown in Fig. 5 ten layers are sewn together and in a crown node sixteen layers);

– for reasons of reduced strength in the seam area, seams in the direction of the greatest main tension are more economical than those running at right angles to it;

– the wastage is generally less in short lengths than in long;

– seams running in the direction of the steepest gradient facilitate the slipping off of snow and the membranes dry more quickly after rain.

Fig. 6 shows some seam patterns for the front joint of a cylindrical membrane. The 'armadillo' pattern is the simplest for production. Exact calculation of the cutting pattern is – apart from specific geometric forms and seam patterns – only possible with an electronic calculator. In Fig. 7 the individual membrane strips are cut out from the surface F through planes E_i. They are approximated by means of very small, flat, triangular elements whose corners lie on the surface of the membrane and on the planes E_i. If the triangles on the broken line K are folded into a

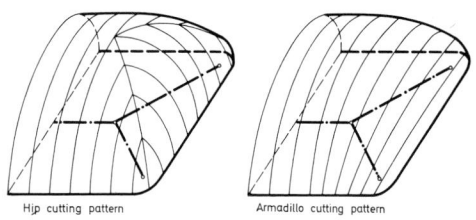

6. Forms for cutting patterns.

Hip cutting pattern   Armadillo cutting pattern

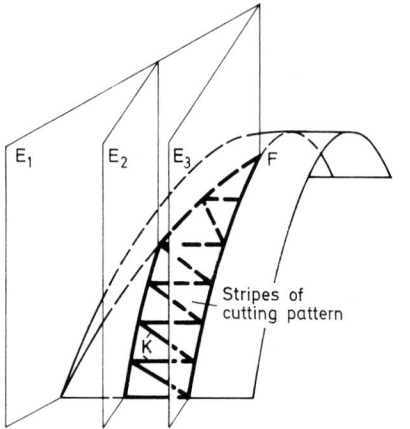

7. Membrane cutting pattern for the hip type.

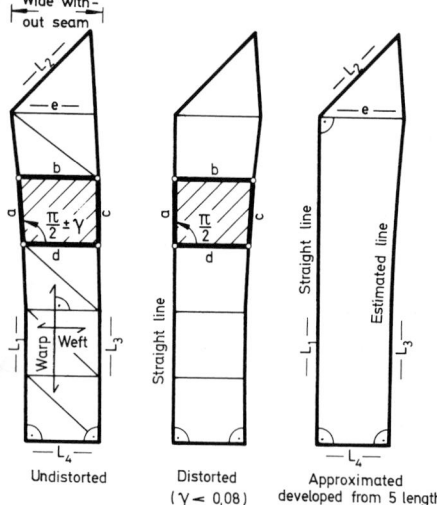

8. Distortion of the cutting pattern.

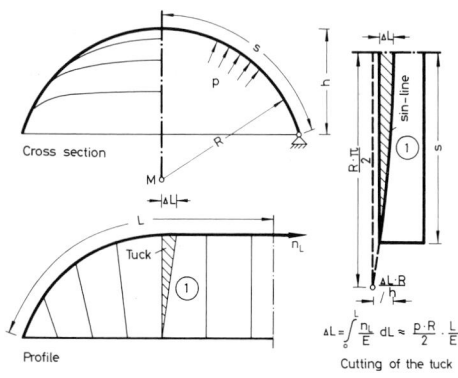

9. Cutting pattern of a strained membrane.

plane, the cutting sector appears. Its form is influenced by the distance and inclination of the planes E_i so that only a low wastage occurs. If one also makes use of the extreme shear flexibility of the material, then more simple cutting patterns are possible. The sections are so distorted without extension of the warp and weft direction, that with equal cutting pattern lengths a straight edge occurs. The permissible shear distortion depends on the material and on the membrane forces. Experience shows that shear strains of 5 to 8% are possible without wrinkles forming.

Approximation methods, as used in tent construction, are suitable for determining the cutting pattern. Commencing with a straight length, the cutting pattern is "developed with the curve template" over the length of the seams and one or two intermediate points (Fig. 8). Small errors are balanced out in the membrane. The use of the approximation method is very simple, although it requires a certain amount of experience.

With larger structures the elongation of the membrane due to the internal pressure (about 1 to 3%) should be taken into account in the cutting pattern. Usually the narrower cutting pattern of individual strips is adequate for adjustment of the membrane elongation (Fig. 9). The distance between the tucks must not be chosen too large because the membrane might be restrained by door and gate constructions in the side of the hall. A tuck width limited to $\Delta l \leq 20$ cm is approved.

6.3. Load assumptions

In pneumatic structures very large deformations can occur which affect wind and snow loading through the changed geometry of the envelope. The load is approximately applied to the undeformed membrane and "affixed" there. Conservative loads, such as deadweight and snow, remain true to direction; internal pressure and wind act always normal to the deformed membrane.

6.3.1. Internal pressure

Air supported halls maintain their form by internal pressure. The size of the internal pressure depends on the geometry of the membrane, loading and permissible deformations. Furthermore the membrane must not wrinkle or flutter in the wind stream. Tests have shown that, depending on the geometry, minimum internal pressures of 50 to 100% of the wind pressure are required for this. More detailed instructions on the minimum internal pressure are contained in the Japanese (Bibl. 32) and East German (Bibl. 158) guidelines. In Table 1 they are compared with the test results according to Bibl. 14 and Bibl. 114.

The minimum internal pressures measured for cylindrical membranes have to be increased by about 15% if one also wishes to avoid wrinkles in the end calottes. In shallow-stretched mem-

branes only suction forces occur so that a low internal pressure is adequate for stabilisation.

In West Germany a minimum internal pressure of $p = 30$ kp/m² is prescribed for envelopes with a height of more than 8 m. This value is not sufficient to stabilise high membranes erected over small ground areas. For any geometry the minimum internal pressure can be determined by restriction of the displacements and by a margin of safety against folding in.

The maximum possible internal pressure is the bursting pressure $p_{Br.}$, which depends on the geometry and on the fabric strength. In the fabrics used today it is generally less than the critical internal pressure p_{crit} at which the membrane suddenly expands and changes to another state of equlibrium (Bibl. 119, p. 64). In Bibl. 53 the formula

$$p_{crit.} = 0.296 \cdot D/R_o \qquad (7)$$

is given for the critical internal pressure of a spherical rubber membrane. Here R_o is the radius of the undeformed sphere and D the tensile stiffness of the membrane (D = E · t, modulus of elasticity E, membrane thickness t). For cylinders the first factor is replaced by the value 0.25. For a coated fabric the equation 7 provides an approximate value.

6.3.2. Deadweight

The specific gravity (γ) of the envelope material lies between 0.8 and 1.6 kp/m² per mm fabric thickness. The fabrics usually used (t ≤ 1 mm, g ≤ 1 kp/m²) are very light, so that the deadload can be ignored in structural calculations. Because of the low weight a pressure difference of 1 to 2 mm water column is sufficient to maintain the envelope in its shape. Even when there are large openings in the envelope the membrane collapses slowly.

Tests are now being run to foam up insulating materials on the inside of the fabric in order to reduce the loss of heat in winter and overheating in summer. A considerably greater dead load must be taken into account in the structural calculation. In cylindrical air supported halls a larger envelope deadweight can also be approximately disregarded if the working pressure exceeds the internal pressure used as a basis for the structural calculation by the value:

Geometry	p/q Building code (Bibl. 158)	p/q Wind tunnel tests (Bibl. 14, Bibl. 114)
Sphere on a tower	1,0	—
Three-quarter sphere	0,85	1,0
Hemisphere	0,65	0,70 ··· 0,74
Cylinder with spherical callottes	0,65	0,60 ··· 0,62
Cylinder over rectangular base	0,55	0,62 [1]

[1] Semicylinder, for a quarter cylinder p/q ≈ 0,5

Table 1. Minimum internal pressure p: wind back pressure q.

$$\Delta p \approx \left[1.1 - \frac{h}{b} \cdot \left(1 + \frac{g}{2\,p} \right) \right] \cdot g \qquad (8)$$

where $h/b \leqslant 0.6$, $\quad g/p \leqslant 0.5$

This increase in pressure only balances out the reduction in membrane forces generated by the deadweight. Likewise in the spherical membrane the increase in pressure

$$\Delta p \approx g \qquad (9)$$

where $h/b \leqslant 1.0$, $\quad g/p \leqslant 0.1$, $\quad g/q \leqslant 0.1$
(q = wind back pressure)

does not lead to any excess in the membrane forces.

6.3.3. Wind load

For the dimensioning of an air supported structure the load condition "internal pressure and wind" is decisive. The membrane forces arising purely from wind pressure are generally greater than those caused by internal pressure. Because of the low mass and great damping no galloping vibrations occur. An approximation of the wind loading is calculated according to the formula

$$w = c \cdot q \qquad (10)$$

where $q = \varrho/2 \cdot v_e^2$ and $\varrho/2 \approx 1/16$
$[kp \cdot s^2/m^4]$.

Here the normalised form function c describes the distribution of wind pressure over the surface, while the back pressure q as a function of the wind speed v_e defines the size of the compression load.

The back pressure q is dependent through the wind speed on the geographical position and height above the ground. Both influences are contained, for example, in the Japanese guidelines (Bibl. 32) of the back pressure distribution (dimensions in kp, m, s):

$$q = \frac{v_e^2}{16} \sqrt[4]{\frac{h}{10}} \geqslant 30 \ kp/m^2 \qquad (11)$$

Because of the low mass of pneumatic structure for v_e the squall speed, with a duration of effect of a few seconds, is critical. Determining factors of the squall speed and its frequency are published in Bibl. 37 for West Germany. According to this the squall speed of $v_e = 35$ m/s lasting 2 to 10 seconds is exceeded only on 2 days a year, if one disregards particularly exposed positions such as high mountain peaks and immediate coastal areas. Special local conditions, such as the nozzle effect in the vicinity of high buildings for example, are not taken into account in the determining factors. In Table 2 the assumed back pressures for various countries are given for comparison. The form function c for the wind distribution can be determined only by tests. It is dependent on the geometry, the deformation and the wind direction. Reynolds Number and the surface roughness of the generally smooth membranes have little influence on the distribution of pressure. Tests on cylindrical membranes gave a greatly varying pressure distribution with distinct suction peaks. Wind directions inclined towards the cylinder axis lead to particularly large wind loadings in the vicinity of the up-

stream calotte. The greatest measured wind pressure distribution (given in Bibl. 114) for the cylinder with spherical calottes affected by transverse and slanting crosswinds is contrasted in Table 3 with the values given in Bibl. 136.
The influence of the deformations on the pressure distribution is slight for a cylinder. The evolved geometry adapts without wrinkles to the support line for the loading, so that even large deformations alter the aerodynamic conditions very little. In contrast to this a spherical membrane takes up randomly distributed wind pressures without appreciable deformations. Only when compressive stresses arise does the membrane bulge locally (snap through). The bulging in the back pressure area influences the form function considerably (Bibl. 14). In Table 4 the form functions c, measured in accordance with Bibl. 114, for different internal pressure conditions are given and compared with the values for the cylinder given in Bibl. 136. When normal to the wind direction the pressure distribution is almost axially symmetrical.
The gradual approximation of the back pressure usual in West Germany is particularly unfavourable. The high back pressure in the apex area coincides with the large ordinates of the form function. The membrane forces calculated according to Bibl. 136 are therefore too large. In comparing Tables 3 and 4 it should be observed that the test results only apply for a constant back pressure distribution.
The internal pressures prescribed for air supported halls are only necessary for stabilisation on a few stormy days in the year. For the rest of the time lower internal pressures are adequate. If the internal pressure is altered to suit every weather condition, then running costs are lower and the life of the envelope is somewhat extended (lower permanent loading). The internal pressure can be easily controlled by means of a wind gauge, whereby the minimum values can still be dependent on the time of year. In Bibl. 61 air valves are suggested to control the internal pressure. The wind presses open the valves as soon as the back pressure exceeds the internal pressure (Fig. 10). Tests (Bibl. 61) and theoretical investigations (Bibl. 56) show that even relatively small valves are sufficient to raise the internal pressure by some $0.7 \cdot q$. Practical testing for the control of varying internal pressures has still to be carried out.

6.3.4. Snow loading

The snow loading of a membrane is dependent on the geometry, the stability of form and the temperature. According to Bibl. 136 no snow will settle on heated membranes with internal temperatures of more than 12 °C. In unheated air supported structures very large deformations occur under snow loading (Figs. 11, 12), which affect snow deposits because of the changed geometry. The shaking effect of the wind combined with the increased snow drift on pneumatic structures usually leads to one-sided snow deposits. The interrelations have been only partly researched.

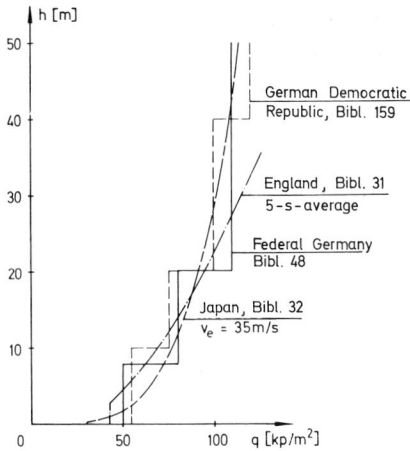

Table 2. Wind back pressures.

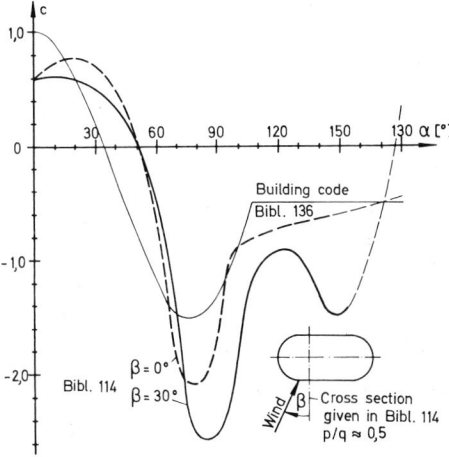

Table 3. Form function c of a semicylinder with quarter sphere, end, test readings (Bibl. 114) – guideline (Bibl. 136).

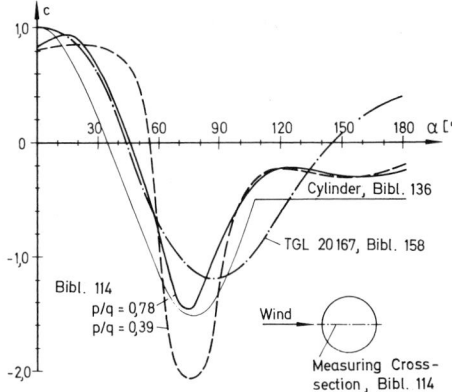

Table 4. Form function c of the hemisphere, test readings (Bibl. 114) – DDR Standard (Bibl. 158).

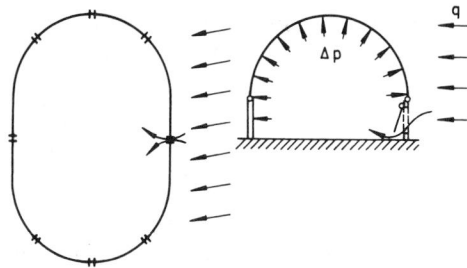

10. Wind valves – see Bibl. 61.

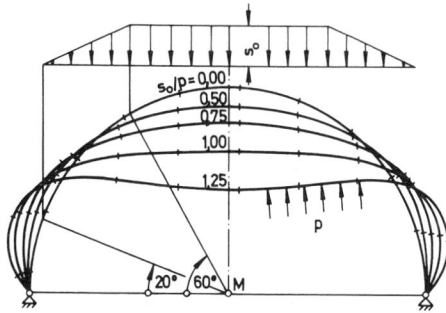

11. Cylinder under full snow loading.

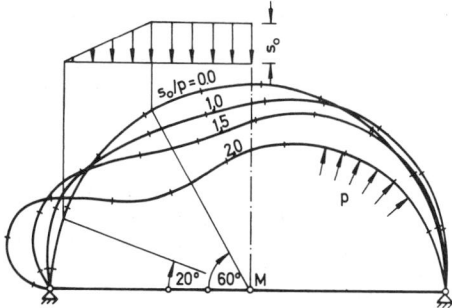

12. Cylinder under one-sided snow loading.

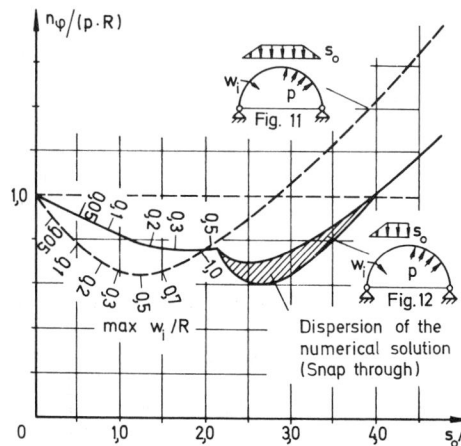

Table 5. Membrane forces of the semicylinder under snow loading.

Snow loading reduces membrane tensions (Table 5) so that splitting of the skin need not be feared. The deformations must be limited according to the use of the structure, so that the envelopes do not "impale themselves" on the installations. Furthermore no pockets of snow should be allowed to arise which could harm the overall stability. As increased snow loading does not increase the risk for the user nor for the structure, the large standard snow loadings (Bibl. 48) based on a 95% probability factor of the annual maximum are not justified. A standard snow load of

$$s_o = 35 \text{ kp/m}^2 \qquad (12)$$

is suggested, as is applied in Bibl. 48 for "buildings for temporary uses". Snow deposits over 10 cm high must be shaken off by varying the internal pressure. Only in regions with great snow falls can the full snow load occur, as here it is often not possible to shake it off. In order to prevent snow pockets arising on cylindrical envelopes and to maintain stability of form, the displacements should not exceed the value

$$\max w \leqslant 0.1 \cdot R. \qquad (13)$$

By comparison in Bibl. 67 $\max w \leqslant 0.03 \ldots 0.05 \cdot R$ is suggested for the semicylinder. The minimum internal pressure is defined by the restriction of deformation (13). Spherical membranes bulge inwards locally under snow load (snap through). The minimum internal pressure which prevents bulging is determined by a folding condition in accordance with Section 6.4.1.

6.4. Calculation procedure

6.4.1. Folding conditions

The principal normal forces of a membrane are calculated by the equation

$$n_{\substack{max \\ min}} = (J_1 \pm \sqrt{(J_1)^2 - 4 \cdot J_2})/2 \qquad (14)$$

with the invariants

$$J_1 = n_{11} + n_{22}, \quad J_2 = n_{11} \cdot n_{22} - (n_{12})^2. \qquad (15)$$

If the membrane forces in the whole envelope fulfil the condition

$$\min n \geqslant o \qquad (16)$$

it is a pure tensile structure. Equivalent to the "folding condition" (16) is the restriction of the invariants

$$J_1 \wedge J_2 \geqslant 0 \qquad (17)$$

or

$$\min n \geqslant J_2/J_1 \geqslant 0. \qquad (18)$$

The very thin membranes do not absorb any compression forces. In order to avoid folding in, the membrane forces actually present must not violate the folding condition (16). However, the actual membrane forces are not known – only known are the calculated forces which can still be influenced by manufacturing accuracy and load scattering. In order to prevent any wrinkle formation, the calculated tensile forces

in the membrane must not become too small. As a criterion for the folding safety under working load the quotient

$$\nu = \frac{\min n (\min p, g, s \vee w)}{\max n (\min p, g)} \geqslant 0.2 \qquad (19)$$

is introduced. In equation 19, $\min n$ is the smallest membrane force determined for the deformed state with the load combinations $(\min p, g, s)$ or $(\min p, g, w)$, and $\min p$ is the smallest working pressure at which the fans start. Folding safety over $\nu = 0.2$ is not generally required. If one disregards the low deadweight of the membrane, equation 19 for the load condition "internal pressure" is converted into condition 6.
The folding safety defined in Bibl. 136 $\bar{\nu} \approx 1/(1 - \nu)$ assumes the superimposition principle and thus a calculation according to the First Order Theory. The definition fails in the case of membranes which can only take up specific loads in the deformed state.

6.4.2. Basic equations of the membrane theory

The basic equations of the membrane theory can be formulated for any membrane forms in tensor notation (sign conventions see Fig. 13). They are given for the linear case in Bibl. 66 and for membranes with large deformations in Bibl. 69. The partial differential equations can be solved for any geometries only numerically. In the following, therefore, only axially symmetrical membranes are investigated for which exact solutions or sufficiently precise approximations are known.
Cylinder co-ordinates as in Fig. 14 are introduced for the axially symmetrical membrane. They run in the direction of the (orthogonal) main lines of curvature. The geometry of a surface element is uniquely described by the following geometric values (Fig. 14):

lengths $\sqrt{a_{11}} = r, \quad \sqrt{a_{22}} = \sqrt{1 + r_{,z}^2}, \qquad (20)$

area $\sqrt{a} = \sqrt{a_{11} \cdot a_{22}}$,

curvatures $b_1^1 = -1/\sqrt{a}, \quad b_2^2 = r^3 \cdot r_{,zz}/a^{3/2}$,

gauss curvature $K = b_1^1 \cdot b_2^2$,

change of lengths $(\sqrt{a_{11}})_{,z} = \Gamma_{12}^1 \sqrt{a_{11}}$,

$$(\sqrt{a_{22}})_{,\varphi} = 0,$$

christoffel symbol $\Gamma_{12}^1 = r_{,z}/r$.

With the specific geometry of the rotational membrane, the physical components of the forces and displacements and the abbreviations,

$$d_1 = \frac{\partial( )}{\sqrt{a_{11}} \ \partial\varphi} , \quad d_2 = \frac{\partial( )}{\sqrt{a_{22}} \ dz} , \qquad (21)$$

$$\Gamma = \Gamma_{12}^1 / \sqrt{a_{22}}$$

the membrane equations derived in Bibl. 66 are converted into
a) equilibrium

$$\begin{bmatrix} d_1 & 2 \ \Gamma + d_2 & o \\ -\Gamma & d_1 & \Gamma + d_2 \\ b_1^1 & o & b_2^2 \end{bmatrix} \cdot \begin{bmatrix} n_{11} \\ n_{12} \\ n_{22} \end{bmatrix} + \begin{bmatrix} p_1 \\ p_2 \\ p_3 \end{bmatrix} = 0, \qquad (22)$$

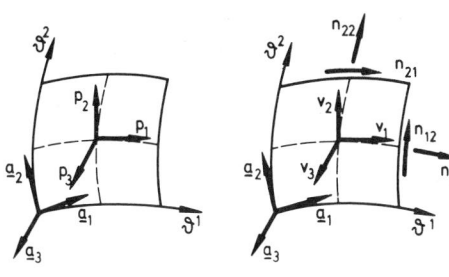

13. Sign convention.

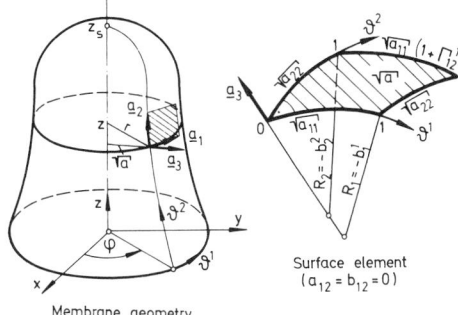

Surface element
$(a_{12} = b_{12} = 0)$

Membrane geometry

14. Rotational membrane.

b) strains

$$
\begin{bmatrix}
d_1 & \Gamma & -b_1^1 \\
d_2 - \Gamma & d_1 & o \\
o & d_2 & -b_2^2
\end{bmatrix}
\cdot
\begin{bmatrix}
v_1 \\
v_2 \\
v_3
\end{bmatrix}
=
\begin{bmatrix}
\alpha_{11} \\
2\,\alpha_{12} \\
\alpha_{22}
\end{bmatrix}
\cdot (23)
$$

The equilibrium conditions (22) are adequate only for calculating the membrane forces when the boundary and transition conditions can also be formulated free of deformation (no cable reinforcements). The displacements then result from the strains and the material laws given in Section 6.4.3. In the case of large deformations the equilibrium conditions depend also on the displacements (Bibl. 69).

For Gauss curvature $K \leqslant 0$ singular solutions ($\alpha_{\alpha\beta} = 0$, $v_i \neq 0$) exist for the strain equations (23). These "stretch-free deformations" do not occur in pneumatic structures with fixed edges. However, the low shear stiffness of coated fabrics (28) can lead to large shear strains ($\alpha_{12} \neq 0$, $\alpha_{11} \wedge \alpha_{22} \approx 0$) and thus to large deformations even with small membrane forces. This effect, which is related to stretch-free deformation, essentially occurs only with very small or negative Gauss curvatures.

Usually the decisive loading for dimensioning of air supported halls – wind loading – is only known in the first approximation. The calculation method may also be equally "rough", as long as it includes the essential loadbearing effects. All the more accurate calculation processes, e.g. the finite element methods, are very expensive numerically. However, they really provide more accurate results only when the wind and snow loading as well as the material laws are known from measurements and tests.

In convex-curved, smooth membranes the deformations remain small as long as folding in is prevented by sufficiently great internal pressure. Alternatively large displacements can occur when
– the load is only transferred by the low shear stiffness of the membrane,
– there are Gauss curvatures $K \leqslant 0$,
– no equilibrium is possible in the undeformed state.

Large deformations have to be taken into account only if stability investigations are required or if the membrane is cable reinforced. Under all other conditions a calculation according to the First Order Theory results in membrane forces being too large. In the deformed state the membrane curves more strongly in the area of heavy load, so that the membrane forces decrease in accordance with the "barrel formula" (Fig. 34). The following rule is valid:
A pneumatic structure calculated by the First Order Theory and dimensioned for the maximum occurring membrane forces is sufficiently safe as long as no instability can occur.

6.4.3. Material laws and dimensioning

The material properties of synthetic films and coated fabrics cover a very wide range. They are not fully described by tensile strength, rupture strains and subsequent tensile strength, even

with the seam strengths determined by short term tests. In particular, aging under ultra violet radiation reduces the material strength. Thus after several years use the tensile strength can decrease by 25%, the subsequent tensile strength by 50% and the water vapour permeability (watertightness) by 65%[1]. Furthermore the long term tensile strength of synthetic materials is relatively low (Table 6). These values show already that dimensioning is problematic using the short term tensile strength as prescribed in Bibl. 136. In order to cover the strength losses caused by aging and permanent loading, large safety factors

$$
v_R \leqslant \frac{n_R}{max\ n} \qquad (24)
$$

are necessary (max n = largest calculated membrane force).

In Bibl. 136 $v_R = 5$ is laid down for the membrane and $v_R = 3.5$ for the seam. As the creep strength is often not known, they suggest a "safety" which is not available in older air supported structures.

Equation 24 only applies for working conditions. As in the linear theory the membrane force is bounded by

$$
perm.\ n = n_R/v_R \geqslant max\ n. \qquad (25)
$$

The coefficient v_R states nothing about a possible increase in load. In an envelope with large deformations the non-proportional growth of the membrane forces can only be determined by a limit design investigation. For proof of loadbearing safety

$$
max\ n\ (v_1 \cdot max\ p,\ v_2 \cdot g,\ v_3 \cdot w) \leqslant n_t \qquad (26)
$$

the v_i fold loadings are applied to the system. Through the time-dependent tensile strength n_t the durability t of the structure is brought into the dimensioning. The following safety factors are recommended:

internal pressure $\qquad\qquad v_1 = 1.2$;

deadweight $\qquad\qquad\qquad v_2 = 1.0$;

wind (Bibl. 143)
– warehouses, agricultural structures $\quad v_3 \geqslant 1.2$
– public buildings $\qquad\qquad\qquad v_3 \geqslant 1.5$
– meeting halls $\qquad\qquad\qquad v_3 \geqslant 1.8$
– structures with particularly great stability,
e.g. radar protection domes $\qquad v_3 \geqslant 2 \ldots 2.5$.

The coefficient v_1 is relatively small as the greatest internal pressure $v_1 \cdot max\ p$ achieved by the fan is generally only a little above the maximum working pressure max p.

[1]All values given here were determined in 5 to 10 year old PVC coated Nylon and Trevira fabrics. They are intended to give no more than a survey over the order of magnitude of the anticipated loss of strength under long term tensioning. For dimensioning reference should be made to the material characteristics given by the manufacturer of the relevant fabric and supported by test certificates.

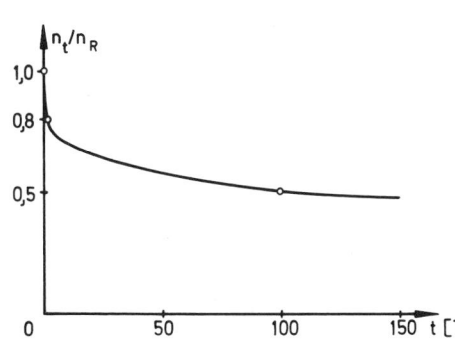

Table 6. Qualitative time-dependent tensile strength for nylon fabrics.

In the case of small strains and equal safety factors ν for all loadings, according to the model law (Bibl. 114) the dimensioning formulae 24 and 26 are equivalent

$$\nu \cdot n (p, g, w) = n (\nu \cdot p, \nu \cdot g, \nu \cdot w). \quad (27)$$

Apart from stability cases, for large deformations and for $\nu_1 < \nu_3$, dimensioning by the limit design theory is more favourable. In buildings for short term use, e.g. exhibition halls, the increased short term tensile strength leads to an especially economic dimensioning.

The material law of the membrane is required for all deformation calculations. In approximation films act isotropically and coated fabrics rectilinearly anisotropically. The material constants can be determined for films only by uniaxial tensile tests. For coated fabrics extensive tests (in accordance with Bibl. 100) under two-dimensional tensioning are required, in order to include the link between the warp and weft direction as well as the non-linear effects related to the material and the fabric structure. The material laws, determined by tests, are very complex as stated in Bibl. 100.

Characteristic is the very weak coupling between the shear strain and the tensile strains in the thread direction as well as the low shear stiffness

$$S \approx 0.05 \ldots 0.1 \cdot D_k \quad (28)$$

compared with the tensile stiffness D_k in the warp direction.

For practical use complicated non-linear material laws are less suited. In the operating condition the errors made by their linearisation are relatively small. Inaccuracies in manufacture, distortions in the cutting pattern, increased seam stiffnesses, as well as wrinkles which only pull smooth by higher internal pressure, often change the stress distribution considerably more. These influences, together with the aging of the membrane, are difficult to determine. Tests have to be made on whether or not very simple material laws, such as for example

$$\begin{bmatrix} \varepsilon_{kk} \\ \varepsilon_{ks} \\ \varepsilon_{ss} \end{bmatrix} = \begin{bmatrix} \dfrac{1}{D_k} & o & -\dfrac{1}{D_k+D_s} \\ o & \dfrac{1}{15\,D_k} & o \\ -\dfrac{1}{D_k+D_s} & o & \dfrac{1}{D_s} \end{bmatrix} \cdot \begin{bmatrix} n_{kk} \\ n_{ks} \\ n_{ss} \end{bmatrix} \quad (29)$$

give sufficiently accurate results. In equation 29 the index k refers to the warp direction and s refers to the weft direction; D is an average secant stiffness taken from the n-ε-diagrams (Table 7). Simple material laws also covering oblique angled fabric directions, creep or plasticising are given in Bibl. 117.

6.4.4. Rotationally symmetric membranes

In convexly-curved membranes (Gauss curvature $K \geqslant 0$) the deformations are small, so that the linear equations 22 and 23 give sufficiently accurate results. If one eliminates n_{11} in equation 22/3, then according to Bibl. 52 the substitution

$$N_{22} = \frac{r^2}{\sqrt{a}} \cdot n_{22}, \quad N_{12} = r^2 n_{12} \quad (30)$$

gives the equations

$$N_{12,z} + r\ r_{,zz} N_{22,\varphi} = -r\sqrt{a}\,p_1 - a\,p_{3,\varphi}, \quad (31)$$
$$N_{12,\varphi} + r^2 N_{22,z} = -r^3 p_2 + r^3 r_{,z}\ p_3.$$

Similarly the elimination of V_3 in equation 23/1 and the substitution

$$V_1 = \frac{v_1}{r}, \quad V_2 = \frac{\sqrt{a}}{r} v_2 \quad (32)$$

leads to the set of equations

$$V_{2,z} + r\ r_{,zz} V_{1,\varphi} = r\ r_{,zz}\,\alpha_{11} + \alpha_{22}, \quad (33)$$
$$V_{2,\varphi} + r^2 V_{1,z} = \sqrt{a}\,\alpha_{12}.$$

The homogenous equations 33 for the strains have the same form as the equilibrium conditions 31. They are solved by the same scheme. For rotationally symmetric loadings the derivatives $(\ )_{,\varphi}$ disappear so that equations 31 are uncoupled. Integration and resubstitution gives

$$n_{22} = \frac{\sqrt{a}}{r^2} \int_{z_s}^{z} (-r \cdot p_2 + r \cdot r_{,z}\,p_3)\,dz + C_1,$$

$$n_{12} = -\frac{1}{r^2} \int_{z_s}^{z} r\sqrt{a}\,p_1\,dz + C_2, \quad (34)$$

$$n_{11} = \sqrt{a}\,p_3 + \frac{r^3 \cdot r_{,zz}}{a}\ n_{22}.$$

For distributed loadings the membrane forces in the zenith remain finite. Equilibrium at the cut-out zenith element provides the integration constants (Fig. 15)

$$C_1 = \sqrt{a} \cdot p_3/2, \quad C_2 = 0. \quad (35)$$

The given surface loads are dispersed in the directions of the moving trihedral $\mathbf{a}_i$.

Internal pressure p:

$$p_1 = p_2 = 0, \quad p_3 = p, \quad (36)$$

Deadweight g:

$$p_1 = 0, \quad p_2 = -r \cdot g/\sqrt{a},$$
$$p_3 = -r \cdot r_{,z} \cdot g/\sqrt{a}. \quad (37)$$

Snow s_0, referring to the basic area x − y:

$$p_1 = 0, \quad p_2 = -r^2 \cdot r_{,z} \cdot s_0/a,$$
$$p_3 = -(r_{,z} \cdot r)^2 \cdot s_0/a. \quad (38)$$

The membrane forces due to internal pressure are given for some membrane forms in Table 8. The evaluation of equations 34 for ellipsoids is shown in Table 18 for internal pressure and in Table 19 for snow load in accordance with Bibl. 136.

In the case of unsymmetrical loads the membrane forces and loadings are developed in Fourier's series:

$$p_1, N_{12} = \sum_{m=0}^{\infty} \binom{m}{} \cdot \sin(m\varphi),$$
$$\quad (39)$$
$$p_2, p_3, N_{11}, N_{22} = \sum_{m=0}^{\infty} \binom{m}{} \cdot \cos(m\varphi).$$

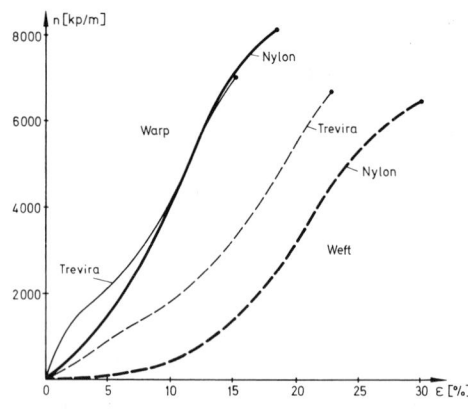

Table 7. n-ε-diagrams for coated fabrics.

15. Equilibrium on the zenith element.

The elimination of the shear force in equations 31 leads to the normal differential equation

$$\overset{m}{N}_{22,zz} + \frac{2\,r_{,z}}{r}\,\overset{m}{N}_{22,z} + \frac{r_{,zz} \cdot m^2}{r}\,\overset{m}{N}_{22} = \overset{m}{F} \quad (40)$$

where

$$\overset{m}{F} = \frac{\sqrt{a}\,m}{r}\,\overset{m}{p}_1 - \frac{a\,m^2}{r^2}\,\overset{m}{p}_3 + \frac{1}{r^2}\left[r^3 \left(-\overset{m}{p}_2 + r_{,z}\,\overset{m}{p}_3 \right) \right]_{,z}.$$
$$\quad (41)$$

The equation leads for the ellipsoid (Bibl. 52)

$$r = \sqrt{R^2 - \alpha \cdot z^2},$$
$$\alpha = R^2/H^2, \quad (42)$$

(see Table 8) with the substitution

$$\omega = \text{artanh}\left(\sqrt{\alpha} \cdot z/R\right) \quad (43)$$

to a differential equation with constant coefficients

$$\overset{m}{N}_{22,\omega\omega} - m^2\,\overset{m}{N}_{22} = \frac{r^4}{R^2\,\alpha}\,\overset{m}{F}. \quad (44)$$

The solution of the differential equation and the resubstitution give for the wind load

a) m = 0:

$$\overset{0}{n}_{22} = -\frac{\sqrt{a}\,\alpha}{r^2} \int_{z_s}^{z} z\,\overset{0}{w}\,dz, \quad (45)$$

b) m = 1:

$$\overset{1}{n}_{22} = \frac{\sqrt{a}}{r^3}\left[-z\int_{z_s}^{z} r\,\overset{1}{w}\,dz + (1-\alpha)\int_{z_s}^{z} r \cdot z \cdot \overset{1}{w}\,dz \right], \quad (46)$$

c) m ≥ 2:

$$\overset{m}{n}_{22} = \frac{\sqrt{a}}{2\,m\sqrt{\alpha}\,R\,r^2}\left[J_1\,e^{-m\omega} + J_2 \cdot e^{m\omega} \right] \quad (47)$$

where

$$J_1 = \int_{z_s}^{\overset{m}{z}} \overset{m}{G}\cdot e^{m\omega}\,dz,$$

$$J_2 = \int_{z_u}^{\overset{m}{z}} \overset{m}{G}\cdot e^{-m\omega}\,dz,$$

$$\overset{m}{G} = (-a\,m^2 - \alpha\,r^2 + 2\,\alpha^2\,z^2)\,\overset{m}{w} - \alpha\,z\,r^2\,\overset{m}{w}_{,z}, \quad (48)$$

$$e^{m\omega} = \left(\frac{R+\sqrt{\alpha}\,z}{R-\sqrt{\alpha}\,z}\right)^{m/2}.$$

Notation		General	Sphere	Cone	Cylinder	Paraboloid	Ellipsoid	Torus
Radius	$r=$	$r(z)$	$\sqrt{R^2-z^2}$	$\mathrm{tg}\,\alpha\cdot z$	R	$\sqrt{2c\cdot z}$ $(c=A^2/2H)$	$\sqrt{R^2-\alpha z^2}$ $(\alpha=R^2/H^2)$	$R+A\cdot\sin\alpha$
Meridian force	$n_z=$	$p\cdot\sqrt{a}\,/2$	$p\cdot R/2$	$p\cdot r/(2\cos\alpha)$	$pR/2$	$p\sqrt{r^2+c^2}/2$	$p\sqrt{r^2+\alpha z^2}/2$	$pA(r+R)/(2r)$
Ring force	$n_\varphi=$	$n_z(2+b_2^2\sqrt{a})$	n_z	$2n_z$	$2n_z$	$n_z\dfrac{2r^2+c^2}{r^2+c^2}$	$n_z\left(2-\dfrac{\alpha R^2}{r^2+\alpha^2 z^2}\right)$	$pA/2$
min $n_\varphi\ge$ / V·max $n_z\le$		$\left(2-V\sqrt{\dfrac{\max a'}{a}}\right)b_2'$ $\le b_2^2$	—	$z\ge V\cdot H$	—	—	$R\le H\sqrt{2-V}$	—

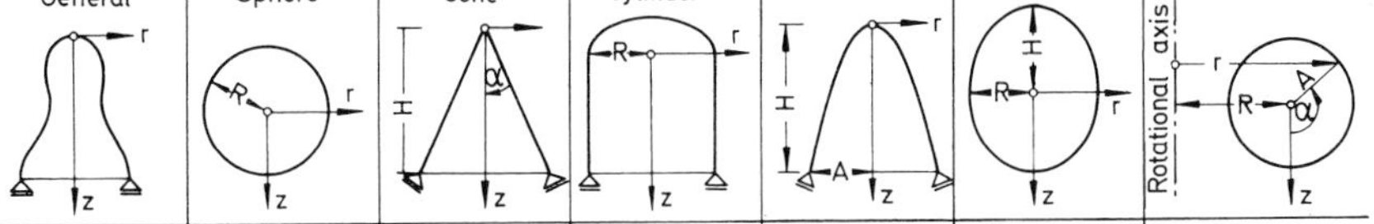

(vertical label between Ellipsoid and Torus columns: "Rotational axis")

Table 8. Membrane forces under internal pressure p, see Bibl. 85.

The integration constants are determined for m = 0,1 by the equilibrium conditions $\Sigma V = 0$, $\Sigma M = 0$ in the zenith (Fig. 15). For the higher series elements $m \ge 2$ residual stress conditions exist, which affect only the form of the membrane force distribution. The integration constants are so adapted that the solution also remains finite in the vertices z_s, z_u. The remaining membrane forces follow from

$$\overset{m}{n}_{12} = \left(-\alpha\cdot z\cdot\overset{m}{p}_3 - \left(\frac{r^2}{\sqrt{a}}\overset{m}{n}_{22}\right)_{,2}\right)/m \quad (m>0),$$

$$\overset{m}{n}_{11} = -\frac{\alpha R^2}{a}\overset{m}{n}_{22} + \sqrt{a}\cdot\overset{m}{w} \quad (\text{all } m). \quad (49)$$

For the ellipsoid the wind loading given for the cylinder in Bibl. 136 is slightly modified (Fig. 16). The integration of the equations 45 to 47 and the superimposition of the series elements are numerically very expensive. The extreme values of the membrane forces which are decisive for dimensioning and for proof of folding safety are given in Tables 20 and 21 for the ellipsoid under constant wind pressure and in Tables 22 to 24 for the sphere under gradually changing wind pressure ($m \le 8$). Numerical results for the spherical membrane under wind loading (Bibl. 158) are also given in Bibl. 62.

The membrane forces for the first two Fourier elements can also be determined by the equilibrium at the dome when cut off at height z (Fig. 15). For the wind load the following equations are obtained for any membrane form:

$$\Sigma V = 0:\ \overset{0}{n}_{22} = -\frac{\sqrt{a_{22}}}{r}\int_{z_s}^{z}\overset{0}{w}\cdot r\cdot r_{,\xi}\,d\zeta,$$

$$\Sigma H = 0:\ \overset{1}{n}_{12} = -\frac{2\sqrt{a_{22}}}{r}\int_{z_s}^{z}\overset{1}{w}\,r\,d\zeta, \quad (50)$$

$$\Sigma M = 0:\ \overset{1}{n}_{22} = -\frac{2\sqrt{a_{22}}}{r^2}\int_{z_s}^{z}\overset{1}{w}\left(r^2 r_{,\xi}+r(\zeta-z)\right)d\zeta$$

where

$$\overset{i}{w} = \frac{1}{2\pi}\int_0^{2\pi} w\cdot\cos(i\varphi)\,d\varphi \quad \text{for } i=0,1. \quad (51)$$

The ring force results from the equilibrium condition 22/3

$$\overset{i}{n}_{11} = \sqrt{a}\,(\overset{i}{w}+b_2^2\cdot\overset{i}{n}_{22}) \quad \text{for } i=0,1. \quad (52)$$

6.4.5. Cylindrical membranes

Cylindrical membranes take up in the ring direction only axially symmetrical loading (barrel formula). All unsymmetrical parts of the load are transferred to the stiff end calottes by shear and longitudinal forces. The low shear stiffness of the membranes leads to large deformations without folds arising in the cylinder. With very short cylinders the First Order Theory provides a good approximation for the membrane forces. For longer cylinders the membrane form changes approximately into the support line for the loading. Both idealisations are further investigated here.

6.4.5.1. Undeformed membranes

For the cylinder the curvature b_2^2 and the Christoffel symbol Γ_{12}^1 disappear, so that the equilibrium conditions 22 are uncoupled and can be directly integrated

$$n_{11} = p_3\cdot R,$$

$$n_{12} = -\int_0^z (p_1+p_{3,\varphi})\,dz + C_1(\varphi), \quad (53)$$

$$n_{22} = -\int_0^z \left(\frac{n_{12,\varphi}}{R}+p_2\right)dz + C_2(\varphi).$$

For symmetrical loadings in the longitudinal direction the constants $C_1(\varphi)$ disappear. The

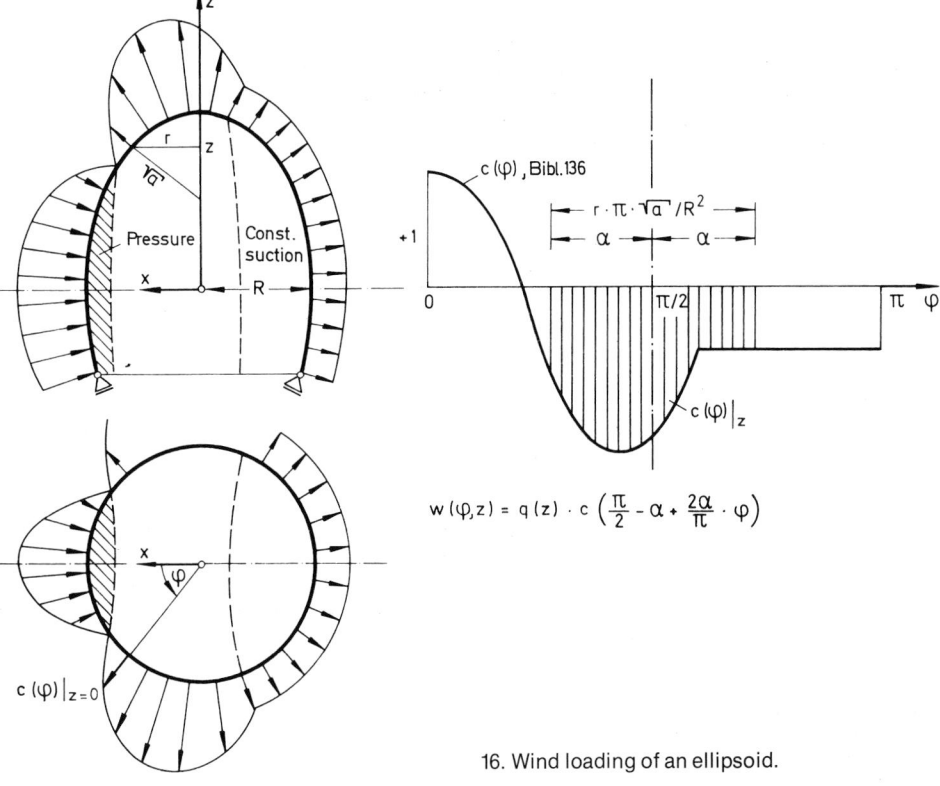

$$w(\varphi,z) = q(z)\cdot c\left(\frac{\pi}{2}-\alpha+\frac{2\alpha}{\pi}\cdot\varphi\right)$$

16. Wind loading of an ellipsoid.

171

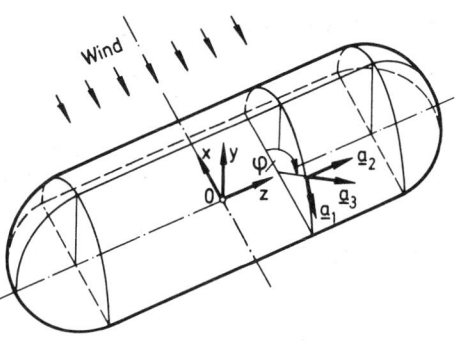

17. Cylinder membrane.

19. Support line – see F. Rudolf (Bibl. 147).

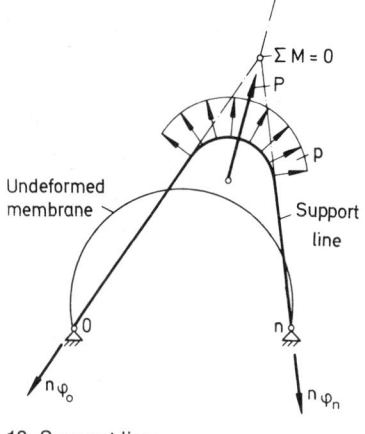

18. Support line.

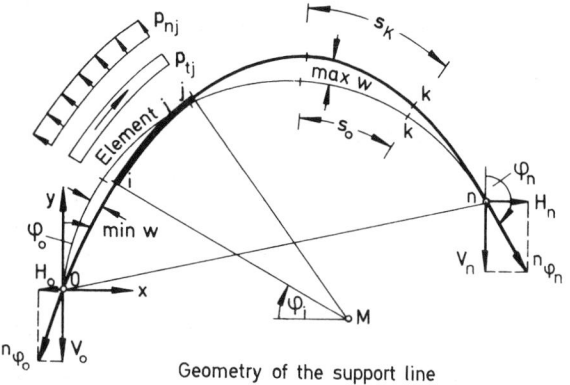

Geometry of the support line

Deformed element

shear forces of the cylinder must be taken up by the end calottes. $C_2(\varphi)$ are the normal forces transferred from the calottes to the cylinder. They are transmitted through the cylinder membrane and are first taken up by the calotte opposite. A detailed calculation of the membrane forces for spherical end calottes is contained in Bibl. 119. Similarly simple formulae for cylinders over rectangular ground areas are not known.

The maximum membrane forces for the dimensioning can be determined without complex calculations. For the load condition internal pressure p and wind w it follows from equation 53/1 that

$$\max n_{11} = (\max w + p) \cdot R. \tag{54}$$

In the longitudinal direction the membrane force is approximately given by

$$\max n_{22} \approx \max n_{11} - p \cdot R/2. \tag{55}$$

These membrane forces, which are calculated by the linear theory, are too large. They are in part considerably reduced by the related changes in curvature.

6.4.5.2. Cylinder as a line of pressure

If one disregards the load distribution in the longitudinal direction, the cylinder can deform without tension. Equilibrium is only possible for

the line of pressure of the loading (Fig. 18). The non-linear differential equation of the line of pressure and its numerical solution is shown in Bibl. 142. F. Rudolf's approximation method (Bibl. 147; Fig. 19) is considerably clearer and simpler. In this the membrane, idealised as a cable, is divided into small sections and constantly loaded element by element.

For the known initial values $n_{\varphi i}$, φ_i, x_i, y_i the variables at the end of the element can be calculated in accordance with Fig. 19:

$$n_{\varphi j} = n_{\varphi i} + p_t \cdot s_o,$$

$$x_j = x_i + 2 R_M \cdot \sin (d \varphi) \cdot \sin (\varphi_i + d \varphi),$$

$$y_j = y_i + 2 R_M \cdot \sin (d \varphi) \cdot \cos (\varphi_i + d \varphi),$$

$$\varphi_j = \varphi_i + 2 \cdot d \varphi \tag{56}$$

where

$$R_M \approx \frac{n_{\varphi i} + n_{\varphi j}}{2 p_n},$$

$$d\varphi = \frac{s_o}{2} \left(\frac{1}{R_M} + \frac{p_n}{D} \right) \tag{57}$$

Starting from one bearing point all the intermediate membrane points are calculated one after another for the prescribed initial values $n_{\varphi o}$, φ_o.

The initial values are then improved iteratively until the calculated end of the cable and the bearing point coincide. At each iteration cycle deadweight and snow loadings must be re-

divided into the components which are tangential and normal to the element.

Tests (Bibl. 14; Bibl. 114) show that even with relatively short cylinders the line of pressure appears as an approximation. Therefore only the smaller membrane forces of the line of pressure are used for dimensioning. For wind and snow load the largest membrane and bearing forces and the largest displacements can be taken from Tables 13 to 17.

6.4.6. Membranes guyed by cables

The cable guys of a pneumatic structure generally only take up the deflection forces from the kink. Variable tangential displacements are scarcely affected by the very pliable connection of the cable with the membrane (Fig. 26). If one disregards even the minimal deadweight, the cable force is

$$S = \text{constant}. \tag{58}$$

The curvatures $\varkappa$ (s) and the torsion ϱ (s) clearly define the cable line. Both quantities can be determined by the membrane geometry (compatibility) and the equilibrium conditions.

6.4.6.1. Compatibility

For a given membrane geometry the cable curve $\mathbf{r}$ (s) is also known (Fig. 20). From the tangential vector of the cable

$$\mathbf{e}_1 = \mathbf{r}_{,s} = \frac{1}{\cos \varphi} \cdot \overset{R}{\mathbf{a}}_3 \times \overset{L}{\mathbf{a}}_3 \tag{59}$$

the curvature $\varkappa$ and torsion ϱ can be determined by means of the Frenet formula

$$\frac{d}{ds} \begin{bmatrix} \mathbf{e}_1 \\ \mathbf{e}_2 \\ \mathbf{e}_3 \end{bmatrix} = \begin{bmatrix} 0 & \varkappa & 0 \\ -\varkappa & 0 & \varrho \\ 0 & -\varrho & 0 \end{bmatrix} \cdot \begin{bmatrix} \mathbf{e}_1 \\ \mathbf{e}_2 \\ \mathbf{e}_3 \end{bmatrix} \tag{60}$$

6.4.6.2. Equilibrium

The deflection forces of the cable must be taken up by the membrane (Fig. 21). The condition leads to the vector equation

$$\overset{L}{n_u} \cdot \overset{L}{\mathbf{u}} + \overset{R}{n_u} \cdot \overset{R}{\mathbf{u}} = \varkappa \cdot S \cdot \mathbf{e}_2. \tag{61}$$

In the longitudinal direction of the cable the equilibrium for S = constant is automatically fulfilled; for the normal direction $\mathbf{e}_2$ and binormal direction $\mathbf{e}_3$, the equilibrium conditions

$$\overset{L}{n_u} \cdot \cos \varphi + \overset{R}{n_u} \cos \varphi = \varkappa \cdot S,$$

$$\overset{L}{n_u} \cdot \sin \varphi - \overset{R}{n_u} \cdot \sin \varphi = 0. \tag{62}$$

are obtained.

The derivation of equation 61 with respect to the arc length s and taking into consideration the Frenet Formula (60) lead to the torsion

$$\varrho = \frac{1}{\varkappa S} \left[\overset{L}{n_{u,s}} \cos \varphi - \overset{R}{n_{u,s}} \cos \varphi \right]. \tag{63}$$

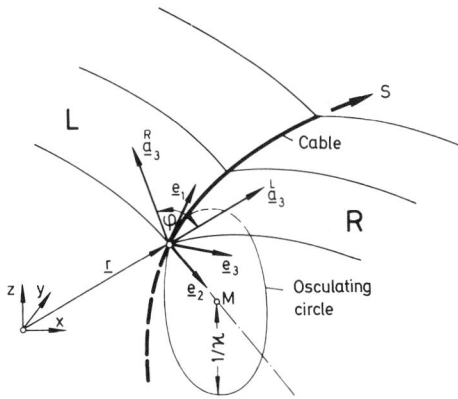

20. Geometry of the cable line.

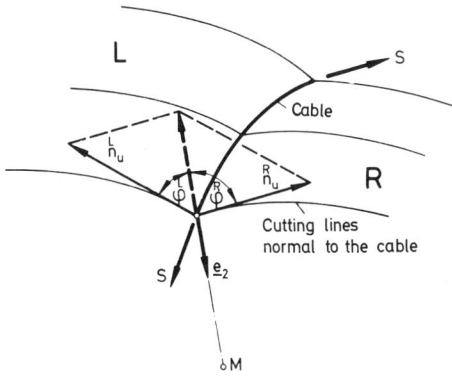

21. Equilibrium on the cable.

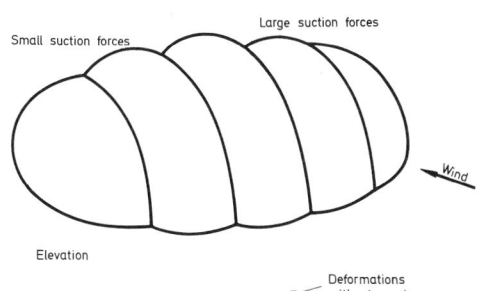

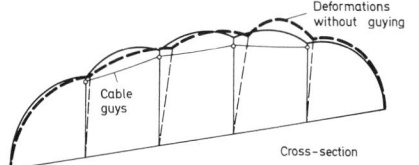

22. Membrane formed with torus zones.

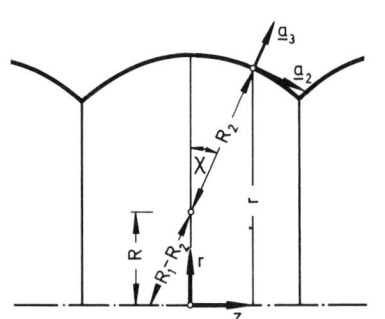

23. Torus geometry.

The actual cable line must fulfil the equilibrium and compatibility conditions. In the undeformed state that is only possible for a definite load condition – usually deadweight and internal pressure – and a geometry selected specifically for it. In all other load conditions large displacements occur, until the cable line which results from the changed geometry can take up the deflection forces from the membrane. Membranes constricted by individual cables must always be calculated by the Second Order Theory. Very large deformations easily lead to "flutter" under wind loading, so that special structural measures for stabilisation are usually necessary.

The membrane formed from torus zones (Fig. 22) fulfils all equilibrium and compatibility conditions in the undeformed state under internal pressure. Unequally distributed wind suction forces in envelopes exposed to cross winds lead to the large deformations illustrated. They are usually prevented by special cable guys.

The membrane forces under internal pressure p are calculated with the geometry quantities (Fig. 23)

$$\sqrt{a_{22}} = 1/\cos\chi, \qquad \sqrt{a} = r/\cos\chi = R_1,$$
$$b_1^1 = -1/R_1, \qquad b_2^2 = -1/R_2 \qquad (64)$$

according to Table 8, Column 1

$$n_{22} = p \cdot r/(2 \cdot \cos\chi), \qquad (65)$$
$$n_{11} = n_{22} \left[2 - \frac{r}{R_2 \cos\chi} \right].$$

In order to get the membrane forces max $n_{11} \geq \nu \cdot$ max n_{22} in the ring direction of the envelope, the folding condition

$$R_2(1 - \nu) \geq R \qquad (66)$$

must be fulfilled.

6.5. Structural indication

6.5.1. Introduction of forces and cable reinforcements

Loops are sewn on to the membrane for anchorage. Steel pipes or cables are pushed through which take up the membrane forces (Bibl. 36). The anchorage using galvanized metal tubes, which can be pushed into each other with 1 to 2 mm clearance, is particularly simple. At the junction the tubes are telescoped and secured against undesired displacements with a small bolt (Fig. 24).

In the rounded-off corner area first the short pipe bend and then the connecting straight anchorage tubes must be threaded and joined. A special anchorage of the pipe bend is not necessary for radii of $R \leq 1.0$ m as the membrane forces are small. Reinforcement of the anchorage in the connection area of the cables and replacements is made possible without difficulty by telescoping one tube into the other. In dimensioning, according to the limit design theory, the bearing capacity of both tubes is fully exploited (Bibl. 42). Galvanized water pipes are only suitable for anchorage when they achieve the minimum material strength. In Federal Germany this is not laid down in DIN 2 440. Moreover the threaded sleeve joint is a very expensive pipe connection which weakens the cross-section. In substructures made of concrete the clamp anchorage is also usual (Fig. 25).

Larger air supported structures are reinforced by individual support cables. The transition to cable net structures, where the membranes only span the individual fields, is fluid. For protection against weather the cables are drawn in on the inside of the envelope, with which they are connected tension-proof (Fig. 26). The mount-

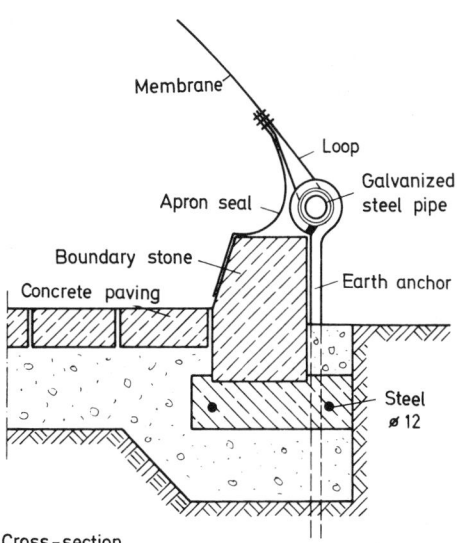

Cross-section

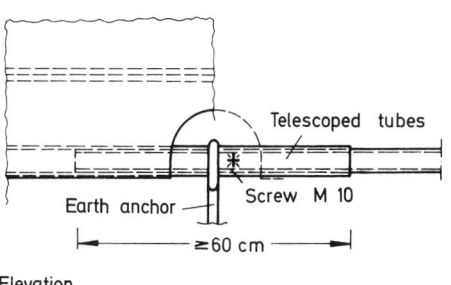

Elevation

24. Loop anchorage with telescope joint.

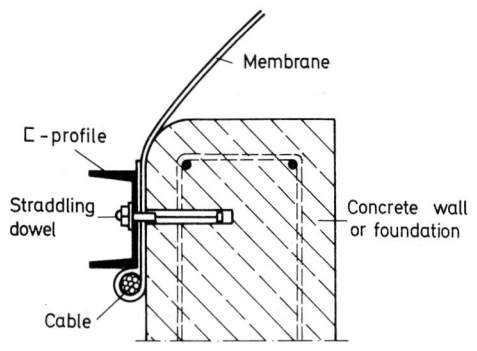

25. Clamp anchorage – see Bibl. 36.

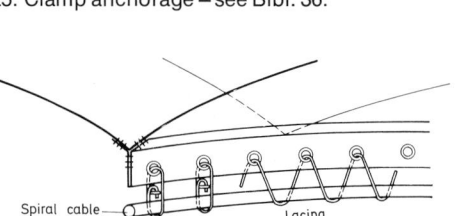

26. Cable/membrane connection.

Type given in Fig. 28	D mm	F_N cm²	G kg/m	S_{Br} [1] Mp	E [2] Mp/cm²
A	8,5	4,31	0,35	6,25	
	9	4,95	0,40	7,18	
	10	6,36	0,52	9,22	
B	11	7,22	0,61	10,45	
	12	8,59	0,72	12,45	≈ 1800
	13	101	0,84	14,65	
	14	117	0,98	16,95	
	16	153	1,28	20,15	
	18	193	1,62	28,05	
	20	239	2,00	34,60	

[1] Tensile strength of the single wire $\sigma_{Br} = 14,5$ Mp/cm²
[2] after stretching

Table 9. Spiral cables – see Bibl. 59.

Type of anchorage	η
Hook or grommet with clamps	0,80
Hook or grommet (spliced)	0,95
Drawing casing, compression socket	0,90
Sealing attachment	1,00

Table 10. η-values for cable anchorages ($\emptyset \leqslant 20$ mm).

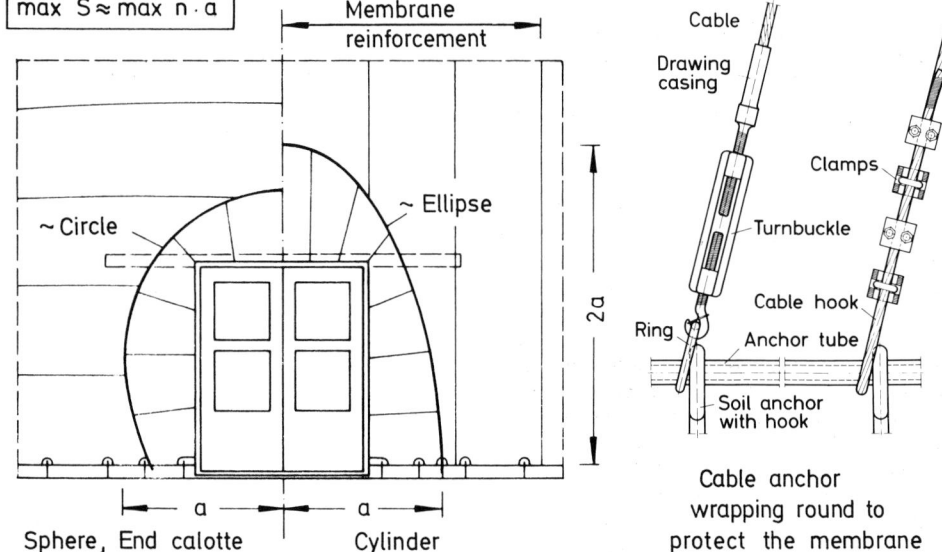

$$\boxed{\max S \approx \max n \cdot a}$$

Membrane reinforcement

~ Circle

~ Ellipse

2a

Sphere, End calotte   Cylinder

a   a

Cable
Drawing casing
Clamps
Turnbuckle
Cable hook
Ring
Anchor tube
Soil anchor with hook

Cable anchor wrapping round to protect the membrane

27. Replacements for spherical and cylinder membranes.
28. Cable forms.

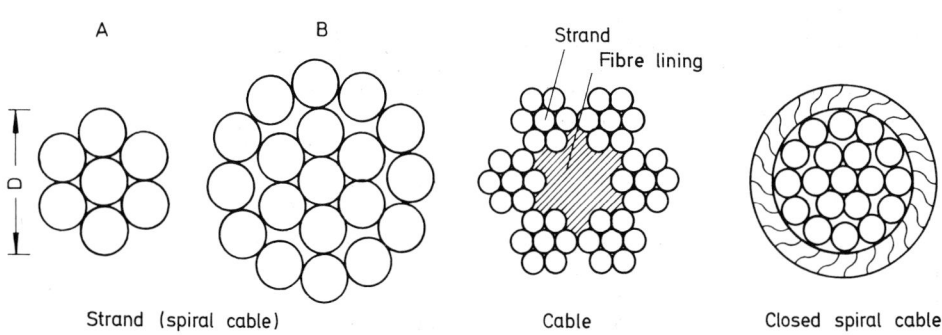

A   B   Strand   Fibre lining

D

Strand (spiral cable)   Cable   Closed spiral cable

ing of the cables in sewn-on loops is very time consuming and only practicable for boundary cables and short gate replacements. When tensions are minimal the cables are also replaced by sewn-on strips of material.

In the case of door and gate replacements boundary cables take up the membrane forces. Thereby additional membrane deformations occur, because parts of the internal pressure will be carried by the gate construction. The geometry of the boundary cable depends only on the membrane forces introduced, whose direction is influenced by the additional deformations. In order to keep deviation from the prescribed membrane form low, the boundary cable for the load condition internal pressure is adapted in approximation to the line of support. Under other load conditions increased membrane forces sometimes occur in the replacement area, as the boundary cable can only take up the forces after large deformations and force rearrangements in the membrane. For cylindrical and spherical membranes the form of the cable line and also an approximation of the largest cable forces are given in Fig. 27.

The support cables usually consist of galvanized steel wires, less frequently of synthetic or organic fibres. The overwhelming choice of steel cables is due to the large E module, the insensibility to environmental influences and the high heat resistance. For pneumatic structures and cable nets spiral strands are mainly used (Fig. 28).

Cables in which several strands are whipped around a fibre centre have a low E module and are more difficult to anchor. They are specially suited to running over a roller. A detailed summary of the different types of strands and cables, their end anchorage and breaking strength is contained in Felten & Guilleaume's catalogue (Bibl. 59). Table 9 is taken from this: from a diameter of 15 mm locked spiral cables are also produced. The cable strength S_{Br} is not the criterion for the dimensioning of the cable, but the, in part, considerably lower strength $V_{Br} = \eta \cdot S_{Br}$ of the end anchorage. The reducing coefficient η is dependent on the type of anchorage and on the cable diameter. With thin cables and a well executed anchorage the η-values given in Table 10 are usually exceeded (Bibl. 13).

6.5.2. Anchorages

Air supported structures can be produced very economically and they have a life of 8 to 15 years. Extensions, alterations or removal to a new site are relatively easy. The membrane anchorage should be just as adaptable and economic. So as not to prejudice later use of the site, the withdrawal or dismantling of the anchorage should not involve any great expense. The membrane forces in pneumatic structures are usually relatively small at 1 to 2 Mp/m. In a direct anchorage of the membrane many anchor points with low strengths are required. They are developed as gravity or earth anchors.

Some gravity anchors are shown in Fig. 29. The anchor forces depend basically on the weight of the anchor only and they disperse very little. For dimensioning of the anchor body two limiting cases have to be investigated (Fig. 30):

a) The horizontal components of the membrane force are taken up by the ground or by an inserted concrete floor. The anchor weight compensates only the vertical component ($N_{Br.} = n_R \cdot$ anchor interval)

$$G \geq V_{Br.} \qquad (67)$$

b) The anchor body turns round and begins to glide unfavourably in the direction of the membrane force. The anchor weight must exceed the value

$$G \geq \frac{N_{Br.}^2}{V_{Br.} + \varrho \cdot H_{Br.}} \qquad (68)$$

(ϱ = sliding friction between anchor and floor). Thereby large displacements occur, which do not affect the loadbearing behaviour of the membrane. They are only of significance with respect to the structural development of the anchor point and the components.

Tests must be made to determine whether the safety factor $v = 1.5$ cannot be reduced in the case of well defined anchorage bodies. The peak values of the membrane forces originate from briefly occurring wind loadings, whose effect is broken down by the mass of the anchor body and by the damping in the ground.

In contrast to gravity anchors, soil anchors are generally easier to place and to remove. Fig. 31 shows a screw anchor and a spread anchor. The screw anchor is especially suitable for lighter soils, while the spread anchor can also be rammed into heavy soils. The lamella of the ground anchor can snap together in highly consistent soils with very great shear strength without a body of earth shearing off. The instability of the lamella is connected with a sudden loss of loadbearing capacity and is prevented by pressing out of the spread anchor foot with concrete.

Under diagonal tension the anchor shaft bends without reducing the bearing capacity significantly. An inclined approach of the anchor or a head plate made of concrete, with which several anchors can be fastened together, prevents the shaft bending.

The loadbearing capacity of the earth anchor depends on the soil physics characteristics – in particular the shear strength – which determine the size of the sheared off earth body. In the area near to the surface drying out, soaking

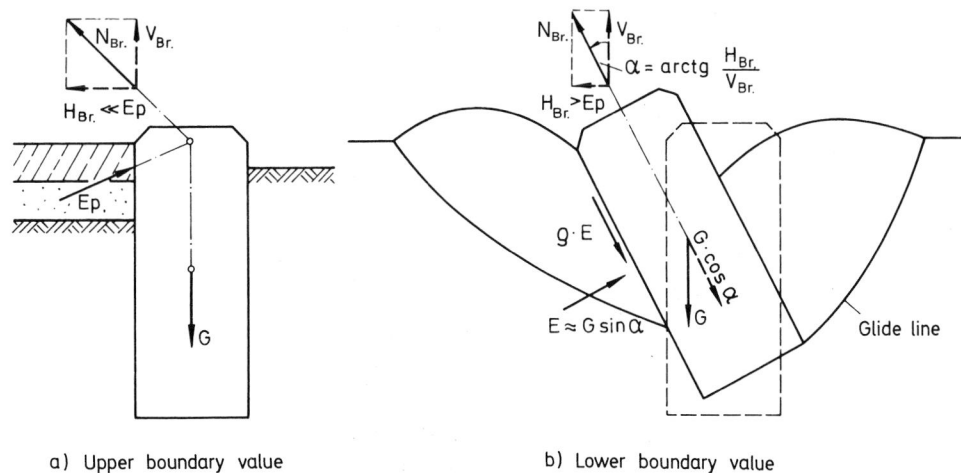

a) Upper boundary value   b) Lower boundary value

30. Limit values of the anchor forces.

or freezing changes the strength properties of the ground. In order to keep the seasonal variations in anchorage capacity as small as possible, the foot of the anchor should lie at least 70 to 100 cm under this zone. Thus one obtains minimum anchor depths of some

$$\min t \approx 1.2 \ldots 1.5 \, (m) \qquad (69)$$

in Federal Germany. In the case of stratified soils the loadbearing capacity of the anchor is dependent mainly on the shear strength of the earth layer directly above the anchor foot. The equation

$$S_{Br.} \approx d \cdot (2.25 \cdot t - 1) \cdot S_o$$

where $1.0 \leq t \leq 2.0 \, [m], \quad 0.1 \leq d \leq 0.35 \, [m] \quad (70)$

gives a cautious indication for the loadbearing capacity of the earth anchor.

Equation 70 only applies for the given anchor depths t and foot diameter d (Fig. 31); the factor S_o can be taken from Table 11. The measured anchor forces disperse over a wide range and in part exceed considerably the values given in equation 70. In the low anchor depths sudden changes in the ground characteristics cannot be excluded. It is therefore suggested that every 5th to 10th anchor in an air supported hall be subjected to a tensile test. In order not to destroy the anchors by the test, only 1.25 times of the working load should be applied.

Larger anchor forces, such as arise in the cable guys of air supported structures and in cable net constructions, are taken up by groups of anchors, reinforced concrete anchor piles or larger gravity anchors. The anchorage costs increase quickly with the size of the force to be anchored, so that large cable forces should be avoided if possible in air supported structures.

6.5.3. Checking of the construction

Air supported halls and their operating installations must be regularly controlled and maintained. A thorough check should be made especially before a large storm gets up. Defects

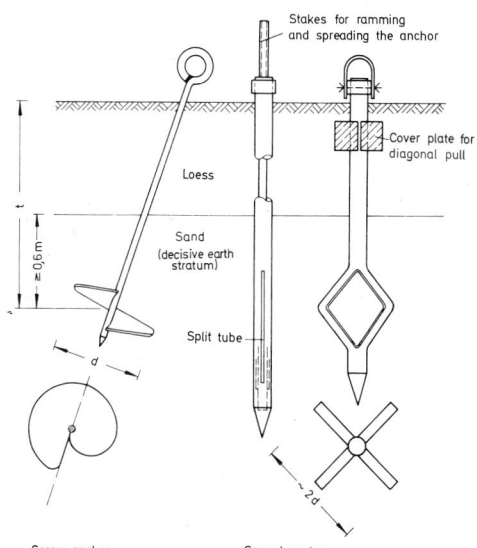

31. Soil anchors.

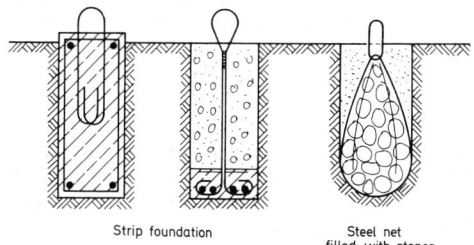

Strip foundation   Steel net filled with stones

29. Gravity anchors.

Type of soil	S_o [Mp/m²]
Rich, well drained clays, mixed with rubbles and stones	8,9
Well graduated gravel-sand mixture, one grain gravel	7,5
Firmly supported rough grained gravel-sand	6,1
Semi-firm, cohesive soils, loam, marl, loess loam	4,7
Filled in, non-compressed soil	3,7

Table 11. Soil coefficients S_o.

No.	Check	References	Date checked		
1	the envelope, by sector, for tears, holes or overstraining				
2	for excessive air leakage around the base				
3	that the envelope is free of tears at the connections with cables, door constructions or components				
4	every old patch for leaks				
5	the proper distance between the envelope and the components or the storage of material	min a =			
6	the firmness of the anchorage, the adjustments and the rust protection				
7	for changed ground conditions (excavations, foundation-trenches)				
8	the ground anchorage of the components				
9	the door linkages and controls for proper operation (doors should not remain open by mistake)				
10	the proper operation of the pressure gauge or manometer				
11	the proper on-off control of the blowers by internal pressure	min p = max p =			
12	the regular servicing of the blowers				
13	the cleanness of the blower intake, the blower and the flexible connection with the envelope – leaves, paper, snow, etc., must be removed				
14	the presence of the warning and direction plates, the snow rope and the repair kit for small leaks				

Table 12. Check list for maintenance – see Bibl. 36.

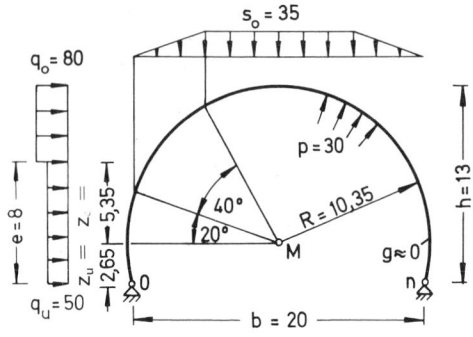

32. Loading scheme. Wind loading – see Bibl. 136.

and deficiencies which are revealed should be repaired immediately. Small tears in the envelope can otherwise quickly tear further and lead to destruction of the whole structure. During a storm the largest permissible internal pressure should be maintained in order to prevent the membrane folding in and the considerably greater wind loadings which follow. A typical check list for maintenance is shown in Table 12, which is taken, somewhat abbreviated, from Bibl. 36.

Alongside this maintenance the operating safety and stability of the air supported structure should be controlled at larger intervals (every 2 years say). Only when the whole structure comes through the next maintenance period with adequate safety will the operating permit be extended. Here a special difficulty of such an inspection becomes clear: a rusted anchor or brittle concrete is easy to recognise, but the residual tensile strength of the aged membrane and seams is not. The owner generally only agrees to sample tests in exceptional cases, as the patches in the membrane stand out very clearly.

A membrane test which satisfies the safety regulations and does not destroy the membrane is offered by "squeezing". In this method of testing, which is taken from pipe construction, the internal pressure on a still day is increased until the membrane force reaches the value

$$\max n \ (p_{test}) \approx 1.1 \cdot \max n \ (p, w). \qquad (71)$$

Factor 1.1 approximately takes in the loss of strength of the membrane in the next maintenance period and the dynamic effect of the wind. The fixed fan installations usually do not attain the test pressure p_{test}. Therefore a specially strong (portable) fan must be connected for the test, which can be borrowed from

the technical inspection authority or from the manufacturer. Only the additional connection nozzles must already be provided on the membrane. Weak points on the membrane are discovered immediately through this "squeezing". Prompt repair avoids greater damage in the next storm.

6.6. Diagrams for membrane calculation

With the following diagrams cylindrical and ellipsoid membranes can be calculated. They are prepared so that the structural quantities necessary for the dimensioning and anchoring of the envelope can easily be read off. The loading corresponds to the data in Bibl. 136. The wind distribution given there for the cylinder is also approximately correct for ellipsoids. The selfweight and the membrane elongations are disregarded. The formulae in Section 6.4 form the basis of the diagrams. Their application is shown in an example in Section 6.7.

In cylinder membranes the deformations must not be ignored. Comparative calculations have shown that only a few parameters considerably influence non-linear behaviour. Those parameters are explicitly contained in the diagrams. The low dispersion of the solution through the unrealised parameters is revealed by the envelope curves of the diagrams. Tables 15 and 16 apply only for constant wind pressure. They may be used in approximation for gradually changing wind pressure.

The membrane deformations of the ellipsoid are very small and can be ignored. Local bulging must be avoided by an adequately large internal pressure. In Tables 19 to 21 only the extreme values of the membrane forces as a function of the z coordinates are shown. The

superposition of the individual load conditions provides the maximum membrane forces for the dimensioning and the minimum for proof of folding safety. Furthermore the extreme values of the membrane are given for the sphere under gradually changing wind pressure.

6.7. Calculation examples

For lack of space the handling of the diagrams given in Section 6.6 is shown here for one load condition only. For a limit load investigation the internal pressure and the back pressure of the wind will be multiplied by the safety factors ν_i, given in Section 6.4.3. The membrane forces, calculated for the ν_i-times of the loading, must not exceed the long term tensile strength of the envelope. In our example a cylinder membrane with spherical calottes will be determined for the load conditions p, w $\vee$ s. The loading scheme is shown in Fig. 32 (dimensions in kp, m). In cylinders with rectangular base the wind load decreases because of higher turbulences. Therefore the membrane forces are smaller than the values determined here.

In the following calculation all the coefficients taken from the diagrams are set in bold type. The reading scheme is marked on the diagrams for example.

6.7.1. Cylinder membrane

6.7.1.1. Input parameters

$h/b = 13/20 = 0,65$,
$e/h = 8/13 = 0.615 \approx 0.6$ (rounded down),
$p/q_u = 30/50 = 0.6$,
$p/q_o = 30/80 = 0.375$,
$p/S_o = 30/35 = 0.857$.

6.7.1.2. Membrane forces

Snow load reduces the membrane forces so that the load condition wind is critical for dimensioning. From Table 13 is taken the membrane force for constant wind pressure and estimated separately for q_u and q_o:

$n_\varphi(q_u) = \mathbf{0.8} \cdot (30 + 50) \cdot 10.35 = 662 \text{ kp/m}$,

$n_\varphi(q_o) = \mathbf{0.765} \cdot (30 + 80) \cdot 10.35 = 872 \text{ kp/m}$.

For gradual wind distribution the membrane force generally lies between these two critical values. Only in the specific input parameters present here is the membrane force $n_\varphi(q_o)$ slightly exceeded. The final membrane force is calculated in accordance with Table 14:

$n_\varphi(q) = -\mathbf{0.025} \cdot 662 + (1 + \mathbf{0.025}) \cdot 872$

$= 877 \text{ kp/m}$.

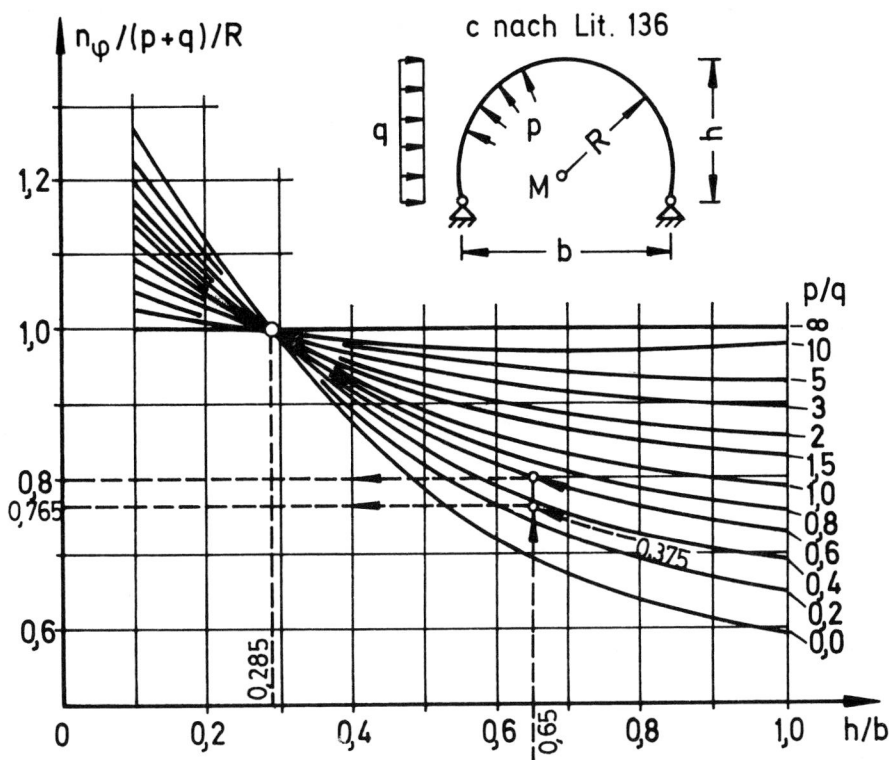

Table 13. Membrane forces of a cylinder under wind loading.

Table 14. Factor λ in gradual wind loading.

$$n_\varphi(q,p) = \lambda \cdot n_\varphi(q_u,p) + (1-\lambda) \cdot n_\varphi(q_0,p)$$

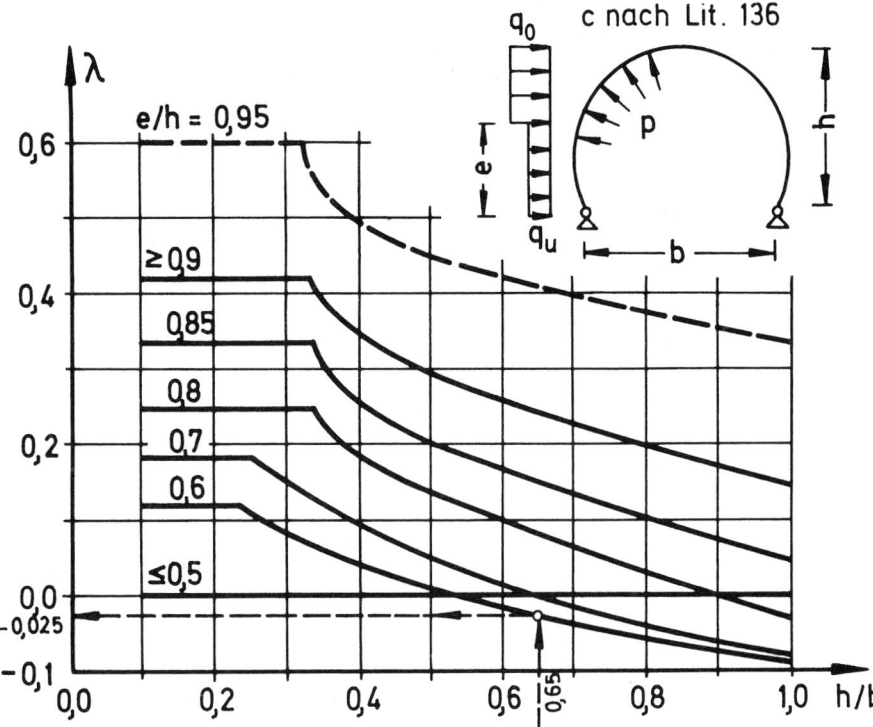

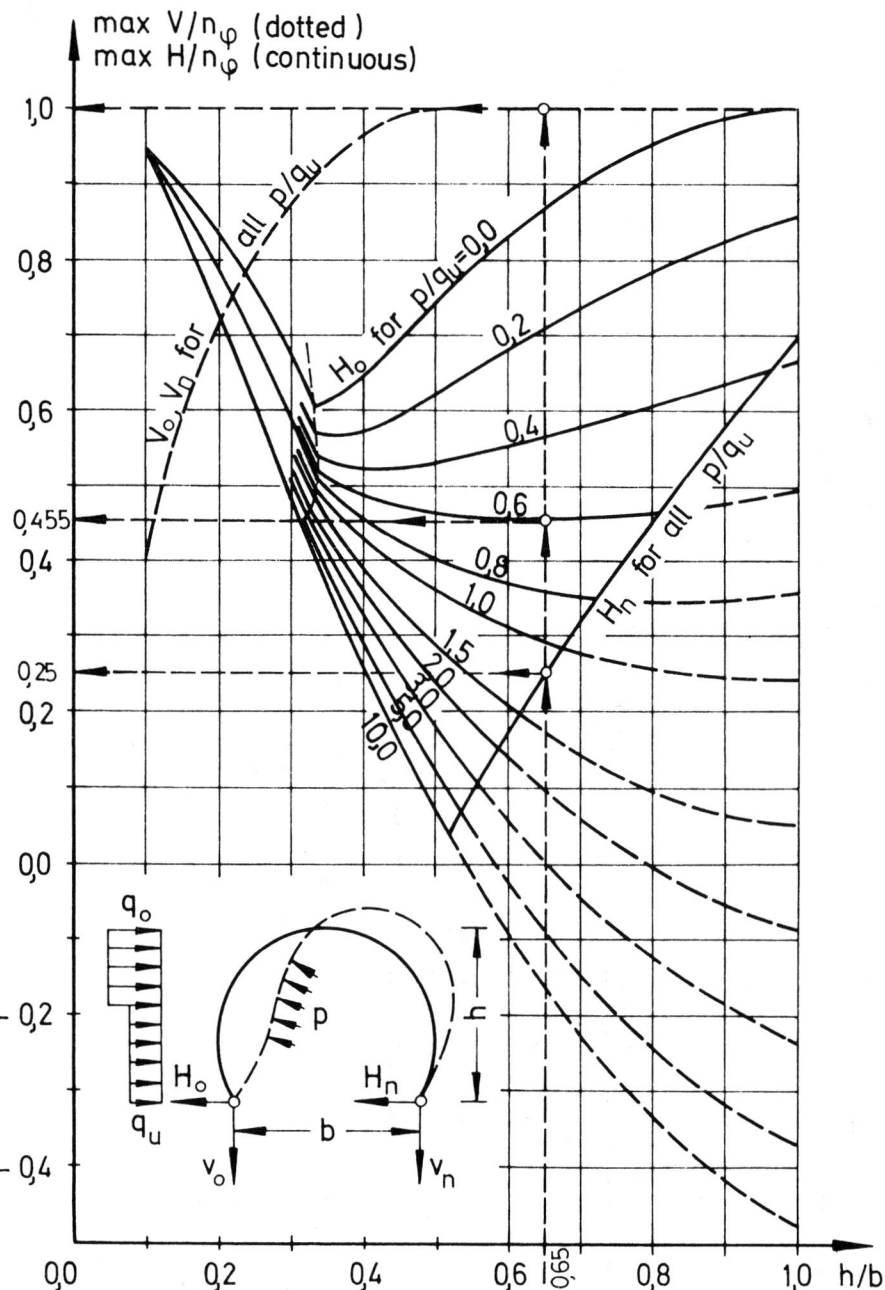

max V/n_φ (dotted)
max H/n_φ (continuous)

V_o, V_n for all p/q_u

H_o for $p/q_u=0{,}0$

0,2

0,4

0,6

0,8

1,0

1,5

2,0

3,0

5,0

10,0

H_n for all p/q_u

q_o

q_u

H_o

H_n

p

b

h

v_o

v_n

0,95

Table 15. Bearing forces of a cylinder under wind loading.

6.7.1.3. Anchorage forces (load condition wind)

In flat cylinder sections when $h/b \leqslant 0.5$ the horizontal anchorage forces $H_o \geqslant H_n$ are always directed outwards. They can change the sign in membranes where $h/b > 0.5$. The anchorage forces are taken from Table 15:

$H_o = \textbf{0.455} \cdot 877 = 400$ kp/m,
$H_n = \textbf{0.25} \cdot 877 = 219$ kp/m,
$V = \textbf{1.0} \cdot 877 = 877$ kp/m.

H_o is directed outwards, H_n inwards.
The maximum forces H and V occur at different bearing points and they are marked off in Table 15 by envelope curves. Therefore

$$n_\varphi (q) \leqslant \sqrt{H^2 + V^2}$$

is valid.

6.7.1.4. Deformation of non-flexible membranes

a) Load condition wind according to Table 16

max $w \approx \textbf{0.2} \cdot 10.35 = 2.07$ m.

The deformation caused by wind must generally be proved for membranes where $h/b \geqslant 0.5$. The greatest displacement occurs in the wind back pressure area.

b) Load condition snow in accordance with Table 17

max $w = \textbf{0.52} \cdot 10.35 = 5.38$ m.

Only in special cases is snow loading taken into consideration for unheated membranes. The one-sided snow loading leads to lower deformations.

The internal pressure must be significantly raised so that the deformations do not exceed the permissible values max $w \leqslant 0.1 \cdot R = 1.035$ m as given in equation 13.

6.7.1.5. Internal pressure for max $w = 0.1 \cdot R$

a) Load condition wind in accordance with Table 16
emp. $p \approx \textbf{1.8} \cdot 50 = 90$ kp/m².

b) Load condition snow in accordance with Table 17
emp. $p = \textbf{2.5} \cdot 3.5 = 87$ kp/m².

Normally one begins with this proof to establish the operating pressure p.

178

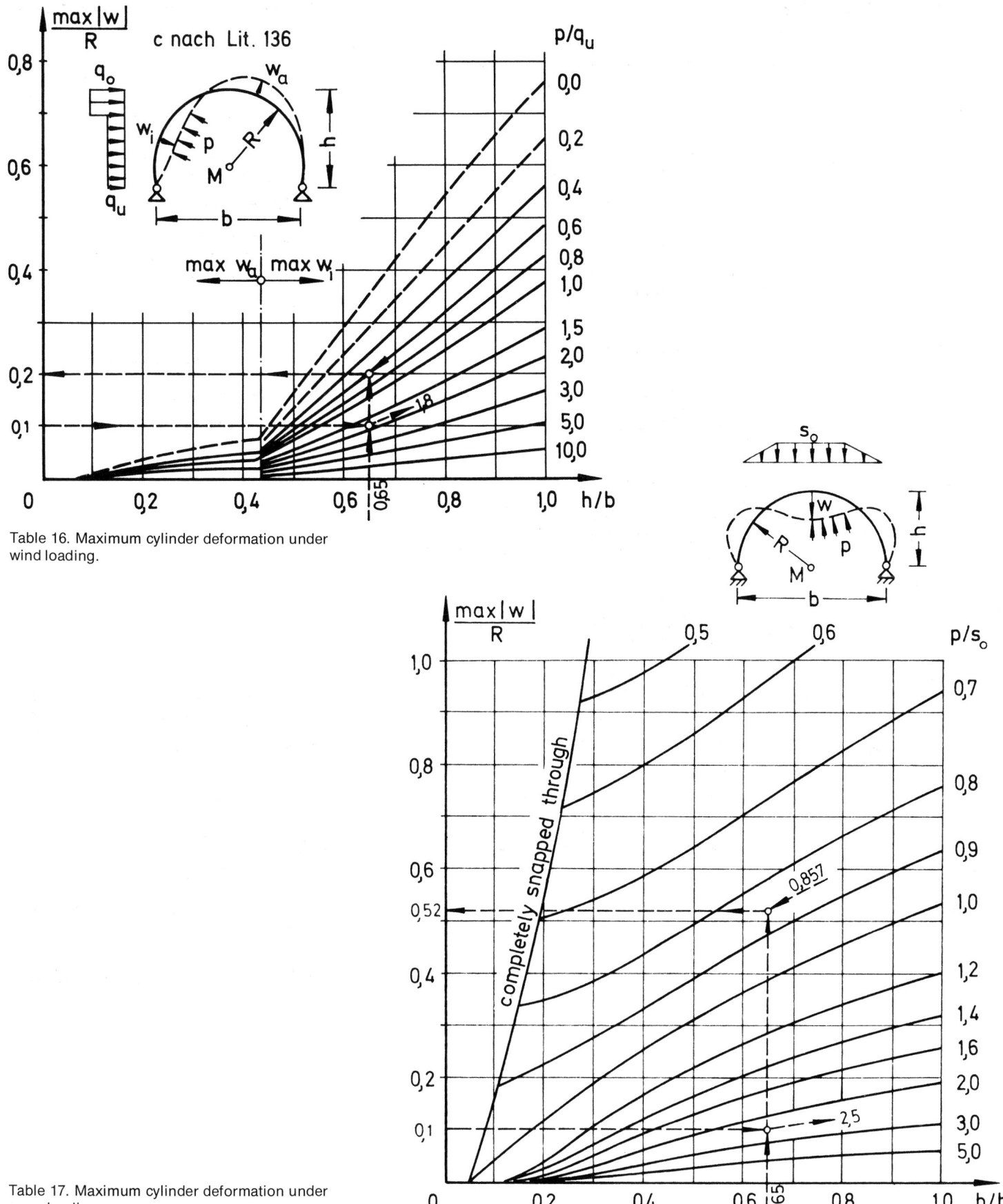

Table 16. Maximum cylinder deformation under wind loading.

Table 17. Maximum cylinder deformation under snow loading.

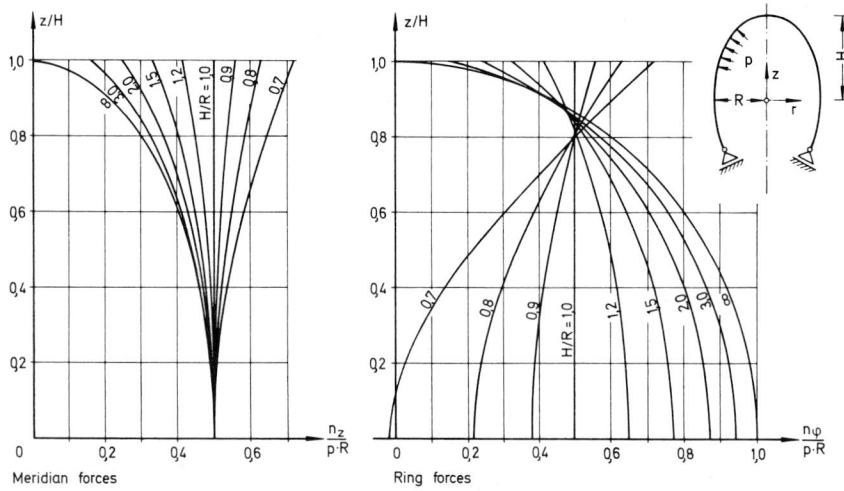

Table 18. Ellipsoid under internal pressure.

Table 19. Ellipsoid under snow loading.

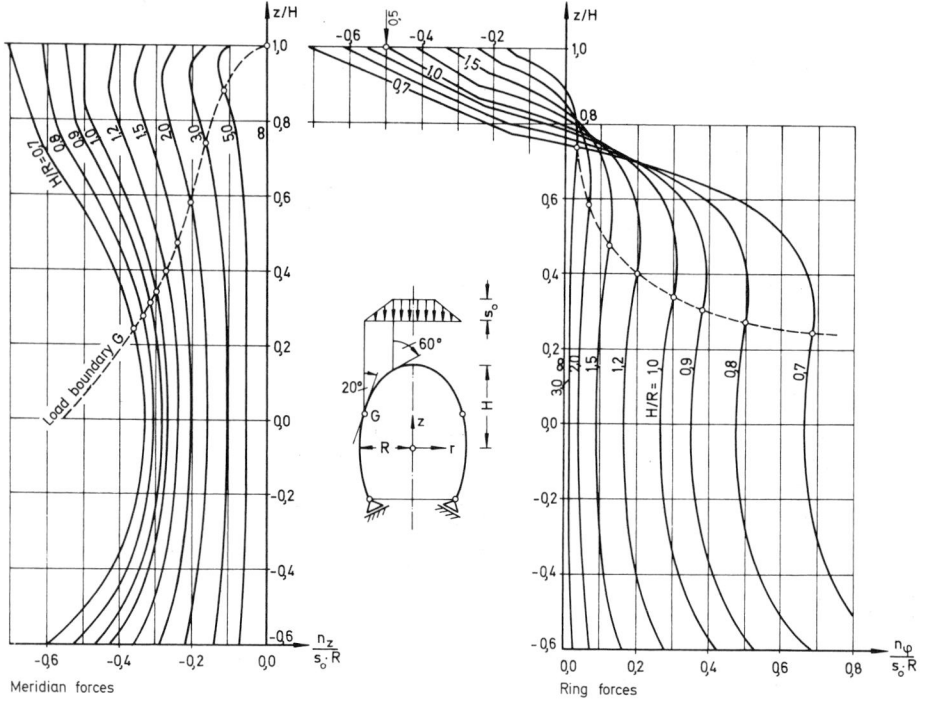

6.7.2. Spherical calottes

The deformations of the double curved end calottes are very small and can be ignored. Thus the superposition principle is maintained. Instead of the deformation restriction of the cylinder membrane folding in must now be avoided by using the necessary margin of safety.

6.7.2.1. Input parameters

$H/R = 1.0$,
$z_u/R = -2.65/10.35 = 0.266 > 0.6$,
$z_o/R = 5.35/10.35 = 0.517$,
$q_u/q_o = 50/80 = 0.625$.

6.7.2.2. Membrane forces

The wind loading is also critical for the dimensioning of spherical membranes. The membrane forces from internal pressure (Table 18) are superimposed with the maximum values from wind (Tables 22, 23):

$$\max n_\varphi = \mathbf{0.5} \cdot 30 \cdot 10.35 + \\ + \mathbf{1.26} \cdot 80 \cdot 10.35 = 1198 \text{ kp/m},$$

$$\max n_z = \mathbf{0.5} \cdot 30 \cdot 10.35 + \\ + \mathbf{1.14} \cdot 80 \cdot 10.35 = 1100 \text{ kp/m}.$$

The membrane forces in the end calottes are 25 to 35% larger than the forces in the cylindrical part (see Section 6.7.3).

The dimensioning forces for a sphere can be very easily determined. For ellipsoids one must plot and superimpose the extreme values of the membrane forces (Tables 18–21) given for the individual load conditions as a function of z. The envelope is dimensioned for the largest membrane forces in the area $z \geqslant z_u$.

6.7.2.3. Anchorage forces

Approximately the anchorage is dimensioned for the maximum membrane forces n_z:

$$\max H \approx \max n_z \cdot z_u/R = -1100 \cdot 0.266 = \\ = -292 \text{ kp/m},$$

$$\max V \approx \max n_z \cdot b/2/R = \\ = 1100 \cdot 20/2/10.35 = 1060 \text{ kp/m};$$

max H points inwards.

6.7.2.4. Formation of wrinkles

a) Load condition wind in accordance with Table 24:

$$\min n_w = -\mathbf{0.43} \cdot 80 \cdot 10.35 = -368 \text{ kp/m}.$$

The minimum membrane force occurs mostly at the jump point of the wind pressure. It does not occur in this order of magnitude in a steady wind distribution.

b) Load condition snow:

$$\min n_s = -\mathbf{0.5} \cdot 35 \cdot 10.35 = 181 \text{ kp/m}.$$

The prescribed internal pressure produces in the end calottes in accordance with Table 18 only the constant membrane force

$$n_{\vartheta} = n_z = \textbf{0.5} \cdot 30 \cdot 10.35 = 158 \; kp/m < |min \; n|.$$

In the end calottes wrinkles occur under wind and snow loading. However, it has not yet been taken into account that the cylinder membranes, by reason of the large deformations, hang on at the stiffer spherical heads.

6.7.2.5. Minimum internal pressure for folding safety ν

Wind loading, which produces the greatest compressive membrane forces, is the critical load condition for calculating the minimum internal pressure.
The minimum internal pressure:

$$min \; p = \frac{|\; min \; n \; (w) \;|}{(1-\nu) \, R/2} = \frac{386}{0.8 \cdot 10.35/2} =$$

$$= 86 \; kp/m^2.$$

arises from equation 19 for folding safety.
In Bibl. 136 the folding safety $\nu = 1.2$ for spherical membranes is prescribed ($\nu \approx (\bar{\nu} - 1)/\bar{\nu} = 0.17$ in accordance with equation 19). The folding safety $\bar{\nu} = 2.0$ ($\nu \approx 0.5$) given there for cylinders with quarto-spherical heads is unrealistic and should be avoided.

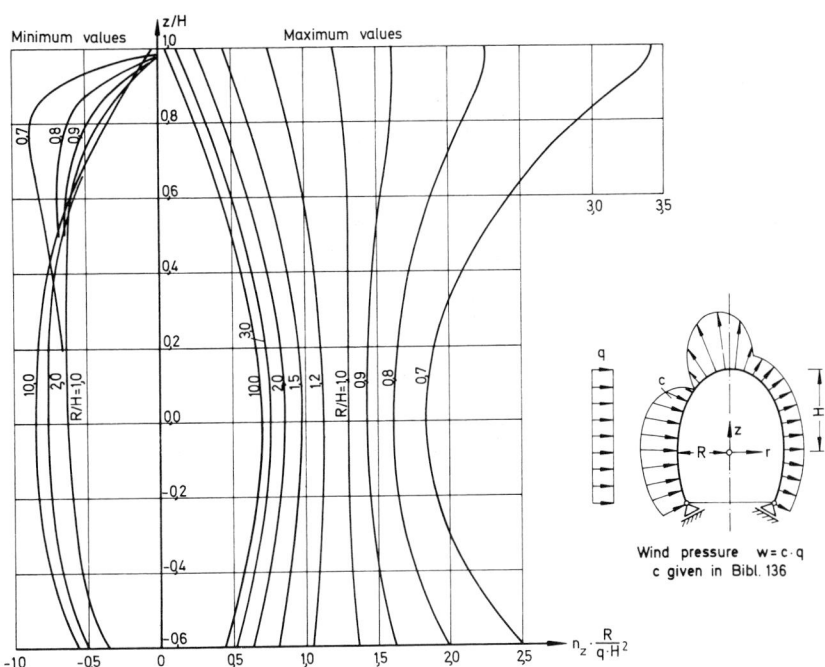

Table 20. Ellipsoid under wind loading, extreme values of the meridian forces.

Table 21. Ellipsoid under wind loading, extreme values of the ring forces.

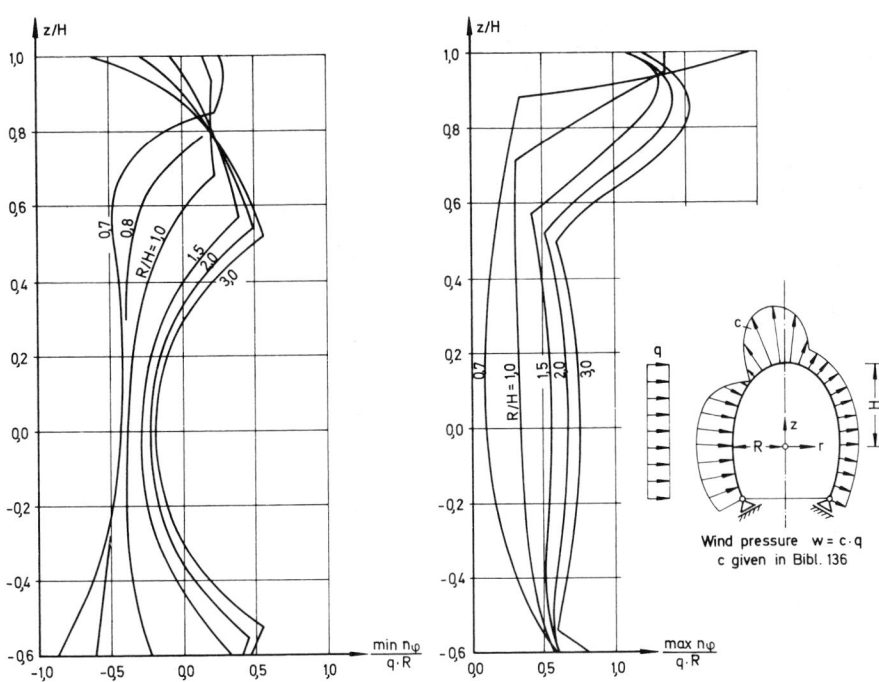

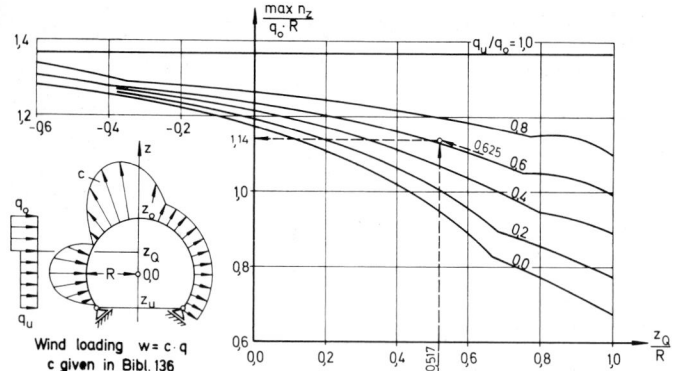

Table 22. Sphere under wind loading, maximum meridian tensile force in the region $-0.6 \leq z/R \leq 1.0$.

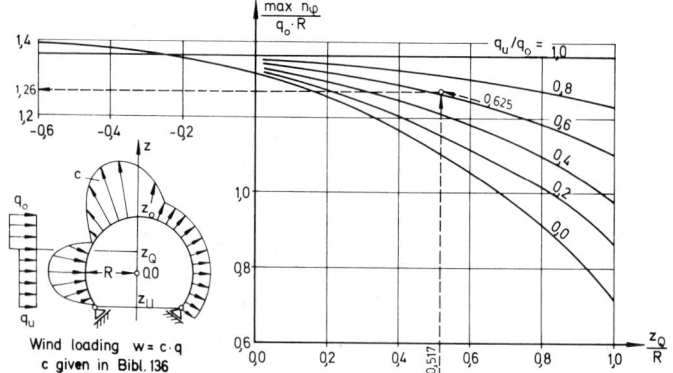

Table 23. Sphere under wind loading, maximum ring tensile force in the region $-0.6 \leq z/R \leq 1.0$.

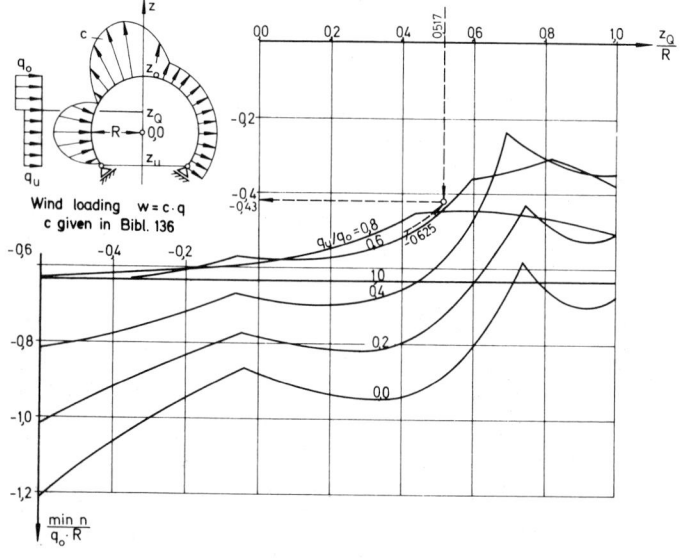

Table 24. Sphere under wind loading, maximum membrane compression force in the region $-0.6 \leq z/R \leq 1.0$.

6.7.3. Discussion of the results

At first glance it is surprising that the maximum membrane forces of the sphere are larger than those of the cylinder. The favourable load-bearing behaviour of the sphere under internal pressure no longer applies for wind loading.

Fig. 33 gives a view of a wind loaded spherical shell. The rings 1, normal to the wind direction, are loaded symmetrically in respect to rotation for the assumed wind distribution c. They bear the wind suction – like the cylinder – mainly in line with the barrel formula. In the wind direction the rings 2 receive pressure and suction so that no large membrane forces can arise which could significantly relieve the load in the transverse direction 1.

In contrast the cylinder adapts better to non-uniform loading. Peaks in the membrane forces such as occur in the sphere are reduced by the deformation of the support line. In the area of higher loading the support line curves more strongly so that only the longitudinal force

$$n_\varphi = \max (p + w) \cdot \min \varrho < \max (p + w) \cdot R$$

arises (Fig. 34).

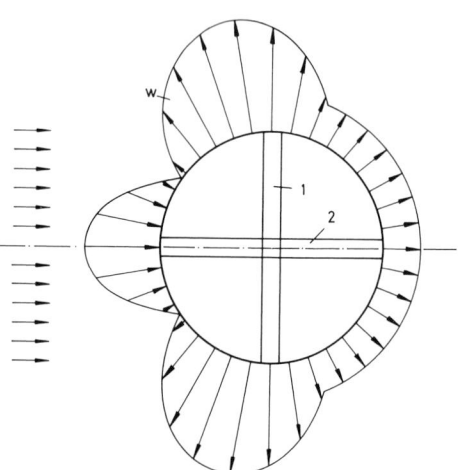

33. View of a wind loaded shell sphere.

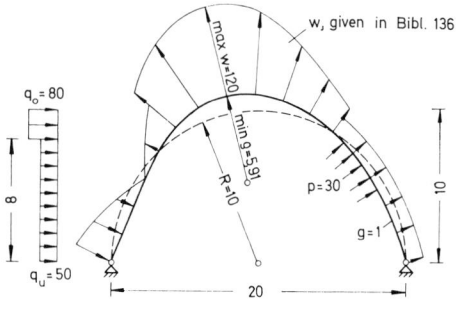

34. Cylinder deformation. Dimensions in kp, m.

7. Excursus: Pneumatic structures as form work for shell construction

As long ago as the early forties the American engineer W. Neff constructed a number of buildings by spraying concrete on to rubber balloons. In the process, however, considerable technical problems arose as there was excessive deformation of the form work due to insufficient internal pressure and furthermore it was not possible to maintain a constant internal pressure during fluctuations in temperature. As a result the concrete shells started cracking. (Bibl. 119, p. 163.)

It was about 20 years before the idea, which was not at all far fetched, reached technical maturity with the Domecrete technique by Haim Heifetz (Fig. 1–4; Bibl. 44, p. 149ff.; Bibl. 70; Bibl. 72). Here a 3 or 4.5 or 6 cm thick coat of high grade concrete is applied or sprayed in layers of 15 mm on to high pressure structures the size of a house. The pneumatic form work can be removed within 90 minutes after application of the concrete. In less than a day a dwelling house including a Vermiculite insulating layer, inside and outside plastering (total wall thickness 10 cm), electricity and water supplies, as well as windows and doors, can be completed.

The production costs are 30 to 56% lower, according to the type of usage, than conventionally produced buildings. Expenses for the special equipment are so moderate that they are usually recovered on the first building.

1

2

3

1–4. Various houses manufactured by the Domecrete technique

4

5. Binishell with fan not yet removed.
6. Diagram of the Binishell process.

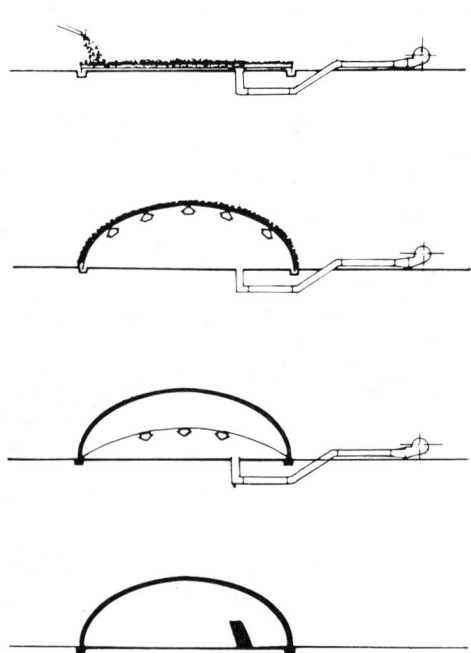

The Bayer Igloo (Figs. 7–10), by Farbenfabriken Bayer AG, was developed as an emergency shelter for use in disaster areas. It is built by spraying polyurethane foam on to a balloon made of PVC film reinforced with fabric. While the pneumatic form work is rotated on a turntable, consisting of a dismountable lightweight steel structure, the spray gun moves up and down on a tripod placed over the form work. The reaction mixture foams up immediately, expands to about 30 times its initial volume and cures after a few seconds. Several layers of spray foam are applied one on top of the other up to a total thickness of 10 cm.

As rigid polyurethane foam also possesses excellent insulating properties, the English firm P. Frankenstein and Sons, Ltd. used this material to develop a shelter for polar regions (Bibl. 44, p. 150). The main difference from the Bayer Igloo is that the air supported structure serving as form work is not only coated from the outside but also from the inside. After the curing of the foam the membrane is no longer structurally necessary, although it cannot be re-used.

A similar process is offered by the American firm Basic Products Development Company. There the inside is also reinforced with steel.

A method combining the above mentioned processes was been developed by the firm of Garteway Towers. Rigidising polyurethane foam is sprayed from the inside on to a pneumatic form work for reinforcement. In addition a steel reinforcement is laid over the outside and covered with concrete by the Torcret technique.

7–10. Various stages in the manufacture of a Bayer igloo.

In the Binishells (Figs. 5, 6; Bibl. 17; Bibl. 44, p. 147 ff.) developed by Dante Bini and manufactured by Du Pont de Nemours International S.A., the form work consists of a flat, 2 to 3 mm thick rubber or neoprene membrane which, because of its high elasticity, can be inflated to hemispherical shape. The setting time of the concrete of which the shell is made is 12 to 48 hours. Removing the form work is particularly simple as the membrane falls back flat again as soon as the tension is slackened.

The reinforcement consists of coiled springs which can adapt well to the arched shape.

The shells can be heat insulated from the inside or the outside. In addition a second concrete shell can later be applied from the outside. Openings are cut out with a saw.

The buildings achieve high strengths and need no scaffolding or cranes in erection. According to the quotations of various firms the price is about 50% below that of a comparable conventional structure.

The rubber form work used by NOE-Schaltechnik, a West German firm, for the production of in-situ concrete conduits (Figs. 11, 12) is particularly economic in the case of large cross-sections of pipes with high loading, but can also be used to manufacture small conduits such as ventilation shafts, chimneys, etc.

First the base is concreted, the external casing erected and, if necessary, reinforcement applied. Then the rubber form work is laid in and inflated.

The form work can also be used in the building of curved pipe conduits; the smallest possible bending radius amounts to about 20 times the tube diameter.

9

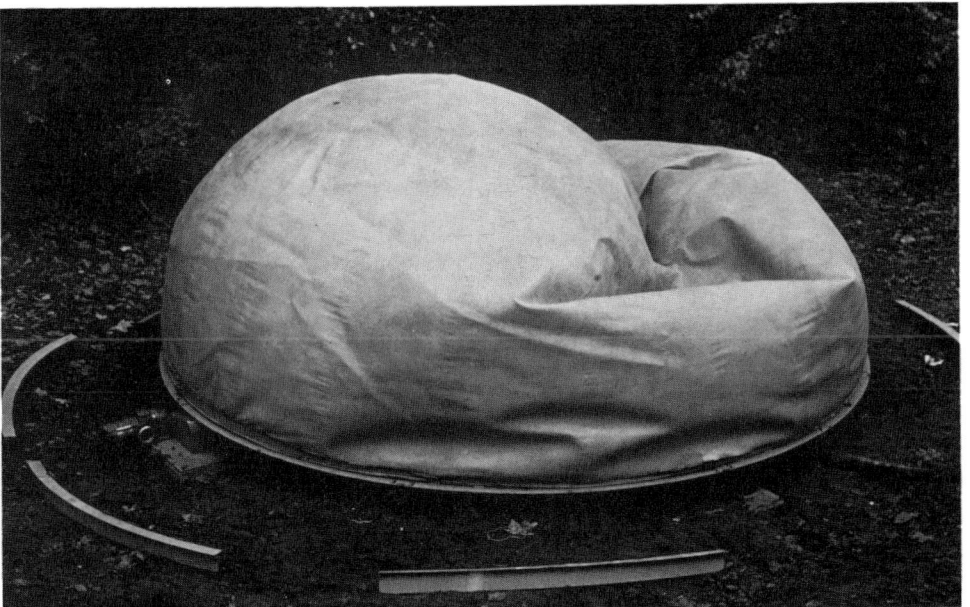

10

11

12

11, 12. Examples of the production of in-situ concrete conduits as developed by the firm of NOE-Schaltechnik.

Bibliography

1. David Allison. "A great balloon for peaceful atoms". *The Architectural Forum* (New York), Nov. 1960.

2. Fred Angerer. *Bauen mit tragenden Flächen.* Munich, 1960.

3. Ant Farm. *Inflatocookbook.* Sausalito, California. 1971.

4. *Architectural Design* (London), 1968, 6.

5. Ballonfabrik—See- und Luftausrüstung GmbH + Co. KG, Augsburg. *Ballonhallen.* Company brochure, Nov. 1970.

6. Ballonfabrik—See- und Luftausrüstung GmbH + Co. KG. *BFA auf einen Blick.* Company brochure, 1972.

7. Reyner Banham. "A Home is not a House". *Art in America* (New York), 1965, 4.

8. *Bauen mit Kunststoffen* (Institut für das Bauen mit Kunststoffen e. V., Darmstadt), 1965, 2.

9. *Bauen mit Kunststoffen,* 1965, 4.

10. *Bauen mit Kunststoffen,* 1965, 5.

11. *Bauen mit Kunststoffen,* 1968, 4.

12. *Bauen mit Plasten.* Seminar report. Technische Hochschule Darmstadt, 1971.

13. J. Beck. *Die Befestigung von Drahtseilen mit aufgezogenen Stahlhülsen.* Thesis. Technische Hochschule Stuttgart, 1940.

14. G. Beger and E. Macher. "Results of Wind Tunnel Tests on some Pneumatic Structures". In: *Proceedings of the 1st International Colloquium on Pneumatic Structures.* Technische Hochschule Stuttgart, 1967.

15. Roland Benesch. "Das Bauen mit Membranen und Netzen". *kib Kunststoffe im Bau* (Heidelberg), 7 (1967).

16. Roland Benesch. "Pneumatische Wände". In: *International Symposium on Pneumatic Structures Delft 1972. Proceedings.* Technische Hogeschool Delft, 1972.

17. Dante Bini. "A New Pneumatic Technique for the Construction of Thin Shells". In: *Proceedings of the 1st International Colloquium on Pneumatic Structures.* Technische Hochschule Stuttgart, 1967.

18. Walter W. Bird. "The Development of Pneumatic Structures. Past, Present, Future". In: *Proceedings of the 1st International Colloquium on Pneumatic Structures.* Technische Hochschule Stuttgart, 1967.

19. Walter W. Bird. "Air structures span space in building revolution". *Rubber World,* 1972, 1, and 1972, 2.

20. Birdair Structures, Inc., Buffalo. N.Y. *Airshelters.* Company brochures, 1971.

21. T. Botschuiver and J. Shaw. "Eventstructures". In: *International Symposium on Pneumatic Structures Delft 1972.* Proceedings. Technische Hogeschool Delft, 1972.

22. C. V. Boys. *Soap bubbles, their formation and the forces that mould them.* London, 1960.

23. *Der Große Brockhaus.* 15th ed., Leipzig, 1928–35.

24. H. Brügge. Zugbeanspruchte Konstruktionen, Marienfeld. *Die Brügge-Innendruck-Lager-Halle.* Company brochure, 1970.

25. Rudolf Brylka. "Belüftung und Beheizung von Traglufthallen". *Haus der Technik — Vortragsveröffentlichungen* (Essen), 124 (1967).

26. Rudolf Brylka. "Zur Konstruktion von Traglufthallen". *Bauen + Wohnen* (Munich), 1968, 6.

27. Rudolf Brylka. "Entwicklung und Anwendungsmöglichkeiten von Traglufthallen". *Der Architekt* (Essen), 1968, 10.

28. Rudolf Brylka. "Traglufthallen heute". *Technische Mitteilungen Krupp* (Fried. Krupp GmbH, Essen), vol. 28 (1970), 2.

29. Rudolf Brylka. "Neue Ausführungsbeispiele und Projekte von pneumatischen Konstruktionen". In *International Symposium on Pneumatic Structures Delft 1972. Proceedings.* Technische Hogeschool Delft, 1972.

30. Rudolf Brylka. "Richtlinien für den Bau und Betrieb von Tragluftbauten in der Bundesrepublik Deutschland". In: *International Symposium on Pneumatic Structures Delft 1972. Proceedings.* Technische Hogeschool Delft, 1972.

31. BSI-Code. 1972. (England.)

32. Building Centre of Japan, Pneumatic Structures Committee. *Pneumatic Structure Design Standard and Commentary.*

33. Philip S. Bulson. "The Behavior of some Experimental Inflated Structures". In: *Proceedings of the 1st International Colloquium on Pneumatic Structures.* Technische Hochschule Stuttgart, 1967.

34. Philip S. Bulson. "3 Metre Diameter Air Bags under Severe Local Loads". In: *International Symposium on Pneumatic Structures Delft 1972. Proceedings.* Technische Hogeschool Delft, 1972.

35. Z. Bychawski. "Rheological Large Deformation of Rotational Membranes". In: *International Symposium on Pneumatic Structures Delft 1972. Proceedings.* Technische Hogeschool Delft, 1972.

36. Canvas Products Association International, Air Structures Division. *Minimum Performance Standard for Single-Wall Air-Supported Structures,* March, 1971.

37. W. Caspar. "Maximale Windgeschwindigkeiten in der Bundesrepublik Deutschland". *Die Bautechnik* (Berlin), vol. 47 (1970), p. 335 ff.

38. Simon Conolly and Mark Fisher. "Auto-mat". *Architectural Design* (London), 1971, 4.

39. Simon Conolly, Mark Fisher and O. W. Neumark. "Experiments and Observations on the Use of Pressure Tensioned Membrane Systems for Locomotion and Flight". In: *International Symposium on Pneumatic Structures Delft 1972. Proceedings.* Technische Hogeschool Delft, 1972.

40. H. J. Cowan and J. G. Pohl. "A Preliminary Investigation into the Load-Bearing Capacity of Open-Ended Cylindrical Columns, Subjected to Internal Fluid Pressure". In: *Proceedings of the 1st International Colloquium on Pneumatic*

Structures. Technische Hochschule Stuttgart, 1967.

41. Justus Dahinden. *Stadtstrukturen für morgen. Analysen, Thesen, Modelle.* Stuttgart, 1971. English: *Urban Structures for the Future.* London and New York, 1971.

42. Deutscher Ausschuß für Stahlbau. *Richtlinien zur Anwendung des Traglastverfahrens im Stahlbau.* May, 1972.

43. Roger N. Dent. *Pneumatic Structures in Architecture with Special Reference to Arctic and Lunar Application.* Thesis. University of Liverpool, 1968.

44. Roger N. Dent. *Principles of Pneumatic Architecture.* London, 1971.

45. Roger N. Dent. "A Unique Pneumatic Application". In: *International Symposium on Pneumatic Structures Delft 1972. Proceedings.* Technische Hogeschool Delft, 1972.

46. *Detail* (Tokyo), April, 1970.

47. Albert G. H. Dietz. "Solar Energy for Air-Supported Buildings". In: *International Symposium on Pneumatic Structures Delft 1972. Proceedings.* Technische Hogeschool Delft, 1972.

48. *DIN 1055. Lastannahmen für Bauten.* Sheet 4, March 1963 issue, and sheet 5, March 1973 draft.

49. *Domus* (Milan), 455 (Oct., 1967).

50. *Domus* (Milan), 467 (Oct., 1968).

51. Milford W. Donaldson and John B. McNicholas. "First Steps in the Development of a Photoelastic Model Technique for the Study of Pneumatic Structures". In: *International Symposium on Pneumatic Structures Delft 1972. Proceedings.* Technische Hogeschool Delft, 1972.

52. H. Duddeck. "Die Biegeberechnung technischer Rotationsschalen mit Rändern entlang den Breitenkreisen". *Der Bauingenieur* (Berlin), vol. 39 (1964), p. 435 ff.

53. E. Dulácska. "Stability of Rubber Ballons". In: *International Symposium on Pneumatic Structures Delft 1972. Proceedings.* Technische Hogeschool Delft, 1972.

54. Dynamit Nobel AG., Troisdorf. *Traglufthallen aus beschichtetem Trevira-hochfest-Gewebe.* Company brochure, 1970.

55. Dynat Gesellschaft für Verschlußtechnik und Feinmechanik mbH, Hildesheim. Company brochure on gas-, water- and pressure-tight fasteners, 1969.

56. G.-A. Euteneuer. "Druckanstieg im Inneren von Gebäuden bei Windeinfall". *Der Bauingenieur* (Berlin), vol. 45 (1970), p. 214 ff.

57. Farbwerke Hoechst AG, Frankfurt/Main. *Information T 21: Vergleichende Prüfung von beschichteten und gummierten Geweben aus Trevira-hochfest auf Schwerentflammbarkeit bzw. Schwerbrennbarkeit.* (Sept. 1967.)

58. E. M. Feldhaus. *Die Technik der Vorzeit, der geschichtlichen Zeit und der Naturvölker.* Leipzig and Berlin, 1914.

59. Felten & Guilleaume Carlswerk AG. *Drahtseile für alle Verwendungszwecke.* Suppliers survey.

60. *Fertigteilbau + Industrialisiertes Bauen* (Waiblingen), 1970, 3.

61. Vladimir Firt. "Air-Supported Structures with Automatic Adjustment of Inner Overpressure". In: *International Symposium on Pneumatic Structures Delft 1972. Proceedings.* Technische Hogeschool Delft, 1972.

62. W. Förster and K.-H. Schüßler. "Der Membranspannungszustand der windbelasteten Kugelschale". *Der Bauingenieur* (Berlin), vol. 42 (1967), p. 21 ff.

63. Bernard Fournier and Donald P. Greenberg. "A Graphical Analysis Approach for Pneumatic Spherical Membrane Structures". In: *International Symposium on Pneumatic Structures Delft 1972. Proceedings.* Technische Hogeschool Delft, 1972.

64. M Fourtané. "Structures Pneumatiques à Géométrie Variable et Structures Mixtes de Grande Portée: Gonflable – Ferme – Cable". In: *International Symposium on Pneumatic Structures Delft 1972. Proceedings.* Technische Hogeschool Delft, 1972.

65. J. E. Gibson, M. Barnes, M. Smolira and W. W. Frischmann, "A Multi Pneumatic Loading System". In: *International Symposium on Pneumatic Structures Delft 1972. Proceedings.* Technische Hogeschool Delft, 1972.

66. A. E. Green and W. Zerna. *Theroretical Elasticity.* London, 1954.

67. J. Grepl. "Die Belastung der Membranen von Überdruckhallen durch Schnee und durch inneren Überdruck". In: *International Symposium on Pneumatic Structures Delft 1972. Proceedings.* Technische Hogeschool Delft, 1972.

68. Blair L. Hamilton. "Pneumatic Structures, Cybernetics and Ecology". In: *International Symposium on Pneumatic Structures Delft 1972. Proceedings.* Technische Hogeschool Delft, 1972.

69. R. Harbord. "Berechnung von Schalen mit endlichen Verschiebungen – Gemischte finite Elemente". *Bericht des Instituts für Statik der Technischen Universität Braunschweig,* 1972, 7.

70. Gershon Har'El and Haim Heifetz. *Realisation of Inflatable Forms Simplifies Hastens and Reduces Cost of Shell Structures.* Publication no. 169 (July, 1971) of the Israel Institute of Technology, Haifa.

71. Eberhard Haug. "Finite Element Analysis of Pneumatic Structures". In: *International Symposium on Pneumatic Structures Delft 1972. Proceedings.* Technische Hogeschool Delft, 1972.

72. Haim Heifetz. "Development in inflatable forms". *Build International* (Barking), Jan./Feb., 1970.

73. Heinz Herlinger. "Über die Herstellung von Chemiefasern". *Süddeutsche Zeitung* (Munich), June 2, 1971.

74. Heinrich Hertel. *Struktur, Form, Bewegung.* Mainz, 1963.

75. Alfons Holslag and Suzan van Westenbrugge. *Pneumatiese Konstrukties,* Technische Hogeschool Delft, 1972.

76. Zdenek Holub. "Neue Lösungen der Stützschlauchkonstruktionen mit höherer Betriebssicherheit". In: *International Symposium on Pneumatic Structures Delft 1972. Proceedings.* Technische Hogeschool Delft, 1972.

77. *IL 2. City in the Arctic.* Institut für leichte Flächentragwerke, Universität Stuttgart, 1971.

78. H. Isler. "Clear – Transparent Roof for a Court". In: *Proceedings of the 1st International Colloquium on Pneumatic Structures.* Technische Hochschule Stuttgart, 1967.

79. H. Isler. "Pneumatic Shape for Concrete Shells". In: *Proceedings of the 1st International Colloquim on Pneumatic Structures.* Technische Hochschule Stuttgart, 1967.

80. Kazuo Ishii. "On Developing of Curved Surfaces of Pneumatic Structures". In: *International Symposium on Pneumatic Structures Delft 1972. Proceedings.* Technische Hogeschool Delft, 1972.

81. Yoshito Isono. "Abstract of 'Pneumatic Structure Design Standard' in Japan". In: *International Symposium on Pneumatic Structures Delft 1972. Proceedings.* Technische Hogeschool Delft, 1972.

82. Yoshito Isono. "The Development of Pneumatic Structures in Japan". In: *International Symposium on Pneumatic Structures Delft 1972. Proceedings.* Technische Hogeschool Delft, 1972.

83. *iup* (Institut für Umweltplanung, Ulm), 7 (1972).

84. *iup,* 9 (1972).

85. V. V. Jermolov and others. *pnevmatičeskie konstrukcii.* Moscow, 1973.

86. Laurent Kaltenbach and Guy Naizot. "Utilité des Gonflables". *Techniques & Architecture* (Paris), 1969, 5.

87. Laurent Kaltenbach, Louis Paul Untersteller and Michel Maillé. "Une Structure Gonflable de 100 m de Portée Résille de Câbles". In: *International Symposium on Pneumatic Structures Delft 1972. Proceedings.* Technische Hogeschool Delft, 1972.

88. Mamoru Kawaguchi and others. *Engineering Problems of Pneumatic Structures.* Lecture manuscript, 1971.

89. Carl Graf von Klinckowstroem. *Knaur Geschichte der Technik,* Munich and Zurich, 1959.

90. Celal N. Kostem. "Thermal Stresses and Deformations in Pneumatic Cushion Roofs". In: *International Symposium on Pneumatic Structures Delft 1972. Proceedings.* Technische Hogeschool Delft, 1972.

91. Fried. Krupp GmbH, Essen. *Untersuchungsbericht 130/70: Leichte Flächentragwerke, Entwürfe und Ideen zur Struktur und Form.*

92. Krupp Universalbau, Essen. Construction drawing on Standard halls, 1970.

93. Krupp Universalbau, Essen. Price lists nos. 1 and 2 for Krupp air-supported halls, Oct., 1970.

94. H. Kurschat. "Zuganker". In: *Räumliche Tragwerke.* Seminar report. Universität Stuttgart, 1968/69.

95. Wolfgang Längsfeld. "Sonsbeek 71". *Süddeutsche Zeitung* (Munich), July 24/25, 1971.

96. N. Laing. "The Use of Solar and Sky Radiation for Air Conditioning of Pneumatic Structures". In: *Proceedings of the 1st International Colloquium on Pneumatic Structures*. Technische Hochschule Stuttgart, 1967.

97. A. Libai. "Pressurized Cylindrical Membranes with Flexible Supports". In: *International Symposium on Pneumatic Structures Delft 1972. Proceedings*. Technische Hogeschool Delft, 1972.

98. Josef Linecker. "Pneumatische Konstruktion für Hallenbäder". In: *International Symposium on Pneumatic Structures Delft 1972. Proceedings*. Technische Hogeschool Delft, 1972.

99. Wolfgang Lochner. *Weltgeschichte der Luftfahrt,* Würzburg, 1970.

100. M. H. Losch. *Bestimmung der mechanischen Konstanten für einen zweidimensionalen, nichtlinearen, anisotropen elastischen Stoff am Beispiel beschichteter Gewebe.* Thesis. Universität Stuttgart, 1971.

101. Victor A. Lundy. "Architectural and Sculptural Aspects of Pneumatic Structures". In: *Proceedings of the 1st International Colloquium on Pneumatic Structures*. Technische Hochschule Stuttgart, 1967.

102. H. L. Malhotra. "Fire Behavior of Single-Skin Air-Supported Structures". In: *International Symposium on Pneumatic Structures Delft 1972. Proceedings*. Technische Hogeschool Delft, 1972.

103. Herbert Mewes. "Beschichtete und gummierte Gewebe aus Trevira-hochfest". *Chemiefasern* (Frankfurt/Main), 1964, 12 u. 1965, 1.

104. Herbert Mewes. "Beschichtete Gewebe aus hochfesten Polyester-Fäden für Traglufthallen und leichte Flächentragwerke". *kib Kunststoffe im Bau* (Heidelberg), 14 (1969).

105. Gernot Minke. *Zur Effizienz von Tragerken.* Stuttgart and Bern, 1970.

106. Gernot Minke. "Pneumatische Konstruktionen". *Bauen + Wohnen* (Munich), 1971, 11.

107. Gernot Minke. "Die konstruktive Effizienz der pneumatischen Stütze". In: *International Symposium on Pneumatic Structures Delft 1972. Proceedings*. Technische Hogeschool Delft, 1972.

108. Gernot Minke. "Übersicht über die Systeme und Typen der pneumatisch stabilisierten Membrantragwerke". In: *International Symposium on Pneumatic Structures Delft 1972. Proceedings*. Technische Hogeschool Delft 1972.

109. C. J. Moore and B. Rawlings. "Inflated Metal Structures – Some Small and Large Scale Tests". In: *International Symposium on Pneumatic Structures Delft 1972. Proceedings*. Technische Hogeschool Delft, 1972.

110. Erwin Mühlestein. "Blow up". *Werk* (Winterthur), 1970, 5.

111. Erwin Mühlestein. "Air Structures Design". *Bauen + Wohnen* (Munich), 1971, 3.

112. Toal'o Muiré. "Instant City Ibiza". *Architectural Design* (London), 1971, 12.

113. Yoshio Nakahara. "The Mechanical Behavior of Circular Membrane Fabric under Uniform Lateral Pressure". In: *International Symposium on Pneumatic Structures Delft 1972. Proceedings*. Technische Hogeschool Delft, 1972.

114. Hans-Jürgen Niemann. "Zur Windbelastung von Traglufthallen". *Konstruktiver Ingenieurbau Berichte* (Ruhruniversität, Bochum), 13.

115. Hans-Jürgen Niemann. "Wind Tunnel Experiments on Aerolastic Models of Air-Supported Structures: Results and Conclusions". In: *International Symposium on Pneumatic Structures Delft 1972. Proceedings*. Technische Hogeschool Delft, 1972.

116. A. Nölker. "Windkanalversuche an Modellen von Traglufthallen: Versuchstechnik und Modellherstellung". In: *International Symposium on Pneumatic Structures Delft 1972. Proceedings*. Technische Hogeschool Delft, 1972.

117. J. T. Oden and W. K. Kubitzka. "Numerical Analysis of Nonlinear Pneumatic Structures". In: *Proceedings of the 1st International Colloquium on Pneumatic Structures*. Technische Hochschule Stuttgart, 1967.

118. Frei Otto. "Bauen im Weltraum". *Bauwelt* (Berlin), 1963, 17.

119. Frei Otto (ed.). *Zugbeanspruchte Konstruktionen.* Vol. 1. Frankfurt/Main and Berlin, 1962. English: *Tensile Structures.* Vol. 1. Cambridge, Mass., and London, 1967.

120. Frei Otto (ed.). *Zugbeanspruchte Konstruktionen.* Vol. 2. Frankfurt/Main and Berlin, 1965. English: *Tensile Structures.* Vol. 2. Cambridge, Mass., and London, 1969.

121. Erasmo Pistolesi and Salvatore Busuttil. *La Colonna Traiana.* Rome, 1846.

122. *Plasticonstruction* (Munich), 1972, 1.

123. *Pneumatische Konstruktionen.* Seminar report. Technische Universität Hanover, 1970.

124. Jens G. Pohl. "Multi-Storey Pneumatic Buildings as a Challenge to the Plastics Industry". *Australian Building Science and Technology* (Sydney), June, 1967.

125. Jens G. Pohl. "Pneumatic Structures". *Architecture in Australia* (Sydney), July, 1968.

126. Jens G. Pohl. "The Structural Pressurized Flexible Membrane Column". *Architectural Science Review* (Sydney), June, 1970.

127. Jens G. Pohl. "The Reinforcement and Bracing of Multi-Storey Pneumatic Buildings". *Architectural Science Review* (Sydney), March, 1971.

128. Jens G. Pohl and H. Cowan. "Multi-Storey Air-Supported Building Construction". *Build International* (Barking), April, 1972.

129. Jens G. Pohl. "Multi-Storey Fluid-Supported Building Systems". In: *International Symposium on Pneumatic Structures Delft 1972. Proceedings*. Technische Hogeschool Delft, 1972.

130. J. M. Prada and R. Aroca. "Some Properties and Possibilities of Third Generation Pneumatic Structures". In: *International Symposium on Pneumatic Structures Delft 1972. Proceedings*. Technische Hogeschool Delft, 1972.

131. Cedric Price, Frank Newby and others. *Air Structures.* London, 1971.

132. August Pütter. "Die Entwicklung des Tierfluges". In: Bernhard Lepsius and Richard Wasmuth (ed.). *Denkschrift der ersten Internationalen Luftschiffahrt-Ausstellung (ILA) zu Frankfurt an Main 1909.* Vol. 1. Berlin, 1910.

133. Wolfgang Rathke. "Pneumatisches Zellentragwerk". In: *International Symposium on Pneumatic Structures Delft 1972. Proceedings*. Technische Hogeschool Delft, 1972.

134. *Règles NV 65.* (France.)

135. G. F. Reitmeier and Milton B. Punnett. "Design Developments in Large Span Cabled Structures". In: *International Symposium on Pneumatic Structures Delft 1972. Proceedings*. Technische Hogeschool Delft, 1972.

136. "Richtlinien für den Bau und Betrieb von Tragluftbauten". July 1971 version. *Ministerialblatt für das Land Nordrhein-Westfalen,* 1971, p. 1658ff.

137. Conrad Roland. *Frei Otto – Spannweiten. Ideen und Versuche zum Leichtbau.* Berlin, Frankfurt/Main and Vienna, 1965. English: *Frei Otto – Structures.* London, 1972.

138. Bernd-Friedrich Romberg. "Lichtdecken über dem Forum Steglitz in Berlin". *Bauwelt* (Berlin), 1970, 19.

139. Bernd-Friedrich Romberg, "Aufgeblasen". *db-Deutsche Bauzeitung* (Stuttgart), 1972, 3.

140. Bernd-Friedrich Romberg. "Konstruktion und Anwendungen von Luftkissen". In: *International Symposium on Pneumatic Structures Delft 1972. Proceedings*. Technische Hogeschool Delft, 1972.

141. Bernd-Friedrich Romberg. "Verfahrbare Raumhüllen aus luftgetragenen Häuten". In: *International Symposium on Pneumatic Structures Delft 1972. Proceedings*. Technische Hogeschool Delft, 1972

142. E. W. Ross. "Large Deflections on an Inflated Cylindrical Tent". *Journal of Applied Mechanics* (New York), 1969, p. 845ff.

143. H. Rühle. "Development of Design and Construction in Pneumatic Structures". In: *Proceedings of the 1st International Colloquium on Pneumatic Structures*. Technische Hochschule Stuttgart, 1967.

144. H. Rühle. *Räumliche Dachtragwerke.* Vol. 2. Cologne, 1970.

145. H. Rühle. "Reale Entwicklungstendenzen pneumatischer Konstruktionen im Bauwesen". In: *International Symposium on Pneumatic Structures Delft 1972. Proceedings*. Technische Hogeschool Delft, 1972.

146. H. Rühle and R. Schulz. "Pneumatische Konstruktionen". *Deutsche Architektur* (East Berlin), 1968, 5.

147. F. Rudolf. "A Contribution to the Design of Air-Supported Structures". In: *Proceedings of the 1st International Colloquium on Pneumatic Structures*. Technische Hochschule Stuttgart, 1967.

148. Seiji Sawada. "Die Kunststoffbauten der Expo '70 verschwinden". *Plasticonstruction* (Munich), 1971, 5.

149. P. R. Smith and Jens G. Pohl. "Pneumatic Construction Applied to Multi-Storey Buildings". *Progressive Architecture* (New York), 1970, 9.

150. *Sport- + Bäderbauten* (Düsseldorf), 1971, 1.

151. Stefan A. Szcelkun. *Shelter, Survival Scrapbook 1,* Brighton and Seattle, 1972.

152. Rudolph Szilard. "Pneumatic Structures for Lunar Bases". In: *Proceedings of the 1st International Colloquium on Pneumatic Structures.* Technische Hochschule Stuttgart, 1967.

153. R. Schulz. "Fire Tests on an Air-Supported Structure". In: *Proceedings of the 1st International Colloquium on Pneumatic Structures.* Technische Hochschule Stuttgart, 1967.

154. Steffens & Nölle GmbH, Berlin. *Das Schwimmbad der Zukunft. Eine neuartige Konstruktion des Allwetterbades. Dargestellt am Beispiel der Gemeinde Unterlüß.* Company brochure, 1972.

155. Graham A. Stevens. "The Development of the Wavetube and Participation as a Working Method". In: *International Symposium on Pneumatic Structures Delft 1972. Proceedings.* Technische Hogeschool Delft, 1972.

156. *Techniques & Architecture* (Paris), 1969, 5.

157. *The Tent. Soft Shell Structures at Expo '70.* Taiyo Kogyo Co., Ltd., Tokyo, 1970.

158. *TGL 10728, Gruppe 20 000: Traglufthallen, Technische Normen, Gütevorschriften, Lieferbedingungen.* (G.D.R.)

159. *TGL 20167, Gruppe 700: Lastannahmen für Bauten.* (G.D.R.)

160. I. Torbe. "Deformation of a Plane Surface of an Inflated Fabric Structure due to Localized Normal Load". In: *International Symposium on Pneumatic Structures Delft 1972. Proceedings.* Technische Hogeschool Delft, 1972.

161. F. J. H. Tutt. "The Effect of Concentrated Loads on Load Bearing Inflatable Structures". In: *International Symposium on Pneumatic Structures Delft 1972. Proceedings.* Technische Hogeschool Delft, 1972.

162. F. J. H. Tutt. "Inflatable Jacking Systems". In: *International Symposium on Pneumatic Structures Delft 1972. Proceedings.* Technische Hogeschool Delft, 1972.

163. F. J. H. Tutt. "A Survey of Inflation Requirements and Methods for Inflatable Structures". In: *International Symposium on Pneumatic Structures Delft 1972. Proceedings.* Technische Hogeschool Delft, 1972.

164. Tekal Vishwanath and P. G. Glockner. "Arbitrarily Large Deformations of Flat Circular Membranes under External Loads and Inflation Pressures". In: *International Symposium on Pneumatic Structures Delft 1972. Proceedings.* Technische Hogeschool Delft, 1972.

165. W. Weisz. "Air-Supported Constructions for Hard Winter Conditions". In: *International Symposium on Pneumatic Structures Delft 1972. Proceedings.* Technische Hogeschool Delft, 1972.

166. Sean R. Wellesley-Miller. "Control Aspects of Pneumatic Structures". In: *International Symposium on Pneumatic Structures Delft 1972. Proceedings.* Technische Hogeschool Delft, 1972.

167. K. L. Wolf. *Tropfen, Blasen und Lamellen.* Berlin, Heidelberg and New York, 1968.

168. Fritz Zwicky. *Entdecken, Erfinden, Forschen im Morphologischen Weltbild.* Munich and Zurich, 1966.

Addendum

169. Henning Drinhausen (editor). *Das Dach von Marl.* 1974. Brochure on the air-cushion roof of the shopping centre in midtown Marl. (The brochure can be ordered free of charge from: Enka Glanzstoff Ag, Ressort Technische Gewebe, D 5600 Wuppertal 1.)

170. *IL 9. Pneus in Nature and Technics.* Institut für leichte Flächentragwerke, Universität Stuttgart, 1976.

171. *IL 12. Convertible Pneus.* Institut für leichte Flächentragwerke, Universität Stuttgart, 1976.

172. *IL 15. Air Hall Handbook.* Institut für leichte Flächentragwerke, Universität Stuttgart, 1976.

173. Frei Otto. "Les pneus – le système des structures pneumatiques". *Techniques et Architecture* (Paris), 304 (1975).

174. David Pelham. *The Penguin Book of Kites.* Harmondsworth, 1976.

Addresses of manufacturers

Air Cruisers Company, P.O.B. 180, Belmar, New Jersey 07719, USA

Air-Tech Industries, Inc., 9 Brighton Road, Clifton, New Jersey 07012, USA

Autoflug GmbH, Industriestraße 10, 2081 Egenbüttel, Western Germany

Badische Anilin- & Soda-Fabrik AG, 6700 Ludwigshafen, Western Germany

Ballonfabrik – See- und Luftausrüstung GmbH + Co. KG, Austraße 35, 8900 Augsburg 3, Western Germany

Barracudaverken, P.O.B. 25, 18251 Djursholm 1, Sweden

Bayer AG, 5090 Leverkusen, Western Germany

Birdair Structures, Inc., 2015 Walden Avenue, Buffalo, New York 14225, USA

H. Brügge Zugbeanspruchte Konstruktionen, 4831 Marienfeld, Western Germany

CIDAIR Structures Company, 130th & Indiana Avenue, Chicago, Illinois 60627, USA

E.I. du Pont de Nemours & Company, Fabrics and Finishes Department, Wilmington, Delaware 19898, USA

Dynamit Nobel Aktiengesellschaft, Postfach 1209, 5210 Troisdorf, Western Germany

Dynat Gesellschaft für Verschlußtechnik und Feinmechanik GmbH, Bergmühlenstraße 10, 3200 Hildesheim, Western Germany

Enka Glanzstoff AG, Ressort Technische Gewebe, Postfach 130120, 5600 Wuppertal 1, Western Germany

Farbwerke Hoechst AG, Verkauf Fasern, Postfach 800320, 6230 Frankfurt (Main) 80, Western Germany

Felten & Guilleaume Carlswerk AG, Schanzenstraße 24, 5000 Köln 80, Western Germany

Firestone Coated Fabrics Company, Firestone Drive, Magnolia, Arkansas 71753, USA

Goodyear Aerospace Corporation, 1210 Massillon Road, Akron, Ohio 44315, USA

Hoogerwerff B.V., Wattstraat 1, Alblasserdam, Holland

KIB Konstruktion und Ingenieurbau GmbH, Flexible Konstruktionen, Frohnhauser Straße 95, 4300 Essen 1, Western Germany

Kléber Renolit Plastiques S.A., 67 Avenue de Verdun, 77470 Trilport, France

Otto Kleyer KG, Lübbecker Straße 12a, 4950 Minden, Western Germany

Krupp Universalbau, Abteilung Flexible Konstruktionen, *since April 1st, 1976* → KIB Konstruktion und Ingenieurbau GmbH, Flexible Konstruktionen

Metzeler AG, Westendstraße 131, 8000 München 2, Western Germany

M.L. Aviation Company, Ltd., Maidenhead, Berkshire, England

NOE-Schaltechnik – Georg Meyer-Keller KG, Kuntzestraße 72, 7334 Süssen, Western Germany

Ogawa Tent Co., Ltd., 28 Fukagawa Fuyukicho, Koto-ku, Tokyo, Japan

Plasteco Milano, Via Ugo Foscolo 13, 20030 Senago, Italy

Polydrom, *taken over in the meantime by* → Smireko AB

RAVEN Industries, Inc., P.O.B. 1007, Sioux Falls, South Dakota 57101, USA

RFD-GQ Limited, Godalming, Surrey, England

J. B. Sanders & Söhne, Postfach 140, 4550 Bramsche, Western Germany

Sarna-Hallen AG, 6078 Lungern, Switzerland

Conrad Scholtz AG, Am Stadtrand 55, 2000 Hamburg 70, Western Germany

Smireko AB, P.O.B. 55, 19060 Bålsta, Sweden

Steffens & Nölle GmbH, Gottlieb-Dunkel-Straße 20/21, 1000 Berlin 42, Germany

L. Stromeyer & Co. GmbH, Stromeyersdorf, 7750 Konstanz, Western Germany

Taisei Construction Co., Ltd., 5, 2-chome, Ginza, Chuo-ku, Tokyo, Japan

Taiyo Kogyo Co., Ltd., 22–1, Higashiyama 3-chome, Meguro-ku, Tokyo 153, Japan

Wülfing und Hauck, Ernst-Abbe-Straße 2, 3504 Kaufungen 1, Western Germany

Index